Waste Valorization
(Volume 1)

Waste Valorization for Value-added Products

Edited by

Vinay Kumar
Department of Community Medicine
Saveetha Medical College & Hospital
Saveetha Institute of Medical and Technical Sciences
(SIMATS)
Chennai, Thandalam-602105, India

Sivarama Krishna Lakkaboyana
Department of Chemistry
Vel Tech Rangarajan Dr. Sagunthala R&D Institute of
Science and Technology
Chennai, Tamil Nadu, India

&

Neha Sharma
Bioprocess Design Laboratory
School of Biotechnology
Jawaharlal Nehru University
New Delhi, India

Waste Valorization

(Volume 1)

Waste Valorization for Value-added Products

Editors: Vinay Kumar, Sivarama Krishna Lakkaboyana and Neha Sharma

ISBN (Online): 978-981-5123-07-4

ISBN (Print): 978-981-5123-08-1

ISBN (Paperback): 978-981-5123-09-8

© 2023, Bentham Books imprint.

Published by Bentham Science Publishers Pte. Ltd. Singapore. All Rights Reserved.

First published in 2023.

need for a court order if at any point you breach any terms of this License Agreement. In no event will any delay or failure by Bentham Science Publishers in enforcing your compliance with this License Agreement constitute a waiver of any of its rights.

3. You acknowledge that you have read this License Agreement, and agree to be bound by its terms and conditions. To the extent that any other terms and conditions presented on any website of Bentham Science Publishers conflict with, or are inconsistent with, the terms and conditions set out in this License Agreement, you acknowledge that the terms and conditions set out in this License Agreement shall prevail.

Bentham Science Publishers Pte. Ltd.
80 Robinson Road #02-00
Singapore 068898
Singapore
Email: subscriptions@benthamscience.net

BENTHAM SCIENCE

CONTENTS

FOREWORD

This book describes the critical and top-priority topic of current research, which is waste valorization for value-added products. The first chapter deals with plant-derived waste utilization. Abundant plant-derived organic wastes can be bio-transformed into bio-fuels like bio-ethanol, bio-butanol, biogas, and hydrogen. They can produce biochemicals like lactic acid, succinic acid, xylose, and xylitol. This chapter discusses some advanced methods for biofuel production. The second chapter discusses the various aspects of lignin bioconversion. Valorization of food waste is another critical area. The organic nature of food waste makes it fit to serve as the raw material for the enzyme industry, bio-fuels, bioactive compounds and bio-degradable plastic. The third chapter discusses this topic. The fourth chapter discusses the use of waste from the olive oil industry. This waste can be used to produce phytochemicals like phenols, flavonoids, and clean energy. Chapter five addresses the use of organic residues present in the waste using manufacturing platform chemicals. Date fruits have earned great importance in human nutrition, owing to their rich content of essential nutrients. Apart from nutraceuticals, a vast and diverse range of biomolecules can be produced, including active pharmaceutical ingredients. Date industry waste can be used for producing a vast array of antibiotics, phenolics, sterols, carotenoids, anthocyanins, flavonoids, different vitamins, economically helpful amino acids, organic acids, bio-surfactants, biopolymers, and exopolysaccharides. Date seeds can be used to produce bio-diesel and biochar and activated carbon. Citrus fruits are equally crucial as dates.

Industrial processing of citrus fruit produces various end-products like juices, concentrated jam, jellies, marmalades, and ice cream. Chapter seven is on the commercial utilization of citrus fruit processing waste to produce various chemicals like essential oils, flavonoids, pectin, enzymes, and methane. The increase in the use of plastic products has caused a significant problem in the disposal of plastic solid waste. The eighth chapter reviews how solid plastic waste can be converted into fuels and other valuable chemicals through thermal degradation, catalytic cracking and gasification, and other novel routes. Chapter nine discusses the lignin structure and the recent significant advancement in different synthesis methods for lignin nanoparticles.

Bio-plastics refer to polymers derived from plants, animals, and microorganisms. The integrated strategy of waste valorization with bio-plastic production is considered a cost-effective and sustainable approach to bio-plastic production and commercialization. Chapter ten describes biotechnological processes for valorizing food waste into commercially important biopolymeric components like Chitosan, polyhydroxyalkanoates, HAp and cellulose-based polymers. Chapter eleven discusses reliable methods for poultry waste management.

Chapter 12 deals with valorization of sugar industry waste for value-added products.

India is the second-largest cultivator of sugarcane. A significant amount of molasses and solid waste, including bagasse and filter cake, are produced every year. Sugarcane industries waste is a rich source of lignocellulosic organic biomass which can be used as a raw material to produce bio-fuel, single-cell proteins, enzymes and organic acids, food additives and nutraceuticals.

During the last century, rapid urbanization, industrialization, and globalization have increased the consumption of resources, polluting the environment. The concept of a circular economy based on restoration and regeneration by creating a connection between technology and biological cycles is gaining ground. Changing the linear economy of the produce-use-throw model into a circular economy can achieve several sustainable development goals. The last chapter discusses how the circular economy brings about transformation in Indian Industries.

This book is a very timely treatise on different valorization processes. Various government policies towards the environment and their implementation have also been discussed. It shall help formulate a business strategy that makes a way for waste valorization and brings actual revenue and tangible benefits to the environment and society.

Katta Venkateswarlu
Laboratory for Synthetic & Natural Products Chemistry
Department of Chemistry
Yogi Vemana University
Kadapa – 516 005,
India

PREFACE

The presented book is a comprehensive compilation of the use of various wastes to produce useful products. The present book contains thirteen chapters. The book highlights the following topics in all the chapters: applications of plant-derived wastes utilization for value-added product formation; lignin valorization for fuels and chemicals production; use of date palm fruit processing wastes to produce high-value products; citrus waste valorization for value-added product production; valorization of sugar industry wastes for value-added products; olive oil wastes valorization for high-value compounds production; food waste bioconversion to high-value products; organic residues valorization for value-added chemicals production; valorization of waste plastics to produce fuels and chemicals and food valorization for bioplastic production and concepts of circular economy in the valorization process. The chapters are written in an organized and strategic manner, which will help the readers gain knowledge related to their subjects. The chapters also include the major research contributions in recent years. It will help researchers advance their knowledge in the areas.

This book covers multidisciplinary concepts, including very recent findings, which will be a great help to the researchers, students, and teachers working in the areas of environmental engineering, waste valorization, agricultural engineering, agricultural biotechnology, nanotechnology, food microbiology, bioremediation, biodegradation, organic chemistry, and agricultural economics. This book will be a great reference for undergraduate, postgraduate, and doctoral students. We are thankful to all the contributing authors for providing their valuable contributions to completing this book. We are thankful to all the reviewers for their valuable suggestions for the improvement of the book.

Vinay Kumar
Department of Community Medicine
Saveetha Medical College & Hospital
Saveetha Institute of Medical and Technical Sciences (SIMATS)
Chennai, Thandalam-602105, India

Sivarama Krishna Lakkaboyana
Department of Chemistry
Vel Tech Rangarajan Dr. Sagunthala R&D Institute of Science and Technology
Chennai, Tamil Nadu, India

&

Neha Sharma
Bioprocess Design Laboratory
School of Biotechnology
Jawaharlal Nehru University
New Delhi, India

List of Contributors

Anjali Khajuria	Department of Zoology, Central University of Jammu, Rahya-Suchani (Bagla), District Samba, J&K, India
Abhinay Thakur	Assistant Professor, PG Department of Zoology, DAV College Jalandhar, (Punjab), India
Adhithya Sankar Santhosh	Department of Life Sciences, CHRIST (Deemed to be University), Bengaluru, Karnataka, India
Barkha Singhal	School of Biotechnology, Gautam Buddha University, Greater Noida (U.P.), India
Charumathi Jayachandran	Department of Biotechnology Bhupat and Jyoti Metha School of Biosciences, Indian Institute of Technology Madras, Chennai, India
Deepika Kumari	Department of Biochemistry, Maharshi Dayanand University, Rohtak, Haryana, India
Debasree Dutta Choudhury	Department of Life Sciences, CHRIST (Deemed to be University), Bengaluru, Karnataka, India
Jithin Thomas	Department of Biotechnology, Mar Athanasius College, Kerala, India
Ketaki Nalawade	Department of Alcohol Technology and Biofuels, Vasantdada Sugar Institute, Manjari (Bk.), Pune, India
Kakasaheb Konde	Department of Alcohol Technology and Biofuels, Vasantdada Sugar Institute, Manjari (Bk.), Pune, India
Lucky Duhan	Department of Biochemistry, Maharshi Dayanand University, Rohtak, Haryana, India
Muskan Pandey	School of Biotechnology, Gautam Buddha University, Greater Noida (U.P.), India
Mridul Umesh	Department of Life Sciences, CHRIST (Deemed to be University), Bengaluru, Karnataka, India
Neha Sharma	Bioprocess Design Laboratory, School of Biotechnology, Jawaharlal Nehru University, New Delhi, India
Neha Kumari	Department of Biotechnology and Bioinformatics, Jaypee University of Information Technology,Waknaghat, Distt. Solan, Himachal Pradesh, India
Pritha Chakraborty	School of Allied Healthcare and Sciences, Jain (Deemed to be) University, Bengaluru, India
Paharika Saikia	Department of Alcohol Technology and Biofuels, Vasantdada Sugar Institute, Manjari (Bk.), Pune, India
Richa Parashar	School of Biotechnology, Gautam Buddha University, Greater Noida (U.P.), India
Rahul Datta	Centre for Agricultural Research and Innovation, Guru Nanak Dev University, Amritsar, Punjab, India
R. Kamatchi	Centre for Biotechnology, Anna University, Chennai, Tamil Nadu, India

Ritu Pasrija	Department of Biochemistry, Maharshi Dayanand University, Rohtak, Haryana, India
Sukhendra Singh	Department of Alcohol Technology and Biofuels, Vasantdada Sugar Institute, Manjari (Bk.), Pune, India
Shuvashish Behera	Department of Alcohol Technology and Biofuels, Vasantdada Sugar Institute, Manjari (Bk.), Pune, India
Sanjay Patil	Department of Alcohol Technology and Biofuels, Vasantdada Sugar Institute, Manjari (Bk.), Pune, India
Sowmiya Balasubramanian	Centre for Biotechnology, Anna University, Chennai, Tamil Nadu, India
Shefali Patel	Department of Biological and Environment Sciences, N. V. Patel College of Pure and Applied Sciences, V. V. Nagar, Gujarat, India
Susmita Sahoo	Department of Biological and Environment Sciences, N. V. Patel College of Pure and Applied Sciences, V. V. Nagar, Gujarat, India
Sivarama Krishna Lakkaboyana	Department of Chemistry, Vel Tech Rangarajan, Dr. Sagunthala R&D Institute of Science and Technology, Chennai, Tamil Nadu, India
Subhrangsu Sundar Maitra	Bioprocess Design Laboraotry, School of Biotechnology, Jawaharlal Nehru University, New Delhi, India
Suma Sarojini	Department of Life Sciences, CHRIST (Deemed to be University), Bengaluru, Karnataka, India
Sapthami Kariyadan	Department of Life Sciences, CHRIST (Deemed to be University), Bengaluru, Karnataka, India
Sruthi Sunil	Department of Biotechnology, Mar Athanasius College, Kerala, India
Saurabh Bansal	Department of Biotechnology and Bioinformatics, Jaypee University of Information Technology, Waknaghat, Distt. Solan, Himachal Pradesh, India
Varsha Sharma	Central Pollution Control Board (CPCB), Waste Management Division-I, East Arjun Nagar, Vishwas Nagar Extension, Vishwas Nagar, Shahdara, Delhi, India
Vinay Kumar	Department of Community Medicine, Saveetha Medical College & Hospital, Saveetha Institute of Medical and Technical Sciences (SIMATS), Chennai, Thandalam, India

CHAPTER 1

Utilization of Plant-derived Wastes For Value Added Product Formation

Ketaki Nalawade[1], Paharika Saikia[1], Sukhendra Singh[1], Shuvashish Behera[1], Kakasaheb Konde[1] and Sanjay Patil[1,*]

[1] *Department of Alcohol Technology and Biofuels, Vasantdada Sugar Institute, Manjari (Bk.), Pune-412307, India*

Abstract: Depletion of fossil fuels and environmental concern has impelled to search for alternative biofuels and biobased chemicals. Biofuels have been considered an alternative clean energy carrier due to their environmentally friendly nature. Recently, research has been focused on finding a readily available, low-cost and renewable lignocellulosic biomass to produce value-added products. In this context, the plant-derived organic wastes can be transformed to produce biofuels (bioethanol, biobutanol, biogas and biohydrogen) and biochemicals (lactic acid, succinic acid, xylose and xylitol). It will be a sustainable effort to reduce the huge amount of plant waste generated. In addition, in the recent decades, several efficient conversion methods have been invented.

During the past few years, a large number of chemical pretreatment methods have also been developed for efficient lignocellulosic conversion. The current chapter discusses the advanced methods for biofuels and biochemicals' production, focusing primarily on different pretreatment methods for effective conversion of plant derived wastes.

Keywords: Anaerobic digestion, Biomass, Biofuels, Bioethanol, Biobutanol, Biogas, Biochemicals, Biohydrogen, Detoxification, Fermentation, Inhibitors, Lignocelluloses, Ligninolytic enzymes, Lactic acid, Plant derived wastes, Pretreatment, Succinic acid, Value added products, Xylitol, Xylose.

INTRODUCTION

Energy plays a crucial role in the socio-economic development of a country. According to the Global Status Report on energy, the major part of energy share of around 78% is obtained from nonrenewable resources (fossil fuels such as petroleum, gases and coal) and only 19% comes from renewable energy resources

* **Corresponding author Sanjay Patil:** Department of Alcohol Technology and Biofuels, Vasantdada Sugar Institute, Manjari (Bk.), Pune-412307, India ; Tel:+91-020-26902341; E-mail: sv.patil@vsisugar.org.in

Vinay Kumar, Sivarama Krishna Lakkaboyana & Neha Sharma (Eds.)
All rights reserved-© 2023 Bentham Science Publishers

(solar, wind, hydropower and biomass) [1, 2]. The fossil fuel reserves are diminishing very rapidly, and also its overuse is creating serious pollution in the environment. Therefore, it is necessary to explore alternative resources of energy to meet the future demand of energy [3]. In this context, plant biomass containing starch and lignocellulose has emerged as a renewable, sustainable and economically feasible source for biofuel production. Scientists and investors have coined a term to the bio-based economy, that is circular bio-economy because of its renewable nature [2 - 4]. In this context, plant biomass containing starch and lignocelluloses can be used to produce value-added products.

Biofuels are classified into primary and secondary biofuels based on the type of biomass used [4]. First-generation biofuels are produced from edible food crops such as starch and sugar containing crops [5]. Since the first-generation biofuels directly compete with the food items, the focus has shifted to second-generation biofuels which are obtained from lignocellulosic materials. Lignocellulosic biomass resources are generally discarded as residual and agricultural wastes. The most significant and abundant renewable biomass resources include crop residues like corn stover, wheat straw, rice straw and sugarcane bagasse [3, 6 - 11]. Due to their abundance and renewable nature, lignocellulosic biomass is considered an excellent alternative substrate for production of several value-added products [12]. Several biofuels and biochemicals can be produced from lignocellulosic biomass [13 - 15].

Lignocellulose is the connecting link between cellulose and lignin. Hemicellulose is present as the matrix surrounding the cellulose skeleton, while lignin is an encrusting material serving hemicelluloses and celluloses as a protective layer [12]. All three components are covalently cross-linked among the polysaccharides and lignin, making biomass a composite material [16, 17]. Therefore, a pretreatment step is mostly required to break these bonds. Pretreatment is an essential pre-requisite to convert lignocellulosic biomass into fermentable sugars with the help of enzymes [18, 19]. Sometimes, these pretreatment strategies further lead to the production of inhibitors such as vanillic acid, uronic acid, 4-hydroxybenzoic acid, phenol, furaldehydes, cinnamaldehyde, and formaldehyde which may intervene with the growth of the fermentative microorganisms. Much advancement has been featured in the field of chemistry which has led to the development of novel processing technologies. These technologies are available at a commercial scale and emerge as promising solutions. In addition, they proved to be low cost at commercial scales [15, 20, 21].

This chapter has been focused on the production of biofuels, and biochemicals. In addition, the nature of inhibitors is also discussed at the end of the chapter.

PLANT BASED BIOMASS PRETREATMENT

Pretreatment is an indispensable step for the preparation of lignocellulosic biomass for its further processing. Pretreatment is essential to weaken the recalcitrant structure of lignocellulose making cellulose, lignin, and hemicellulose more accessible for enzymes or chemicals. Moreover, pretreatment is followed by the removal of lignin, degradation of hemicellulose, reduction in cellulose crystallinity, and an increase of surface porosity [2, 22]. Pretreatment is considered as the most expensive step in the entire biomass processing. Therefore, necessary efforts should be made to lower the operating costs, and increase the process effectiveness, and recovery of lignocellulosic components [23]. The critical factors for biomass pretreatment that should be considered are: (1) The possibility of large-scale feedstock processing; (2) High yields regardless of the type and origin of biomass; (3) Reducing the waste and inhibitors; (4) Compatibility of the pretreatment with further processing; (5) Efficient recovery of lignin; and (6) Reducing equipment and energy cost. Pretreatment methods of plant based biomass are classified into three basic categories: physical, chemical and biological [1, 24, 25].

Physical pretreatment consists of an increase in temperature and/or pressure, which causes structural changes in the biomass. Chemical treatment is characterized by the use of organic or inorganic compounds, which disrupts the lignocellulosic structure [2, 23]. Although individual pretreatment methods are effective, but their combination has higher efficacy. Biological pretreatment includes the microorganisms and enzymes for the hydrolysis of lignocellulosic polymers into their monomers [2]. An extensive number of the research papers concerning plant based biomass pretreatment have been published in the last decade focusing on the strengths and weaknesses of various technologies to get a competent pretreatment suitable for an eco-friendly cost effective process. The schematic route of pretreatment is shown in Fig. (**1**).

DIFFERENT PRETREATMENT METHODS

Physical

Plant materials require a rigorous method to break them into components. There are several physical methods available for plant-based biomass pretreatment. Mechanical, microwave, ultrasound, and hydrodynamic cavitation are the most common techniques used for plant-based biomass pretreatments [23, 26].

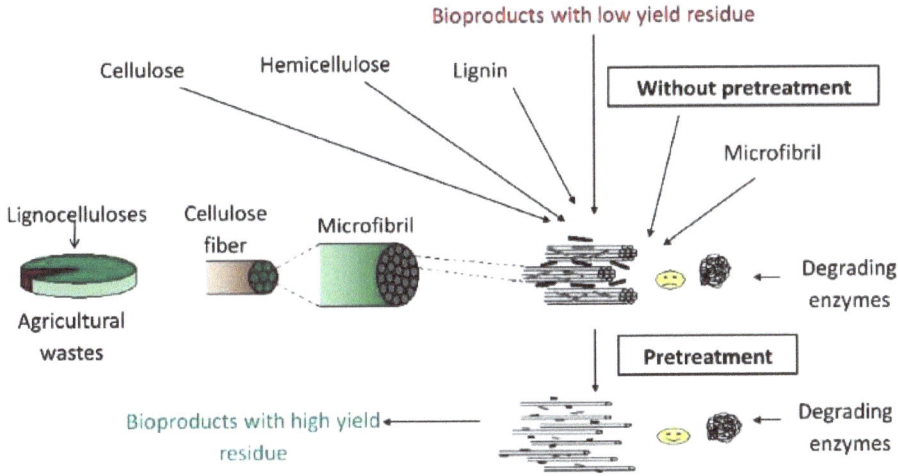

Fig. (1). Schematic configuration showing composition of lignocellulosic biomass before and after the pretreatment.

Milling

Milling is the first step of pretreatment of a plant-based biomass. The milling process reduces the crystallinity of plant biomass cellulose and the degree of polymerization [18]. Milling requires high energy consumption. To achieve the high yield, biomass has to be ground into fine particles which is the first step in the pretreatment process [27]. Ball milling, roll milling, hammer milling, colloid milling, knife milling and disk milling are several types of millings used in the pretreatment of plant-based biomass. Zhang *et al.* [28] studied the knife milling of poplar wood with size reduction to increase the sugar yield. They reported that the poplar wood particles with a size of 4 mm provided higher sugar yield than 1 mm and 2 mm particles.

Ultrasound

Ultrasound pretreatment is used to break and remove the hemicellulose from lignin and cellulose [29]. Ultrasound disrupts the α-O-4 and β-O-4 linkages in lignin which leads to the formation of small cavitation bubbles by splitting structural polysaccharides and lignin fractions [13 - 30]. Ultrasound is considered as an effective technology because it generates a high pressure and temperature [31]. This method is efficient in the production of high cellulose content and less undesired components such as lignin and hemicellulose [32]. Ultrasonication has been reported to increase biohydrogen production rate in pulp and paper mill effluent [33].

Microwave

Microwave (MW) irradiation is widely used for plant-based biomass pretreatment It has added advantages such as smooth operation, low energy requirement, high heating capacity and less inhibitor formation [34]. It is effective alone and in combination with other methods [13, 35]. When cellulose is treated by microwave irradiation in the presence of ionic liquid, the solubility of cellulose increases. This also decreases the degree of polymerization. Therefore, the rate of enzymatic hydrolysis of cellulose increases to manifold [36]. Keshwani *et al.* [37] studied the microwave-assisted alkali pretreatment of switch-grass and coastal bermudagrass for improving the enzymatic hydrolysis. They obtained glucose and xylose yield after subsequent enzymatic hydrolysis was as high as 82% and 63% for swichgrass and 89% and 59% bermudagrass, respectively. Li *et al.* [38] performed the microwave assisted KOH pretreatment of bamboo biomass. They reported that the yields of glucose (20.87%) and xylose (63.06%) were increased up to 8.7-fold and 20.5-fold, respectively after pretreatment with a cellulase.

Chemical

Chemical pretreatment is a crucial method in terms of recovery of sugar monomers. Acid and alkali-based hydrolyses are the most commonly used chemical pretreatments [39].

Acidic

Acid hydrolysis is the most commonly used method for pretreatment. The most commonly used acids are sulphuric, hydrochloric, nitric, phosphoric and peracetic acid. Acid pretreatment is generally carried out with concentrated or diluted acids. Concentrated acid hydrolysis is performed at a low-temperature range (30-60°C). The use of higher acid concentration yields more sugar [40]. The process can be made effective by the acid recovery. Diluted acid hydrolysis is the most efficient treatment in hemicellulose. It increases the biomass porosity and cellulose hydrolysis. It is performed at high temperature ranging from 120-170°C and 0.5-6.0% acids with a variable time frame. Dilute acid hydrolysis weakens the glycosidic bond in the lignocellulosic biomass. Dilute acid treatment increases the plant biomass porosity. High hydrolysis yields have been obtained when plant-based biomass was pretreated with dilute sulfuric acid as compared to hydrochloric, phosphoric, and nitric acid. Acid pretreatment forms the furan and short-chain aliphatic acid derivatives, which act as potent inhibitors in the microbial fermentation process [39].

Sulfuric (H_2SO_4) and phosphoric (H_3PO_4) acids are commonly used for acid pretreatment because they are relatively inexpensive and give efficient hydrolysis of plant-based biomass. HCl has better penetration efficiency in the biomass. It is more volatile in nature and the recovery of HCl is easier than H_2SO_4 [40]. Peracetic acid (PAA) is a strong oxidizing agent, which helps to remove lignin. At a substrate concentration of 2% and PAA concentration of 20% (for 120 min at 120°C temperature), it gives a high yield of reducing sugars [41]. De Vasconcelos *et al.* [42] optimized dilute phosphoric acid pretreatment for sugarcane bagasse and found 0.2% acid concentration and 186°C as very effective for hemicellulose solubilisation. Moraes *et al.* [43] studied the effects of sugarcane bagasse with dilute mixture of sulfuric acid and acetic acid pretreatment. It efficiently hydrolysed the hemicellulose with a removal efficiency of 90%. In addition, cellulose degradation was observed to be below 15% corresponding to low crystallinity fraction. On the other hand, Singh *et al.* [44] investigated a research on the sulfuric acid pretreatment of rice straw, corncorb, barley straw and wheat bran. They found promising results for rice straw as compared to other plants' biomass taken.

Alkaline

Alkali pretreatment can be carried out using sodium hydroxide (NaOH), potassium hydroxide (KOH), calcium hydroxide ($CaOH_2$), ammonium hydroxide (NH_4OH) and ammonia (NH_3). Alkali treatment is used to disrupt the hemicellulose and lignin binding. It can increase the digestibility of hemicellulose which helps to promote the enzyme access to cellulose [45]. Lignin decomposition is the cleavage of a-aryl ether bonds from its polyphenolic monomers, whereas hemicellulose dissolution and cellulose swelling are due to hydrogen bond weakening [1, 46, 47].

Alkaline pretreatment is the de-lignification process which disrupts the cell wall by dissolving hemicellulose, lignin and cellulose. This process gives a liquid fraction (containing hemicellulose and lignin) and solid fraction (cellulose) [46]. Physical structure and chemical composition of the substrate are critical factors in the effective alkaline pretreatment. Sodium hydroxide gives the most significant lignin degradation as compared to other alkalis, such as sodium carbonate, ammonium hydroxide, calcium hydroxide, and hydrogen peroxide [40]. NaOH can be used in the pretreatment of different plant-based biomasses to remove lignin under appropriate conditions. Wunna *et al.* [48] reported 83.7% and 87.3% as the maximum removal of lignin from sugarcane bagasse.

Steam Explosion

Steam explosion is also known as auto-hydrolysis or steam disruption. It is a physiochemical process carried out by steaming with or without explosion. Steam explosion is typically initiated at 160-260°C temperature with a corresponding pressure of 0.69-4.83 MPa for several seconds to few minutes before exposure to the atmospheric pressure. High temperature and pressure cause degradation of hemicellulose and lignin. It further increases the cellulose hydrolysis potential of plant-based biomass. Steam explosion is generally used for ethanol and biogas production by using plant-based biomass. Saturated high-pressure steam is used for the pretreatment of plant-based biomass [13]. Boboescu *et al.* [49] employed steam-treatments on whole sweet sorghum biomass under certain severity factors which resulted in proper de-lignification and hydrolysis of lignocellulose fibres [49]. Walker *et al.* [50] optimized the steam explosion pretreatment method with phosphoric acid for four feedstocks wheat straw, corn stover, Miscanthus, and willow for xylose release. They reported that the maximum xylose recovery of 90% and >1000 L of wheat straw hydrolysate could be achieved at optimized conditions.

Ammonia Fibre Explosion (AFEX)

Ammonia fibre explosion (AFEX) is a physicochemical pretreatment carried out by variation of water loading, ammonia loading, reaction time and temperature. This process is advantageous because it doesn't require a small particle size of biomass as well as there is no formation of inhibitors during the process. For high lignin content biomass, this process is less effective [51]. The AFEX process is very similar to the steam explosion pretreatment. AAFEX process for plant biomass disruption includes 1-2 kg of liquid ammonia/kg of dry biomass at 90°C temperature and 30 min residence time. AFEX gives a rapid expansion of the liquid ammonia, which causes swelling and physical disruption of lignocellulosic biomass fibres and partial de-crystallisation of cellulose. AFEX process can reduce or modify cellulose crystallinity and lignin fraction of the plant-based biomass. During the AFEX process, deacetylation process causes the removal of the least acetyl groups, which results in increasing digestibility of plant-based biomass [52].

Carbon Dioxide Explosion

In this process, CO_2 is used as a supercritical fluid under high pressure, where CO_2 penetrates the plant biomass resulting in the increased digestibility. Supercritical carbon dioxide is the promising green solvent in hydrothermal treatment as well as in the organosolv de-lignification, in single or multistage processes [53]. It is believed that once CO_2 dissolves in water, it forms carbonic acid, which helps in the hydrolysis of hemicellulose. The yields of CO_2 explosion are lower than those obtained by steam or ammonia explosion, but the yield is higher as compared to that without pretreatment. Park *et al.* [53] studied simultaneous pretreatment by CO_2 explosion with enzymatic hydrolysis and obtained 100% glucose yield.

Biological Pretreatment

Biological plant biomass pretreatment is carried out by using a wide variety of naturally found microorganisms or enzymes. The microorganisms secrete the hydrolytic enzymes such as hydrolases and ligninolytic enzymes which degrade or depolymerize the lignin [35, 54]. In the process, the cell wall structure is disrupted and subsequently results in the hydrolysis of cellulose and hemicelluloses. In biological plant biomass, pretreatment is affected by the processes such as temperature, moisture, substrate size, aeration, pH, and structural complexity [13]. Biological pretreatment is environment-friendly and sustainable. This is because it requires very less energy, is chemical-free, does not release toxic products and can be performed in mild conditions. However, as compared to other processes, this process is very convenient [27].

Some bacteria (*Azospirillum* sp.*, Bacillus* sp., *Cellulomonas* sp., *Clostridium* sp., *Pseudomonas* sp., *Streptomyces* sp. and *Thermomonospora* sp.) and several fungi (*Aspergillus* sp., *Fusarium* sp. *Neurospo*ra sp. and *Trichoderma* sp.) are reported to hydrolyze plant-based biomass [26, 35, 55]. *Streptomycetes* is one of the most important genera of ligninolytic microorganisms which produce several oxidative enzymes. These enzymes include peroxidases and laccases [9, 56]. It has been observed that *Pseudomonas* sp. and *Actinobactor* sp. removed 52% and 57% of lignin from poplar wood in a 30 day pretreatment [57, 58]. Other bacterial species such as *Streptomyces cyaneous* and *Thermomonospora mesophila* can degrade 50% of lignin from barley straw. There are several predominant rumen cellulolytic bacteria (for example *Fibrobactor succinogenes*, *Ruminococcus flavefaciens*, and *Roseovarius albus*) and anaerobic bacteria (*Clostridium thermocellum* and *Bacteroides cellulosolvens etc.*) that produce cellulases with high enzyme activity [26, 54]. Bacterial species such as *Azospirillum lipoferum*, and *Bacillus subtilis* have been studied for bacterial laccase, which depolymerizes the lignin [1, 59]. Therefore, more research work is required in the field of bacterial lignin degradation from plant-based biomass. Nanoparticle-mediated methods can be used to produce value-added products [60 - 63]. The lignocellulolytic fungi include species from the ascomycetes (*e.g. Aspergillus* sp., *Penicillium* sp., *Trichoderma reesei*), basidiomycetes including white-rot fungi (*e.g. Schizophyllum* sp., *P. chrysosporium*), brown-rot fungi (*e.g. Fomitopsis palustris*) and few anaerobic species (*e.g. Orpinomyces* sp.) [26]. In recent years, a number of research studies have been performed using fungal strains for the pretreatment of plant-based biomass [2, 64, 65]. These fungi secrete the ligninolytic enzymes such as laccase, magnease peroxidase and lignin peroxidase, which help in depolymerization and mineralization of lignin [66].

PRODUCT FORMATION

Plant-based biomass can be used to produce biofuels and biochemicals. Biofuels include bioethanol, biobutanol, biogas, and biohydrogen. Biochemicals include lactic acid, succinic acid, xylitol and xylose.

Biofuels

Bioethanol

Bioethanol is widely known as a renewable and sustainable liquid fuel with economic, environmental and strategic attributes [67]. It can be blended with petrol to increase the octane number. In addition, it can be used directly to reduce greenhouse gas emissions up to 80% as compared to gasoline. Therefore it is considered as a cleaner fuel for future [68, 69]. There are different generations of bioethanol production based on different substrates These include 1st generation (produced from food grade material), 2nd generaion (produced from agro residues) and 3rd generation (produced from microalgae). Bioethanol can be produced from a variety of plant-based biomasses including rice straw, wheat straw, corn stover, sugarcane bagasse and microalgae [70 - 77].

In the current scenario, most of the bioethanol is produced from 1st generartion feedstock *i.e.* sugarcane and maize. However, many scientific and industrial efforts are going on to produce 2nd generation bioethanol to make the process economic. In India, around 120-160 MMT of surplus biomass is available which has the potential to produce 3000 crore litres of bioethanol annually [78]. Bioethanol production from lignocellulosic biomass (2G ethanol) generally consists of four sequential steps *i.e.* pre-treatment, hydrolysis, fermentation and distillation. 2G ethanol is generally produced through separate hydrolysis and fermentation (SHF) or simultaneous saccharification and fermentation (SSF) [79, 80]. *Zymomonas mobilis* and *Saccharomyces cerevisiae* are the most commonly used microorganisms for the production of ethanol from hexose sugars (glucose) [80]. However, these two microorganisms are incapable of fermenting the pentose sugar (xylose), and therefore, yeast strains such as *Candida shehatae*, *Pichia stipitis*, and *Pachysolen tannophilus* are being used for the xylose-based ethanol production [80]. Liu and coworkers [81] subjected the alkali-pretreated sugarcane bagasse for ethanol production at high solid loading (30%, w/v) in a fed batch with simultaneous saccharification and fermentation. They achieved 66.92 g/L of ethanol with a conversion efficiency of 72.89% after 96 h of fermentation. Some of the recent studies related to ethanol production from pretreated biomass are mentioned in Table **1**.

Table 1. Production of ethanol from plant-based biomass after different pretreatments

Raw Material	Pretreatment	Titer	Yield	Reference
Sugarcane Bagasse	Alkali	66.92 g/L	-	[81]
	Acid-alkali	31.5 g/L	0.34 g/g	[82]
	Hydrodynamic cavitation (HC)-assisted alkaline hydrogen peroxide (AHP)	28.44 g/L	0.43 g/g	[83]
Wheat Straw	Alkaline	54.5 g/L	0.23 g/g	[84]
	Dilute H_2SO_4	36.0 g/L	0.29 g/g	[85]
	$H_3PO_4 + H_2O_2$	71.2 g/L	0.11 g/ g	[86]
	Sodium carbonate	99.4 g/L	-	[87]
	Choline chloride/glycerol	36.7 g/L	-	[88]
Corn Stover	$H_2SO_4 + NaOH$	115.3 g/L	0.09 g/g	[72]

Biobutanol

Butanol is considered a promising gasoline substitute to ethanol. For gasoline blending, four-carbon alcohol (butanol) is highly desirable due to its higher energy density, lower hygroscopicity, lower vapour pressure, blending ability, and use in conventional combustion engines without engine modification [89]. Biobutanol production using renewable feedstocks through the ABE (acetone–butanol–ethanol) fermentation process involving microorganisms is a sustainable approach [89, 90]. However, there are certain challenges in the ABE fermentation process. These include strict anaerobic conditions, rapid pH shift, low butanol titer, low solvent yield, high recovery cost, and low solvent tolerance by the microorganisms [91]. Alternative plant-based materials can be used to produce economic biobutanol [92]. These plant-based biomasses need the process of pretreatment, enzymatic saccharification and further ABE fermentation for the production of solvents. Strains like *C. acetobutylicum* and *C. beijerinckii* are mostly studied microorganisms for the production of butanol with the involvement of two phases *i.e.* acidogenesis and solventogenesis [92].

Amiri *et al.* [93] followed organosolv pretreatment of rice straw for the production of butanol and obtained 10.5 g/L of ABE. Li *et al.* [94] compared the pretreatment of sugarcane bagasse using diluted acid, and aqueous ammonia and their combinations for enzymatic hydrolysis, structural characterization and ABE fermentation. Diluted acid and oxidative ammonolysis in combination could improve the digestibility. It further enhanced butanol production to 12.12 g/L in ABE fermentation. Qi *et al.* [95] performed pretreatment of wheat straw by ammonium sulfite for enhanced production of acetone-butanol-ethanol (ABE).

Fermentation of corn stover (CS) hydrolysate could produce 50.14 g/L of ABE with a yield of 0.43 and a productivity of 0.70 g/L/h [96]. Butanol production from plant-based biomass after pretreatments from some recent studies is shown in Table **2**.

Table 2. Production of butanol from plant-based biomass after pretreatment.

Raw material	Pretreatment	Titer	Yield	References
Sugarcane bagasse	Dilute acid (HCl/ H_2SO_4)	15 g/L	0.3 g/g	[97]
	Alkali	21.11 g/L	0.15 g/g	[98]
	Hydrothermal	14.5 g /L	0.29 g/g	[99]
Rice Straw	Foraceline+ sodium carbonate	9.5 g/L	0.25 g/g	[100]
	liquid ammonia	7.0 g/L	0.36 g/g	[101]
	Alkali pretreatment	6.9 g/L	-	[102]
	H_2SO_4	4.9 g /L	-	[101]
Soybean Hull	Dilute acid (HCl/ H_2SO_4)	15 g/L	0.3 g/g	[97]
Cotton Stalk	Dilute acid (HCl/ H_2SO_4)	15 g/L	0.3 g/g	[97]

Biogas

Plant-based biomass with high organic content can be converted into another form of energy such as biogas *via* anaerobic digestion (AD). Anaerobic digestion is a biochemical process which converts organic substrates into methane-rich biogas by sequential stages including hydrolysis, acidogenesis, acetogenesis and methanogenesis [103]. The four steps of AD process are (i) Hydrolysis of proteins, lipids, carbohydrates, (ii) Conversion of hydrolysis products and monomers into volatile fatty acids (VFAs), (iii) Conversion of VFAs into acetate, carbon dioxide, and hydrogen, and (iv) Methane formation by methanogenesis [104]. Biogas composition slightly varies with the different feedstocks used in the anaerobic digestion and mainly composed of CH_4 (40-75%) and CO_2 (25-60%), with H_2S, NH_3 in minor amounts [105].

Mechanical pretreatment is the most significant process to increase methane yield, but it is not able to remove lignin which acts as a barrier in the bioavailability of carbohydrates [106 - 108]. During chemical pretreatment, the cost of reagents, operations (neutralization step) and the requirement for corrosion-resistant reactors are the known limitations. Thus, achieving higher efficiency and lowering the formation of inhibitory compounds by combining lower concentration of chemical reagents with other pretreatments can help to reduce the cost [108, 109]. For the substitution of natural gas and medium calorific value gases, the methane production using agricultural residues *via* the anaerobic digestion process is an effective method. Hashemi *et al.* [110] reported that the pretreatment of sugarcane bagasse by ammonia improved the biogas production up to 299 mL/g VS. Mancini *et al.* [111] employed the organosolv pretreatment method with ethanol for wheat straw and improved the biogas production up to 274 mL/g VS [111]. Rajput *et al.* [112] studied the effect of thermal pretreatment (180°C) on wheat straw in anaerobic batch digestion reaction and showed 615 mL/g VS of biogas yield and 69% of volatile solids reduction [112]. The biogas production from different biomass after pretreatment is shown in Table **3**.

Table 3. Production of biogas from plant-based biomass after different pretreatments.

Pretreatment	Substrate/ Biomass	Methane yield (Untreated) mL/g VS	Methane yield (Treated) mL/g VS	Reference
Mechanical, Multistage Milling	Wheat straw	167.8	245.6	[113]
Mechanical, Milling	Rice straw	58.1	65.7	[114]
Chemical, Alkali	Wheat straw	274	315	[111]
Physicochemical, Hydrothermal	Rice straw	92	280	[115]
Physicochemical	Sugarcane bagasse	105.6	299.3	[110]
Physicochemical, Hydrothermal	Wheat straw	388.9	488.3 -611.7	[112]
Physicochemical, Steam Explosion	Reed	188	226 & 355	[116]
Biological	Sawdust	89.9	155.2	[117]
Chemical, Acid	Wheat straw	159.1	163.7	[118]
Physicochemical, Steam Explosion	Corn stover	285	348	[119]
Physicochemical, Alkaline and Hydrothermal	Sugarcane bagasse	215	318	[120]

Biohydrogen

Hydrogen is considered a potential clean fuel because of its carbon-free nature, and it oxidizes to water as a combustion product. Conventional method of hydrogen production is usually based on fossil fuels, but due to high energy requirement and CO_2 emission, it causes a greenhouse effect which is not considered environment friendly [121]. Biohydrogen can be produced through two biological routes *i.e.* light-dependent photo fermentation and light-independent dark fermentation which has several advantages in comparison to the conventional method [45]. Several microorganisms such as *Chlamydomonas reinhardtii, Platymonas subcordiformis, Chlorella fusca, Scenedesmus obliquus, Chlorococcum littorale, Rhodopseudomonas palustris,* and *Rhodobacter* *sphaero*ides are used for light dependent biohydrogen production [122]. Facultative (*E. coli, Enterobacter* sp.) and obligate microorganisms (*Clostridia* and rumen bacteria) are used for the dark fermentation process of biohydrogen production [123]. The cost of the hydrogen production process can be lowered by using starch and lignocelluloses-based renewable raw materials [124].

In a study, biohydrogen production from rice straw through combined pretreatments, resulted in a higher biohydrogen yield (129 mL/g COD) [45]. Moodley *et al.* [125] compared the acidic pretreatment (HCl, H_2SO_4 and HNO_3) of sugarcane leaf wastes for the production of hydrogen. The HCl based pretreatment provided 160% more sugar and gave a yield of 18.6 ml H_2/g fermentable sugar. Gonzales *et al.* [126] performed acidic pretreatment of rice husk and could get 1860 ± 245 mL H_2/L/day hydrogen yield by optimizing the downstream of fermentable sugar. Mirza *et al.* [127] carried out photofermentative biohydrogen production in a batch process using raw sugarcane bagasse (SCB) for the production of 148-513 mL/L of H_2 with purple non-sulfur bacteria (PNSB). Biohydrogen production from various types of plant-based biomass using different pretreatments is shown in Table **4**.

Table 4. Production of biohydrogen from plant-based biomass after pretreatments.

Raw Material	Pretreatment	Yield	Reference
Empty palm fruit bunch	Acid	1.27 mol H_2/ mol of sugar	[126]
Poplar leaves	Alkali	44.92 mL H_2/g sugar	[128]
Pine tree wood	Acid + Alkali	179 mL H_2/g sugar	[126]
Waste sorghum leaves	Physicochemical	213.14 mL H_2/g sugar	[129]
Sugarcane leaf waste	Physicochemical	18.6 ml H_2/g sugar	[125]

Biochemicals

Lactic Acid Production

Lactic acid (LA) is a 2-hydroxypropanoic acid and it is used in the various industries like food, pharmaceutical, cosmetic, chemical, leather tanning and to produce biodegradable polymers [130, 131]. Global lactic acid market size is expected to rise to USD 8.77 billion by 2025 with an increase in the CAGR of 18.7%. The global demand for lactic was 1220.0 kt in 2016 but with increasing annual demand of about 16.2%, it will reach up to 1960.1 kt in 2025 [132]. It is present naturally in two optical isomeric forms *i.e* L (+)-lactic acid and D (-)-lactic acid. Generally, lactic acid is produced through microbial fermentation and chemical synthesis. Chemical synthesis of lactic acid produces a racemic mixture of two optical isomers, whereas microbial fermentation can give racemic mixtures as well as optically pure isomers depending on microorganisms used for synthesis and fermentation conditions.

Lactic acid production *via* microbial synthesis is currently driven by several microbes such as lactic acid bacteria (LAB), Bacilli and genetically-modified strains, including *E. coli* and *Coryne-bacterium* sp [133, 134]. Most of these microorganisms are mesophilic or neutrophilic and their conditions are ideal for the survival of commonly found contaminating microorganisms which adversely affect the fermentation process [135]. Batch, fed-batch, and continuous fermentation are commonly used processes in the production of lactic acid. Higher lactic acid titre is generally obtained by using batch and fed-batch cultures rather than continuous culture [136]. Lactic acid can also be produced from fungal and bacterial cells by solid state fermentation. *Rhizopus sp.* is the most commonly used fungal species [137] and in case of bacterial cultures, *Lactobacillus* sp. is widely used [138].

Various factors such as nitrogen sources and neutralizing agents affect lactic acid production [139]. There are cheaper alternative sources of plant-based biomass. These include starchy raw materials such as corn, maize, cassava, barley, potato, rice, rye, and wheat [140 - 144]. Plant-based wastes include sugarcane bagasse, wheat straw, rice straw, corn stover, banana stalk, corn cobs, and sweet sorghum [98, 145 - 147] for lactic acid production. Lactic acid production from plant-based biomass is based on four major steps *i.e.* pretreatment, sachharification, fermentation and purification [75, 148, 149].

Nalawade *et al.* [134] reported that 2G lactic acid was produced using *Bacillus coagulans* NCIM 5648 from alkali pre-treated sugarcane bagasse. Optically pure L-lactic acid (52.5 g/L) was obtained within 45 to 55 h of fermentation time [134]. Wishral *et al.* [150] performed SSCF of pre-treated hemicellulosic hydrolysate and partially delignified cellulignin (PDCL) from sugarcane bagasse by using *L. pentosus* ATCC 8041. The study reported 64.8 g/L lactic acid with productivity of 1.01 g/L/h. Dilute acid pre-treatment usually generates hemicellulose hydrolysate (HH) fraction which is rich in xylose. Xylose fermentation is more difficult as compared to the fermentation of glucose. One of the microorganisms *L. pentosus* assimilate xylose during the fermentation process [151]. Several techniques are used to purify lactic acid. These include liquid-liquid extraction, diffusion dialysis, and bipolar membrane electrodialysis [131, 152]. Although, lactic acid purification is not an economic process. However, plant waste based substrate used for the production will definitely meet the current demand of chemicals with an affordable price.

Succinic Acid

Succinic acid is a naturally-occurring chemical, and is considered a promising candidate for industrial applications [153]. Succinic acid can be used for manufacturing of lacquers, resins, and other coating chemicals as well as a flavour additives in the food and beverage industry [132]. The annual production of succinic acid worldwide is about 16, 000tons approximately and the price ranges from $6-9 per kilogram depending on its purity [132]. It has four existing markets; (1) Surfactant, additives, foaming agents and detergent, (2) Iron chelators, (3) The food market, and (4) Pharmaceutical industry [154].

Succinic acid can be produced chemically through the techniques of paraffin oxidation, catalytic hydrogenation, electro-reduction of maleic acid or maleic anhydride. However, these techniques are not considered as environment-friendly and cause serious pollution [132]. Succinic acid can be produced by fermentation of sugars from renewable feedstocks such as plant biomass. It can be produced from bio-based techniques as a building block for commodity and high value chemicals [155]. The recent developments in the production of succinic acid have been focused on biotechnological alternatives that include processes such as microbial transformation. Lactic acid can be produced from sugarcane bagasse [156], corn stalk [157], sweet sorghum [158], agave [159], cassava bagasse [160], rice straw [161], and wheat straw [162]. The succinic acid pathway partly or wholly were observed in the following bacterial species: *Actinobacillus succinogenes* [157], *Anaerobiospirillum, succiniciproducens* [163], *Corynebacterium crenetum* [164], *Corynebacterium glutamicum* [165] *Escherichia coli* [166], *Fibrobacter succinogenes* [162], and *Mannheimia succiniciproducens* [165].

Borges *et al.* [156] optimized nutrients concentration of batch fermentation to produce succinic acid from sugarcane bagasse using *A. succinogenes* CIP 106512. They reported 22.5 g L^{-1} of succinc acid at optimized conditions. Shi *et al.* [160] reported an enhanced production of succinic acid with a concentration of 22.5 g L^{-1} and productivity of 0.42 g L^{-1} h^{-1} using immobilized cells of *Corynebacterium glutamicum* in batch fermentation. The study produced 35 g L^{-1} of glucose cassava bagasse. In another study, Salvach *et al.* [153] reported a yield of 0.69 g/g and a productivity of 0.43 g $L^{-1}h^{-1}$. In the study, *B. succiniciproducens* was used on corn stover in a 0.5 L reactor. Lo *et al.* [158] reported the production of 17.8 g L^{-1} of succinic acid by *Actinobacillus succinogenes* from 28.9 g L^{-1} of cellulosic glucose which was obtained by enzymatic hydrolysis of acid pretreated with sweet sorghum bagasse.

Although biological techniques for succinate production are considered an eco-friendly route as compared to petrochemical-derived succinate, yet it is not an economic option. This is due to some drawbacks, such as the high cost of feedstocks, low concentration of products, the co-production of low-value acid by-products and difficulties in the recovery of product [167]. Future work should be focused on the metabolic engineering to maximize product yield. To minimize the problems such as production inhibition, processes such as in-situ product removal can be performed.

Xylitol

Xylitol is a five-carbon sugar alcohol with its commercial uses in food, confectionary industries, different healthcare sectors and most specifically as an alternative sweetener for diabetic patients [168]. Xylitol has created an attractive global demand mainly due to its insulin-independent metabolism, anticariogenic properties, and pharmacological properties [169, 170]. Based on its utility as a building block, it can also be used as an important intermediate source and a heat-transferring agent for many other processes such as the production of polyester resins, PET bottling, hydraulic fluids, paintings, coatings and de-icing fluids used in aircraft [171]. Its market is tremendously rising and is estimated to be over US$ 340 million/year and priced at US$ 6-7 per kg [171].

Xylitol can be synthesized either by chemical hydrogenation or by enzymatic bio-transformation of purified substrates. The purified substrates are quite expensive due to some limitations and high purification cost [172]. The chemical method required for the manufacturing of xylitol is very laborious, energy consuming and cost-intensive [173]. However, it can also be produced by the microbial fermentation process from xylose sugar [174]. The plant biomass for xylitol production includes corn cob [175], wheat straw [176], corn waste [177], wheat bran [178] and miscanthus [179]. Xylitol can be produced by xylose-assimilating microorganisms such as *Candida parapsilosis* [180], *Candida tropicalism* [181], *Candida guilliermondii* [182], *Pachysolentannophilus* [183], *Corynebacterium* sp [184], *Petromyces albertensis* [185], *etc*. The biotechnological route for xylitol production offers several advantages as compared to the chemical route. These advantages include (1) Xylose purification and crystallization is not required as it is a biologically selective process; (2) It requires milder temperature and pressure conditions; (3) Some organic and inorganic impurities of the hemicellulosic hydrolysate are partially degraded or used as nutrients by the microorganism, making xylitol purification convenient and less expensive; (4) Exhaustion of xylose and glucose during the fermentation process favours xylitol purification, despite the presence of other contaminant sugars such as galactose and arabinose [186].

Xylitol is industrially produced by xylose reduction. The chemical process for the xylitol production comprise multiple steps. These include (i) Hydrolysis of plant-based biomass by acid pre-treatment; (ii) Purification and separation of the hydrolysate to obtain a pure form of xylose; (iii) Catalytic reduction of xylose to xylitol; and (iv) Crystallization and separation of xylitol [187]. The hydrolysate generally contains a variety of sugars such as xylose, arabinose, glucose, galactose and mannose in different proportions depending on the type of biomass used [188]. A two-step biotransformation step is involved in xylitol production from glucose. The first step involves glucose conversion into D-arabitol by yeast. In the second step, D-arabitol is consumed by *Gluconobacter* strains to produce xylitol [189]. The production of xylitol from D-arabitol depends on the oxidation of D-arabitol into xylulose by D-arabitol dehydrogenase. This is a high efficiency reaction (>75%) and the xylulose reduction to xylitol by xylitol dehydrogenase is a low efficiency reaction [190].

Strategies have been explored to overcome the limitations associated with a microbial conversion of xylitol into xylose. Metabolic pathway engineering can be enhanced for xylose transport by a heterologous expression of a transporting gene, changes in cofactor dependency and the use of enzyme technology to increase xylitol production [191, 192]. Dasgupta *et al.* [193] improved xylitol production in the native yeast strain of *Kluyveromyces marxianus* IIPE453 by the over-expression of an endogenous d-xylose reductase gene. They observed a 2.1-fold enhanced xylose reductase with a 1.62-fold increase in the overall xylitol yield in the modified strain. The study reported an excess of 58.62 ± 0.15 g xylitol production from 2 kg of fermentable sugars derived from sugarcane bagasse. Ethanol production by this strain was unaffected through this strain engineering approach. They observed a 1.6 fold increase in xylitol yield as compared to the native strain. In addition, an excess of 58.62 ± 0.15 g of xylitol production from 2 kg of fermentable sugars derived from sugarcane bagasse was recorded. Ko *et al.* [190] performed the gene disruption of xylitol dehydrogenase enzyme of *Candida tropicalis* for enhancement in xylitol production using D-xylose as a substrate and glycerol as a co-substrate. They observed xylitol volumetric productivity of 3.23 g L^{-1} h^{-1}, and specific productivity of 0.76 g g^{-1} h^{-1}. However, studies also reported that the use of engineered strains delivered some challenges which are needed to be overcome such as an increase in their metabolic stability.

Xylose

Xylose is a five-carbon sugar obtained from the hemicelluloses and is used as a substrate for the production of ethanol, furfural, xylitol, green surfactants and furan resins [194]. It has been used by several industries such as xylitol is used as a flavour, fragrance, food-beverage, and pet food [195]. Xylose is produced by industries in two forms: first as raw material grade and refine grade. Raw grade xylose is commonly used by xylitol and glycoside industry because of its low price. Whereas the refine grade xylose is mainly used by the flavour and fragrance industry, food and beverage industry and pet food industry and other industries [186]. According to the study, the current worldwide market for xylose is 250 million USD and it is expected that it would reach 270 million USD in 2024.

Xylose is commercially produced by the hydrolysis of xylan by diluted acid at a high temperature of 140-180°C [196]. Xylans are heteropolymers mainly consisting of xylose and arabinose which can be obtained from plant biomass rich in hemicelluloses, such as sawdust, sugarcane bagasse, rice straw, rice husk, and corn cobs [197]. The hydrolysis of hemicelluloses can be enhanced by using conventional acid catalysts. After the pretreatment and hydrolysis of plant-based biomass by enzymes, glucose and xylose are found as major sugars. Microorganisms can readily utilize glucose for the production of valuable products such as ethanol, lactic acid, acetic acid, *etc* [19]. However, xylose is leftover in the mother liquor, which can be recovered for other biotransformation processes. The utilization of xylose present in hydrolyzed liquor could improve the process economy.

Several process strategies have been proposed for efficient separation and recovery of xylose from the mother liquor. A xylose recovery method from the spent liquor was employed having 30-60% liquor, using a nano-filtration system to get xylose in the crystallized form with 85-95% of xylose [198]. A screw-steam-explosive extrusion process was employed for the pretreatment and hydrolysis of corncob at 1.55 MPa with 9 mg of sulphuric acid/g of corncob concentration for xylose recovery. In this process, about 3.575 kg of crystals xylose was produced and recovered from 22 kg of corncob [199]. Chen *et al.* [200] proposed a chromatography technique where a cation exchange resin (Amberlite IRP69 (Ca^{2+})) was used for the recovery of monosaccharide from pine biomass hydrolysate with 88% recovery of xylose. In another study, about 13.8 g of xylose was produced and recovered from 100 g of rice straw. In this process, rice straw was firstly pre-impregnated with nitric acid at with 0.5% HNO_3 and further hydrolysed by using 2.5% (w/v) NaOH [201]. Loow *et al.* [202] proposed a xylose recovery method from stalks of oil palm fronds using inorganic salts including $FeCl_3$ and $CuCl_2$. Table **5** shows the production parameters of some important biochemicals from plant-based biomass.

INHIBITOR FORMATION

Inhibitory compound formation is strongly based on the type of feedstock and the type of pretreatment used in the process. Pretreatment generates inhibitory products mostly due to severe conditions like long duration, pH, high temperature and pressure [217].Various inhibitors such as phenolics, furfural, 5-hydroxymethyl furfural, *etc.* are formed during pretreatment [39]. Inhibitors negatively affect the microbial cell growth and fermentation which reduce productivity and the yield of the end product. A small inhibitory compound easily penetrates through the cell membrane and disrupts the cell structure. Hydrolysis efficiency is reduced in the presence of inhibitors which induce protein precipitation. Phenolics with higher carbonyl content inhibit the cellulase hydrolysis activity.

Table 5. Production of different biochemicals from plant-based biomass after different pretreatments.

Product	Substrate	Microorganism	Pretreatment	Titer (g/L)	Yield (g/g)	Productivity (g/L/h)	Reference
Lactic acid	Sugarcane bagasse	*Lactobacillus pentosus*	Steam-Acid	72.75	0.61	1.01	[203]
	Sugarcane bagasse	*B. coagulans* DSM 2314	Acid-Steam explosion	70.4	0.9	1.14	[204]
	Sugarcane bagasse	*Bacillus* sp. P38	Acid-alkaline	185	0.99	1.93	[205]
	Corn stover	Engineered *Pediococcus Acidilactici*	Acid	104.4	0.72	1.45	[206]
	Corncob	*Lactobacillus pentosus* CECT 4023T	Acid	26	0.53	0.34	[207]
Xylitol	Sugarcane bagasse	*C. tropicalis*	Autohydrolysis	32	0.46	0.27	[208]
	Sugarcane bagasse	*Debaryomyceshansenii* NRRL Y-7426	Acid	10.54	0.71	0.22	[209]
	Rice straw	*C. tropicalis* JH030	Acid	31.1	0.71	0.44	[210]
	Corn cob	*C. tropicalis* W103	Acid	68.4	0.7	0.95	[211]
Succinic acid	Corn stover	*B. succiniciproducens* (CCUG57335)	Deacylated –ACID	30	0.69	0.43	[153]
	Cotton stalk hydrolysate	*A. succinogenes* 130ZT	Steam-Alkali-oxidation	63 0.	-	1.17	[212]
	Wheat bran hydrolysate	*C. crenatum*	Acid	43.6	1.03	4.36	[213]
	Wheat bran hydrolysate	*A. succinogenes*	Milling	62.1	1.02	0.91	[214]

(Table 5) cont.....

Product	Substrate	Microorganism	Pretreatment	Titer (g/L)	Yield (g/g)	Productivity (g/L/h)	Reference
Xylose	Sugarcane bagasse	-	Acid	-	0.96	-	[215]
	Oil palm empty fruit bunches	-	Acid	44.94	0.53	-	[216]

To overcome the inhibitory effect on fermentation process, either the hydrolysate needs to be detoxified or microorganisms can be adapted to these inhibitory compounds. Detoxification of lignocellulosic hydrolysate for the removal of inhibitors can be carried out using different methods to reduce the inhibitory effect on fermentation and enzyme hydrolysis. Detoxification of hydrolysates includes physicochemical methods (membrane filtration, evaporation, biochar adsorption, alkaline detoxification and ion exchange process) and biological methods (enzymatic and microbial methods) [23]. Phenolic inhibitors formed during pretreatment can be removed by washing, but it can lead to a 15% loss of sugars. Lignocellulosic hydrolysates which are in acidic conditions need to neutralize before fermentation. Several alkali solutions used for neutralization of hydrolysates are NH_4OH, $CaOH_2$ and $NaOH$. The main problems related to this processes are toxicity, cost and recycling of solvents [23, 39]. Biological detoxification methods are more substrate-specific and eco-friendly in nature. Laccase and peroxidase are widely used for enzymatic detoxification and they are distributed in plants, fungi, bacteria, algae and insects. Laccase activity depends on structural variations in plant-based biomass. Enzymes production cost is the main problem in enzymatic detoxification which can be minimized by binding them with the carrier.

COMMERCIAL PLANTS

The details of commercial plants based on the plant-derived feedstocks are mentioned in Table **6**.

Table 6. Details of commercial plants based on plant-derived feedstocks

S. No.	Product	Substrate	Plant Detail	Place
1	Ethanol	Corn stover	DuPont Danisco Cellulosic Ethanol plant	Nevada, Iowa, US
2	Biogas	Agricultural Residue	Verbio India Pvt. Ltd.	Village Bhutalkalan, Lehragaga, Sangrur, Punjab, India

(Table 6) cont.....

S. No.	Product	Substrate	Plant Detail	Place
3	Biogas	Agricultural Waste	Shri Govardhannathji Energies LLP	Nadiad, Kheda, Gujarat, India

CONCLUSION

The significance of various pre-treatments for the production of biofuels is discussed in this chapter. Pretreatment varies with the desired end-products like biogas, bioethanol, biobutanol, biohydrogen and other value-added products. As discussed in the chapter, pretreatment is the most energy-consuming and costlier process in biofuel production. Harsh operating conditions generate some inhibitors which affect the downstream processing of the product and also reduce the yield and productivity of the specific product. However, while choosing the pretreatment method, cost analysis study should be carried out to get a cost-effective process. Furthermore, proper selection of pretreatment is based on the energy requirement, solvent (chemical) recycling and effect on the environment. Due to high availability of plant-based biomass, the sustainability norms should comply with the pretreatment methods to signify eco-friendly vicinity with great industrial applications. As per the discussion reviewed about varied pretreatments, researchers have to keep in mind that the overall process for biofuel and biochemical production has to be cost-effective, energetically viable, environment-friendly and less inhibitory for the industrial avenues.

ACKNOWLEDGMENTS

This research was financially supported by the Department of Biotechnology (DBT, India) to VSI, Puneunder the Indo-UK Industrial Waste Challenge 2017 project. The authors are thankful to Mr. Shivajirao Deshmukh, Director General, VSI for providing necessary facilities to complete this work and his constant motivation.

REFERENCES

[1] Mahmood H, Moniruzzaman M, Iqbal T, Khan MJ. Recent advances in the pretreatment of lignocellulosic biomass for biofuels and value-added products. Curr Opin Green Sustain Chem 2019; 20: 18-24.
[http://dx.doi.org/10.1016/j.cogsc.2019.08.001]

[2] Kumar B, Bhardwaj N, Agrawal K, Chaturvedi V, Verma P. Current perspective on pretreatment technologies using lignocellulosic biomass: An emerging biorefinery concept. Fuel Process Technol 2020; 199106244.
[http://dx.doi.org/10.1016/j.fuproc.2019.106244.]

[3] Singh R, Kumar S. A review on biomethane potential of paddy straw and diverse prospects to enhance its biodigestibility. J Clean Prod 2019; 217: 295-307.
[http://dx.doi.org/10.1016/j.jclepro.2019.01.207]

[4]   Hans M, Kumar S, Chandel AK, Polikarpov I. A review on bioprocessing of paddy straw to ethanol using simultaneous saccharification and fermentation. Process Biochem 2019; 85: 125-34.
        [http://dx.doi.org/10.1016/j.procbio.2019.06.019]

[5]   Raud M, Kikas T, Sippula O, Shurpali NJ. Potentials and challenges in lignocellulosic biofuel production technology. Renew Sustain Energy Rev 2019; 111: 44-56.
        [http://dx.doi.org/10.1016/j.rser.2019.05.020]

[6]   Moretti MMS, Perrone OM, Nunes CCC, *et al.* Effect of pretreatment and enzymatic hydrolysis on the physical-chemical composition and morphologic structure of sugarcane bagasse and sugarcane straw. Bioresour Technol 2016; 219: 773-7.
        [http://dx.doi.org/10.1016/j.biortech.2016.08.075] [PMID: 27578061]

[7]   Kaur K, Phutela UG. Enhancement of paddy straw digestibility and biogas production by sodium hydroxide-microwave pretreatment. Renew Energy 2016; 92: 178-84.
        [http://dx.doi.org/10.1016/j.renene.2016.01.083]

[8]   Liu L, Zhang Z, Wang J, *et al.* Simultaneous saccharification and co-fermentation of corn stover pretreated by H_2O_2 oxidative degradation for ethanol production. Energy 2019; 168: 946-52.
        [http://dx.doi.org/10.1016/j.energy.2018.11.132]

[9]   Tsegaye B, Balomajumder C, Roy P. Optimization of microwave and NaOH pretreatments of wheat straw for enhancing biofuel yield. Energy Convers Manage 2019; 186: 82-92.
        [http://dx.doi.org/10.1016/j.enconman.2019.02.049]

[10]   Wang Z, He X, Yan L, *et al.* Enhancing enzymatic hydrolysis of corn stover by twin-screw extrusion pretreatment. Ind Crops Prod 2020; 143111960
        [http://dx.doi.org/10.1016/j.indcrop.2019.111960]

[11]   Yuan H, Song X, Guan R, Zhang L, Li X, Zuo X. Effect of low severity hydrothermal pretreatment on anaerobic digestion performance of corn stover. Bioresour Technol 2019; 294122238
        [http://dx.doi.org/10.1016/j.biortech.2019.122238.] [PMID: 31610486]

[12]   Adeeyo O, Oresegun OM, Oladimeji TE. Compositional analysis of lignocellulosic materials: Evaluation of an economically viable method suitable for woody and non-woody biomass. Am J Eng Res 2015; 4(4): 14-9. [AJER].

[13]   Kumar AK, Sharma S. Recent updates on different methods of pretreatment of lignocellulosic feedstocks: a review. Bioresour Bioprocess 2017; 4(1): 7.
        [http://dx.doi.org/10.1186/s40643-017-0137-9] [PMID: 28163994]

[14]   Gaurav N, Sivasankari S, Kiran GS, Ninawe A, Selvin J. Utilization of bioresources for sustainable biofuels: A Review. Renew Sustain Energy Rev 2017; 73: 205-14.
        [http://dx.doi.org/10.1016/j.rser.2017.01.070]

[15]   Hou W, Kan J, Bao J. Rheology evolution of high solids content and highly viscous lignocellulose system in biorefinery fermentations for production of biofuels and biochemicals. Fuel 2019; 253: 1565-9.
        [http://dx.doi.org/10.1016/j.fuel.2019.05.136]

[16]   Payne CE, Wolfrum EJ. Rapid analysis of composition and reactivity in cellulosic biomass feedstocks with near-infrared spectroscopy. Biotechnol Biofuels 2015; 8(1): 43.
        [http://dx.doi.org/10.1186/s13068-015-0222-2] [PMID: 25834638]

[17]   Karimi K, Taherzadeh MJ. A critical review of analytical methods in pretreatment of lignocelluloses: Composition, imaging, and crystallinity. Bioresour Technol 2016; 200: 1008-18.
        [http://dx.doi.org/10.1016/j.biortech.2015.11.022] [PMID: 26614225]

[18]   Maurya DP, Singla A, Negi S. An overview of key pretreatment processes for biological conversion of lignocellulosic biomass to bioethanol. 2015.
        [http://dx.doi.org/10.1007/s13205-015-0279-4]

[19]   Kucharska K, Rybarczyk P, Hołowacz I, Łukajtis R, Glinka M, Kamiński M. Pretreatment of lignocellulosic materials as substrates for fermentation processes. Molecules 2018; 23(11): 2937. [http://dx.doi.org/10.3390/molecules23112937] [PMID: 30423814]

[20]   Ferreira JA, Taherzadeh MJ. Improving the economy of lignocellulose-based biorefineries with organosolv pretreatment. Bioresour Technol 2020; 299122695 [http://dx.doi.org/10.1016/j.biortech.2019.122695.] [PMID: 31918973]

[21]   Islam MK, Wang H, Rehman S, *et al.* Sustainability metrics of pretreatment processes in a waste derived lignocellulosic biomass biorefinery. Bioresour Technol 2020; 298122558 [http://dx.doi.org/10.1016/j.biortech.2019.122558.] [PMID: 31862395]

[22]   Yan HL, Li Z-K, Wang Z-C, *et al.* Characterization of soluble portions from cellulose, hemicellulose, and lignin methanolysis. Fuel 2019; 246: 394-401. [http://dx.doi.org/10.1016/j.fuel.2019.03.019]

[23]   Bhatia SK, Jagtap SS, Bedekar AA, *et al.* Recent developments in pretreatment technologies on lignocellulosic biomass: Effect of key parameters, technological improvements, and challenges. Bioresour Technol 2020; 300122724 [http://dx.doi.org/10.1016/j.biortech.2019.122724.] [PMID: 31926792]

[24]   Sankaran R, Parra Cruz RA, Pakalapati H, *et al.* Recent advances in the pretreatment of microalgal and lignocellulosic biomass: A comprehensive review. Bioresour Technol 2020; 298122476 [http://dx.doi.org/10.1016/j.biortech.2019.122476.] [PMID: 31810736]

[25]   Awasthi SK, Sarsaiya S, Kumar V, *et al.* Processing of municipal solid waste resources for a circular economy in China: An overview. Fuel 2022; 317123478 [http://dx.doi.org/10.1016/j.fuel.2022.123478.]

[26]   Sharma HK, Xu C, Qin W. Biological pretreatment of lignocellulosic biomass for biofuels and bioproducts: an overview. Waste Biomass Valoriz 2019; 10(2): 235-51. [http://dx.doi.org/10.1007/s12649-017-0059-y]

[27]   Rooni V, Raud M, Kikas T. Technical solutions used in different pretreatments of lignocellulosic biomass: a review. Agron Res (Tartu) 2017; 15(3): 848-58.

[28]   Zhang M, Song X, Deines T W, Pei Z J, Wang D. Biofuel manufacturing from woody biomass: effects of sieve size used in biomass size reduction, Journal of Biomedicine and Biotechnology 2012; 2012

[29]   Luo J, Fang Z, Smith RL Jr. Ultrasound-enhanced conversion of biomass to biofuels. Pror Energy Combust Sci 2014; 41: 56-93. [http://dx.doi.org/10.1016/j.pecs.2013.11.001]

[30]   Shirkavand E, Baroutian S, Gapes DJ, Young BR. Combination of fungal and physicochemical processes for lignocellulosic biomass pretreatment – A review. Renew Sustain Energy Rev 2016; 54: 217-34. [http://dx.doi.org/10.1016/j.rser.2015.10.003]

[31]   Kunaver M, Jasiukaitytė E, Čuk N. Ultrasonically assisted liquefaction of lignocellulosic materials. Bioresour Technol 2012; 103(1): 360-6. [http://dx.doi.org/10.1016/j.biortech.2011.09.051] [PMID: 22029956]

[32]   Esfahani MR, Azin M. Pretreatment of sugarcane bagasse by ultrasound energy and dilute acid. Asia-Pac J Chem Eng 2012; 7(2): 274-8. [http://dx.doi.org/10.1002/apj.533]

[33]   Hay JXW, Wu TY, Juan JC, Jahim JM. Improved biohydrogen production and treatment of pulp and paper mill effluent through ultrasonication pretreatment of wastewater. Energy Convers Manage 2015; 106: 576-83. [http://dx.doi.org/10.1016/j.enconman.2015.08.040]

[34]   Yu T, Deng Y, Liu H, *et al.* Effect of alkaline microwaving pretreatment on anaerobic digestion and

biogas production of swine manure. Sci Rep 2017; 7(1): 1668.
[http://dx.doi.org/10.1038/s41598-017-01706-3] [PMID: 28490754]

[35] Xu R, Zhang K, Liu P, *et al.* Lignin depolymerization and utilization by bacteria. Bioresour Technol 2018; 269: 557-66.
[http://dx.doi.org/10.1016/j.biortech.2018.08.118] [PMID: 30219494]

[36] Bichot A, Lerosty M, Radoiu M, *et al.* Decoupling thermal and non-thermal effects of the microwaves for lignocellulosic biomass pretreatment. Energy Convers Manage 2020; 203112220
[http://dx.doi.org/10.1016/j.enconman.2019.112220.]

[37] Keshwani DR, Cheng JJ. Microwave-based alkali pretreatment of switchgrass and coastal bermudagrass for bioethanol production. Biotechnol Prog 2010; 26(3): 644-52.
[http://dx.doi.org/10.1002/btpr.371] [PMID: 20039265]

[38] Li Z, Jiang Z, Yu Y, Cai Z. Effective of microwave-KOH pretreatment on enzymatic hydrolysis of bamboo," Journal of Sustainable Bioenergy Systems, Volume 2, 2012; pp 104-107 2012; 2: 104-7.
[http://dx.doi.org/10.4236/jsbs.2012.24015]

[39] Behera S, Arora R, Nandhagopal N, Kumar S. Importance of chemical pretreatment for bioconversion of lignocellulosic biomass. Renew Sustain Energy Rev 2014; 36: 91-106.
[http://dx.doi.org/10.1016/j.rser.2014.04.047]

[40] Sabiha-Hanim S, Halim NAA. Sugarcane bagasse pretreatment methods for ethanol production.Fuel Ethanol Production from Sugarcane. IntechOpen 2018.

[41] Sabiha-Hanim S, Abd Halim NA. Sugarcane bagasse pretreatment methods for ethanol production.Fuel Ethanol Production from Sugarcane. IntechOpen 2018.

[42] de Vasconcelos SM, Santos AMP, Rocha GJM, Souto-Maior AM. Diluted phosphoric acid pretreatment for production of fermentable sugars in a sugarcane-based biorefinery. Bioresour Technol 2013; 135: 46-52.
[http://dx.doi.org/10.1016/j.biortech.2012.10.083] [PMID: 23186685]

[43] Jackson de Moraes Rocha G, Martin C, Soares IB, Souto Maior AM, Baudel HM, Moraes de Abreu CA. Dilute mixed-acid pretreatment of sugarcane bagasse for ethanol production. Biomass Bioenergy 2011; 35(1): 663-70.
[http://dx.doi.org/10.1016/j.biombioe.2010.10.018]

[44] Singh DP, Trivedi RK. Acid and Alkaline pretreatment of lignocellulosic biomass to produce ethanol as biofuel. Int J Chemtech Res 2013; 5(2): 727-34.

[45] Sivagurunathan P, *et al.* A critical review on issues and overcoming strategies for the enhancement of dark fermentative hydrogen production in continuous systems. 2016.
[http://dx.doi.org/10.1016/j.ijhydene.2015.12.081]

[46] Lorenci Woiciechowski A, Dalmas Neto CJ, Porto de Souza Vandenberghe L, *et al.* Lignocellulosic biomass: Acid and alkaline pretreatments and their effects on biomass recalcitrance – Conventional processing and recent advances. Bioresour Technol 2020; 304122848
[http://dx.doi.org/10.1016/j.biortech.2020.122848.]

[47] Kim JS, Lee YY, Kim TH. A review on alkaline pretreatment technology for bioconversion of lignocellulosic biomass. Bioresour Technol 2016; 199: 42-8.
[http://dx.doi.org/10.1016/j.biortech.2015.08.085] [PMID: 26341010]

[48] Wunna K, Nakasaki K, Auresenia J, Abella L, Gaspillo P. Effect of alkali pretreatment on removal of lignin from sugarcane bagasse. Chem Eng Trans 2017; 56: 1831-6.

[49] Boboescu IZ, Damay J, Chang JKW, *et al.* Ethanol production from residual lignocellulosic fibers generated through the steam treatment of whole sorghum biomass. Bioresour Technol 2019; 292121975
[http://dx.doi.org/10.1016/j.biortech.2019.121975.] [PMID: 31445238]

[50]   Walker DJ, Gallagher J, Winters A, Somani A, Ravella SR, Bryant DN. Process optimization of steam explosion parameters on multiple lignocellulosic biomass using Taguchi method—a critical appraisal. Front Energy Res 2018; 6: 46.
[http://dx.doi.org/10.3389/fenrg.2018.00046]

[51]   Hernández-Beltrán JU, Hernández-De Lira IO, Cruz-Santos MM, Saucedo-Luevanos A, Hernández-Terán F, Balagurusamy N. Insight into pretreatment methods of lignocellulosic biomass to increase biogas yield: current state, challenges, and opportunities. Appl Sci (Basel) 2019; 9(18): 3721.
[http://dx.doi.org/10.3390/app9183721]

[52]   Blümmel M, Teymouri F, Moore J, *et al.* Ammonia Fiber Expansion (AFEX) as spin off technology from 2nd generation biofuel for upgrading cereal straws and stovers for livestock feed. Anim Feed Sci Technol 2018; 236: 178-86.
[http://dx.doi.org/10.1016/j.anifeedsci.2017.12.016]

[53]   Park CY, Ryu YW, Kim C. Kinetics and rate of enzymatic hydrolysis of cellulose in supercritical carbon dioxide. Korean J Chem Eng 2001; 18(4): 475-8.
[http://dx.doi.org/10.1007/BF02698293]

[54]   Bayer E, Shoham Y, Lamed R. Lignocellulose-decomposing bacteria and their enzyme systems. 2013.
[http://dx.doi.org/10.1007/978-3-642-30141-4_67]

[55]   Tu WC, Hallett JP. Recent advances in the pretreatment of lignocellulosic biomass. Curr Opin Green Sustain Chem 2019; 20: 11-7.
[http://dx.doi.org/10.1016/j.cogsc.2019.07.004]

[56]   Tuncer M, Kuru A, Isikli M, Sahin N, Celenk FG. Optimization of extracellular endoxylanase, endoglucanase and peroxidase production by Streptomyces sp. F2621 isolated in Turkey. J Appl Microbiol 2004; 97(4): 783-91.
[http://dx.doi.org/10.1111/j.1365-2672.2004.02361.x] [PMID: 15357728]

[57]   Odier E, Janin G, Monties B. Poplar lignin decomposition by gram-negative aerobic bacteria. Appl Environ Microbiol 1981; 41(2): 337-41.
[http://dx.doi.org/10.1128/aem.41.2.337-341.1981] [PMID: 16345706]

[58]   Salvachúa D, Karp EM, Nimlos CT, Vardon DR, Beckham GT. Towards lignin consolidated bioprocessing: simultaneous lignin depolymerization and product generation by bacteria. Green Chem 2015; 17(11): 4951-67.
[http://dx.doi.org/10.1039/C5GC01165E]

[59]   Miron J, Ben-Ghedalia D, Morrison M. Invited review: adhesion mechanisms of rumen cellulolytic bacteria. J Dairy Sci 2001; 84(6): 1294-309.
[http://dx.doi.org/10.3168/jds.S0022-0302(01)70159-2] [PMID: 11417686]

[60]   Vallinayagam S, *et al.* Recent developments in magnetic nanoparticles and nano-composites for wastewater treatment. J Environ Chem Eng 2021; 9(6)106553
[http://dx.doi.org/10.1016/j.jece.2021.106553.]

[61]   Lakkaboyana SK, Khantong S, Asmel NK, *et al.* Indonesian Kaolin supported nZVI (IK-nZVI) used for the an efficient removal of Pb(II) from aqueous solutions: Kinetics, thermodynamics and mechanism. J Environ Chem Eng 2021; 9(6)106483
[http://dx.doi.org/10.1016/j.jece.2021.106483.]

[62]   Lakkaboyana SK, Soontarapa K, Asmel NK, *et al.* Synthesis and characterization of Cu(OH)2-NW--PVA-AC Nano-composite and its use as an efficient adsorbent for removal of methylene blue. Sci Rep 2021; 11(1): 5686.
[http://dx.doi.org/10.1038/s41598-021-84797-3] [PMID: 33414495]

[63]   Lakkaboyana SK, Soontarapa K, Vinaykumar , Marella RK, Kannan K. Preparation of novel chitosan polymeric nanocomposite as an efficient material for the removal of Acid Blue 25 from aqueous environment. Int J Biol Macromol 2021; 168: 760-8.

[http://dx.doi.org/10.1016/j.ijbiomac.2020.11.133] [PMID: 33232701]

[64]   Zabed HM, Akter S, Yun J, *et al.* Recent advances in biological pretreatment of microalgae and lignocellulosic biomass for biofuel production. Renew Sustain Energy Rev 2019; 105: 105-28.
[http://dx.doi.org/10.1016/j.rser.2019.01.048]

[65]   Troiano D, Orsat V, Dumont MJ. Status of filamentous fungi in integrated biorefineries. Renew Sustain Energy Rev 2020; 117109472
[http://dx.doi.org/10.1016/j.rser.2019.109472.]

[66]   Zhanga F, *et al.* Regulation and production of lignocellulolytic enzymes from Trichoderma reesei for biofuels production. Advances in Bioenergy 2019; p. 79.
[http://dx.doi.org/10.1016/bs.aibe.2019.03.001]

[67]   Ayodele BV, Alsaffar MA, Mustapa SI. An overview of integration opportunities for sustainable bioethanol production from first-and second-generation sugar-based feedstocks. J Clean Prod 2019.118857

[68]   Mithra MG, Jeeva ML, Sajeev MS, Padmaja G. Comparison of ethanol yield from pretreated lignocellulo-starch biomass under fed-batch SHF or SSF modes. Heliyon 2018; 4(10)e00885
[http://dx.doi.org/10.1016/j.heliyon.2018.e00885.] [PMID: 30417150]

[69]   Saini JK, Saini R, Tewari L. Lignocellulosic agriculture wastes as biomass feedstocks for second-generation bioethanol production: concepts and recent developments. 2015.

[70]   Takano M, Hoshino K. Bioethanol production from rice straw by simultaneous saccharification and fermentation with statistical optimized cellulase cocktail and fermenting fungus. Bioresour Bioprocess 2018; 5(1): 16.
[http://dx.doi.org/10.1186/s40643-018-0203-y]

[71]   Wang L, Littlewood J, Murphy RJ. Environmental sustainability of bioethanol production from wheat straw in the UK. Renew Sustain Energy Rev 2013; 28: 715-25.
[http://dx.doi.org/10.1016/j.rser.2013.08.031]

[72]   Li WC, Zhang S-J, Xu T, *et al.* Fractionation of corn stover by two-step pretreatment for production of ethanol, furfural, and lignin. Energy 2020; 195117076
[http://dx.doi.org/10.1016/j.energy.2020.117076.]

[73]   Singh S, Chakravarty I, Pandey KD, Kundu S. Development of a process model for simultaneous saccharification and fermentation (SSF) of algal starch to third-generation bioethanol. Biofuels 2020; 11(7): 847-55.

[74]   Awasthi MK, Kumar V, Yadav V, *et al.* Current state of the art biotechnological strategies for conversion of watermelon wastes residues to biopolymers production: A review. Chemosphere 2022; 290133310
[http://dx.doi.org/10.1016/j.chemosphere.2021.133310.] [PMID: 34919909]

[75]   Kumar V, Sharma N, Umesh M, *et al.* Emerging challenges for the agro-industrial food waste utilization: A review on food waste biorefinery. Bioresour Technol 2022; 362127790
[http://dx.doi.org/10.1016/j.biortech.2022.127790.] [PMID: 35973569]

[76]   Liu H, Kumar V, Jia L, *et al.* Biopolymer poly-hydroxyalkanoates (PHA) production from apple industrial waste residues: A review. Chemosphere 2021; 284131427
[http://dx.doi.org/10.1016/j.chemosphere.2021.131427.] [PMID: 34323796]

[77]   Awasthi SK, Kumar M, Kumar V, *et al.* A comprehensive review on recent advancements in biodegradation and sustainable management of biopolymers. Environ Pollut 2022; 307119600
[http://dx.doi.org/10.1016/j.envpol.2022.119600.] [PMID: 35691442]

[78]   Saravanan AP, Mathimani T, Deviram G, Rajendran K, Pugazhendhi A. Biofuel policy in India: A review of policy barriers in sustainable marketing of biofuel. J Clean Prod 2018; 193: 734-47.
[http://dx.doi.org/10.1016/j.jclepro.2018.05.033]

[79]   Singh S, Chakravarty I, Kundu S. Mathematical modelling of bioethanol production from algal starch hydrolysate by Saccharomyces cerevisiae. Cell Mol Biol 2017; 63(6): 83-7.
       [http://dx.doi.org/10.14715/cmb/2017.63.6.17] [PMID: 28968215]

[80]   Aditiya HB, Mahlia TMI, Chong WT, Nur H, Sebayang AH. Second generation bioethanol production: A critical review. Renew Sustain Energy Rev 2016; 66: 631-53.
       [http://dx.doi.org/10.1016/j.rser.2016.07.015]

[81]   Liu Y, Xu JX, Zhang Y, *et al.* Improved ethanol production based on high solids fed-batch simultaneous saccharification and fermentation with alkali-pretreated sugarcane bagasse. BioResources 2016; 11(1): 2548-56.
       [http://dx.doi.org/10.15376/biores.11.1.2548-2556]

[82]   de Araujo Guilherme A, Dantas PVF, Padilha CEA, dos Santos ES, de Macedo GR. Ethanol production from sugarcane bagasse: Use of different fermentation strategies to enhance an environmental-friendly process. J Environ Manage 2019; 234: 44-51.
       [http://dx.doi.org/10.1016/j.jenvman.2018.12.102] [PMID: 30599329]

[83]   Terán Hilares R, Dionízio RM, Sánchez Muñoz S, *et al.* Hydrodynamic cavitation-assisted continuous pre-treatment of sugarcane bagasse for ethanol production: Effects of geometric parameters of the cavitation device. Ultrason Sonochem 2020; 63104931
       [http://dx.doi.org/10.1016/j.ultsonch.2019.104931.] [PMID: 31945566]

[84]   Yuan Z, Li G, Hegg EL. Enhancement of sugar recovery and ethanol production from wheat straw through alkaline pre-extraction followed by steam pretreatment. Bioresour Technol 2018; 266: 194-202.
       [http://dx.doi.org/10.1016/j.biortech.2018.06.065] [PMID: 29982039]

[85]   Saha BC, Nichols NN, Qureshi N, Kennedy GJ, Iten LB, Cotta MA. Pilot scale conversion of wheat straw to ethanol *via* simultaneous saccharification and fermentation. Bioresour Technol 2015; 175: 17-22.
       [http://dx.doi.org/10.1016/j.biortech.2014.10.060] [PMID: 25459799]

[86]   Qiu J, Ma L, Shen F, *et al.* Pretreating wheat straw by phosphoric acid plus hydrogen peroxide for enzymatic saccharification and ethanol production at high solid loading. Bioresour Technol 2017; 238: 174-81.
       [http://dx.doi.org/10.1016/j.biortech.2017.04.040] [PMID: 28433905]

[87]   Molaverdi M, Karimi K, Mirmohamadsadeghi S. Improvement of dry simultaneous saccharification and fermentation of rice straw to high concentration ethanol by sodium carbonate pretreatment. Energy 2019; 167: 654-60.
       [http://dx.doi.org/10.1016/j.energy.2018.11.017]

[88]   Kumar AK, Parikh BS, Shah E, Liu LZ, Cotta MA. Cellulosic ethanol production from green solvent-pretreated rice straw. Biocatal Agric Biotechnol 2016; 7: 14-23.
       [http://dx.doi.org/10.1016/j.bcab.2016.04.008]

[89]   Karimi K, Tabatabaei M, Sárvári Horváth I, Kumar R. Recent trends in acetone, butanol, and ethanol (ABE) production. Biofuel Research Journal 2015; 2(4): 301-8.
       [http://dx.doi.org/10.18331/BRJ2015.2.4.4]

[90]   Dürre P. Fermentative production of butanol—the academic perspective. Curr Opin Biotechnol 2011; 22(3): 331-6.
       [http://dx.doi.org/10.1016/j.copbio.2011.04.010] [PMID: 21565485]

[91]   Behera S, Sharma NK, Kumar S. Prospects of solvent tolerance in butanol fermenting bacteria.Biorefining of biomass to biofuels. Springer 2018; pp. 249-64.
       [http://dx.doi.org/10.1007/978-3-319-67678-4_11]

[92]   Behera S, Kumar S. Potential and Prospects of Biobutanol Production from Agricultural Residues. Liquid Biofuel Production 2019; pp. 285-318.

[http://dx.doi.org/10.1002/9781119459866.ch9]

[93]   Amiri H, Karimi K, Zilouei H. Organosolv pretreatment of rice straw for efficient acetone, butanol, and ethanol production. Bioresour Technol 2014; 152: 450-6.
[http://dx.doi.org/10.1016/j.biortech.2013.11.038] [PMID: 24321608]

[94]   Li H, Xiong L, Chen X, *et al.* Enhanced enzymatic hydrolysis and acetone-butanol-ethanol fermentation of sugarcane bagasse by combined diluted acid with oxidate ammonolysis pretreatment. Bioresour Technol 2017; 228: 257-63.
[http://dx.doi.org/10.1016/j.biortech.2016.12.119] [PMID: 28081523]

[95]   Qi G, Huang D, Wang J, Shen Y, Gao X. Enhanced butanol production from ammonium sulfite pretreated wheat straw by separate hydrolysis and fermentation and simultaneous saccharification and fermentation. Sustain Energy Technol Assess 2019; 36100549.
[http://dx.doi.org/10.1016/j.seta.2019.100549]

[96]   Qureshi N, Cotta MA, Saha BC. Bioconversion of barley straw and corn stover to butanol (a biofuel) in integrated fermentation and simultaneous product recovery bioreactors. Food Bioprod Process 2014; 92(3): 298-308.
[http://dx.doi.org/10.1016/j.fbp.2013.11.005]

[97]   Li J, Du Y, Bao T, *et al.* n-Butanol production from lignocellulosic biomass hydrolysates without detoxification by Clostridium tyrobutyricum Δack-adhE2 in a fibrous-bed bioreactor. Bioresour Technol 2019; 289121749
[http://dx.doi.org/10.1016/j.biortech.2019.121749] [PMID: 31323711]

[98]   Pang Z-W, Lu W, Zhang H, *et al.* Butanol production employing fed-batch fermentation by Clostridium acetobutylicum GX01 using alkali-pretreated sugarcane bagasse hydrolysed by enzymes from Thermoascus aurantiacus QS 7-2-4. Bioresour Technol 2016; 212: 82-91.
[http://dx.doi.org/10.1016/j.biortech.2016.04.013] [PMID: 27089425]

[99]   Zetty-Arenas AM, Alves RF, Portela CAF, *et al.* Towards enhanced n-butanol production from sugarcane bagasse hemicellulosic hydrolysate: Strain screening, and the effects of sugar concentration and butanol tolerance. Biomass Bioenergy 2019; 126: 190-8.
[http://dx.doi.org/10.1016/j.biombioe.2019.05.011]

[100] Xing W, Xu G, Dong J, Han R, Ni Y. Novel dihydrogen-bonding deep eutectic solvents: Pretreatment of rice straw for butanol fermentation featuring enzyme recycling and high solvent yield. Chem Eng J 2018; 333: 712-20.
[http://dx.doi.org/10.1016/j.cej.2017.09.176]

[101] Dalal J, Das M, Joy S, Yama M, Rawat J. Efficient isopropanol-butanol (IB) fermentation of rice straw hydrolysate by a newly isolated Clostridium beijerinckii strain C-01. Biomass Bioenergy 2019; 127105292.
[http://dx.doi.org/10.1016/j.biombioe.2019.105292]

[102] Kiyoshi K, Furukawa M, Seyama T, Kadokura T, Nakazato A, Nakayama S. Butanol production from alkali-pretreated rice straw by co-culture of Clostridium thermocellum and Clostridium saccharoperbutylacetonicum. Bioresour Technol 2015; 186: 325-8.
[http://dx.doi.org/10.1016/j.biortech.2015.03.061] [PMID: 25818258]

[103] Hosseini Koupaie E, Dahadha S, Bazyar Lakeh AA, Azizi A, Elbeshbishy E. Enzymatic pretreatment of lignocellulosic biomass for enhanced biomethane production-A review. J Environ Manage 2019; 233: 774-84.
[http://dx.doi.org/10.1016/j.jenvman.2018.09.106] [PMID: 30314871]

[104] Barrios-Pérez J, Sepúlveda-Gálvez A, Carrillo-Reyes J, Buitrón-Méndez G, Vargas-Casillas A. Effect of the variation of the operating parameters in the production of methane from lignocellulosic waste. IFAC-PapersOnLine 2018; 51(13): 639-43.
[http://dx.doi.org/10.1016/j.ifacol.2018.07.352]

[105] Mokomele T, da Costa Sousa L, Balan V, van Rensburg E, Dale BE, Görgens JF. Incorporating

anaerobic co-digestion of steam exploded or ammonia fiber expansion pretreated sugarcane residues with manure into a sugarcane-based bioenergy-livestock nexus. Bioresour Technol 2019; 272: 326-36.
[http://dx.doi.org/10.1016/j.biortech.2018.10.049] [PMID: 30384207]

[106] Tsapekos P, Kougias PG, Angelidaki I. Mechanical pretreatment for increased biogas production from lignocellulosic biomass; predicting the methane yield from structural plant components. Waste Manag 2018; 78: 903-10.
[http://dx.doi.org/10.1016/j.wasman.2018.07.017] [PMID: 32559985]

[107] Abraham A, Mathew AK, Park H, *et al.* Pretreatment strategies for enhanced biogas production from lignocellulosic biomass. Bioresour Technol 2020; 301122725
[http://dx.doi.org/10.1016/j.biortech.2019.122725.] [PMID: 31958690]

[108] Edwiges T, Bastos JA, Lima Alino JH, d'avila L, Frare LM, Somer JG. Comparison of various pretreatment techniques to enhance biodegradability of lignocellulosic biomass for methane production. J Environ Chem Eng 2019; 7(6)103495
[http://dx.doi.org/10.1016/j.jece.2019.103495.]

[109] Eskicioglu C, Monlau F, Barakat A, *et al.* Assessment of hydrothermal pretreatment of various lignocellulosic biomass with CO_2 catalyst for enhanced methane and hydrogen production. Water Res 2017; 120: 32-42.
[http://dx.doi.org/10.1016/j.watres.2017.04.068] [PMID: 28478293]

[110] Sajad Hashemi S, Karimi K, Majid Karimi A. Ethanolic ammonia pretreatment for efficient biogas production from sugarcane bagasse. Fuel 2019; 248: 196-204.
[http://dx.doi.org/10.1016/j.fuel.2019.03.080]

[111] Mancini G, Papirio S, Lens PNL, Esposito G. Increased biogas production from wheat straw by chemical pretreatments. Renew Energy 2018; 119: 608-14.
[http://dx.doi.org/10.1016/j.renene.2017.12.045]

[112] Rajput AA, Zeshan , Visvanathan C. Effect of thermal pretreatment on chemical composition, physical structure and biogas production kinetics of wheat straw. J Environ Manage 2018; 221: 45-52.
[http://dx.doi.org/10.1016/j.jenvman.2018.05.011] [PMID: 29793209]

[113] Dell'Omo P, Froscia S. Enhancing anaerobic digestion of wheat straw through multistage milling. Modelling, Measurement and Control C 2018; 79(3): 127-32.
[http://dx.doi.org/10.18280/mmc_c.790310]

[114] Chandra R, Takeuchi H, Hasegawa T, Vijay V. Experimental evaluation of substrate's particle size of wheat and rice straw biomass on methane production yield. Agric Eng Int CIGR J 2015; 17(2)

[115] Luo T, Huang H, Mei Z, *et al.* Hydrothermal pretreatment of rice straw at relatively lower temperature to improve biogas production *via* anaerobic digestion. Chin Chem Lett 2019; 30(6): 1219-23.
[http://dx.doi.org/10.1016/j.cclet.2019.03.018]

[116] Lizasoain J, Rincón M, Theuretzbacher F, *et al.* Biogas production from reed biomass: Effect of pretreatment using different steam explosion conditions. Biomass Bioenergy 2016; 95: 84-91.
[http://dx.doi.org/10.1016/j.biombioe.2016.09.021]

[117] Ali SS, Abomohra AEF, Sun J. Effective bio-pretreatment of sawdust waste with a novel microbial consortium for enhanced biomethanation. Bioresour Technol 2017; 238: 425-32.
[http://dx.doi.org/10.1016/j.biortech.2017.03.187] [PMID: 28458176]

[118] Pellera FM, Gidarakos E. Chemical pretreatment of lignocellulosic agroindustrial waste for methane production. Waste Manag 2018; 71: 689-703.
[http://dx.doi.org/10.1016/j.wasman.2017.04.038] [PMID: 28456458]

[119] Lizasoain J, Trulea A, Gittinger J, *et al.* Corn stover for biogas production: Effect of steam explosion pretreatment on the gas yields and on the biodegradation kinetics of the primary structural compounds. Bioresour Technol 2017; 244(Pt 1): 949-56.
[http://dx.doi.org/10.1016/j.biortech.2017.08.042] [PMID: 28847085]

[120]   Mustafa AM, Li H, Radwan AA, Sheng K, Chen X. Effect of hydrothermal and Ca(OH)$_2$ pretreatments on anaerobic digestion of sugarcane bagasse for biogas production. Bioresour Technol 2018; 259: 54-60.
[http://dx.doi.org/10.1016/j.biortech.2018.03.028] [PMID: 29536874]

[121]   Sindhu R, Binod P, Pandey A, Gnansounou E. Agroresidue-based biorefineries.Refining Biomass Residues for Sustainable Energy and Bioproducts. Elsevier 2020; pp. 243-58.
[http://dx.doi.org/10.1016/B978-0-12-818996-2.00011-9]

[122]   Sun Y, He J, Yang G, Sun G, Sage V. A review of the enhancement of bio-hydrogen generation by chemicals addition. Catalysts 2019; 9(4): 353.
[http://dx.doi.org/10.3390/catal9040353]

[123]   Rittmann S, Herwig C. A comprehensive and quantitative review of dark fermentative biohydrogen production. Microb Cell Fact 2012; 11(1): 115.
[http://dx.doi.org/10.1186/1475-2859-11-115] [PMID: 22925149]

[124]   Ntaikou I, Antonopoulou G, Lyberatos G. Biohydrogen production from biomass and wastes *via* dark fermentation: a review. Waste Biomass Valoriz 2010; 1(1): 21-39.
[http://dx.doi.org/10.1007/s12649-009-9001-2]

[125]   Moodley P, Kana EG. Comparative study of three optimized acid-based pretreatments for sugar recovery from sugarcane leaf waste: A sustainable feedstock for biohydrogen production. 2018.

[126]   Gonzales RR, Kumar G, Sivagurunathan P, Kim S-H. Enhancement of hydrogen production by optimization of pH adjustment and separation conditions following dilute acid pretreatment of lignocellulosic biomass. 2017.
[http://dx.doi.org/10.1016/j.ijhydene.2017.05.021]

[127]   Mirza SS, Qazi JI, Liang Y, Chen S. Growth characteristics and photofermentative biohydrogen production potential of purple non sulfur bacteria from sugar cane bagasse. Fuel 2019; 255115805
[http://dx.doi.org/10.1016/j.fuel.2019.115805.]

[128]   Cui M, Yuan Z, Zhi X, Wei L, Shen J. Biohydrogen production from poplar leaves pretreated by different methods using anaerobic mixed bacteria. Int J Hydrogen Energy 2010; 35(9): 4041-7.
[http://dx.doi.org/10.1016/j.ijhydene.2010.02.035]

[129]   Rorke D, Kana EG. Biohydrogen process development on waste sorghum (Sorghum bicolor) leaves: Optimization of saccharification, hydrogen production and preliminary scale up. 2016.

[130]   Komesu A, Oliveira JAR, Martins LHS, Wolf Maciel MR, Maciel Filho R. Lactic acid production to purification: a review. BioResources 2017; 12(2): 4364-83.
[http://dx.doi.org/10.15376/biores.12.2.Komesu]

[131]   Abdel-Rahman MA, Tashiro Y, Sonomoto K. Recent advances in lactic acid production by microbial fermentation processes. Biotechnol Adv 2013; 31(6): 877-902.
[http://dx.doi.org/10.1016/j.biotechadv.2013.04.002] [PMID: 23624242]

[132]   Alves de Oliveira R, Komesu A, Vaz Rossell CE, Maciel Filho R. Challenges and opportunities in lactic acid bioprocess design—From economic to production aspects. Biochem Eng J 2018; 133: 219-39.
[http://dx.doi.org/10.1016/j.bej.2018.03.003]

[133]   Abdel-Rahman MA, Tashiro Y, Zendo T, Sonomoto K. Improved lactic acid productivity by an open repeated batch fermentation system using Enterococcus mundtii QU 25. RSC Advances 2013; 3(22): 8437-45.
[http://dx.doi.org/10.1039/c3ra00078h]

[134]   Nalawade K, Baral P, Patil S, *et al.* Evaluation of alternative strategies for generating fermentable sugars from high-solids alkali pretreated sugarcane bagasse and successive valorization to L (+) lactic acid. Renew Energy 2020; 157: 708-17.
[http://dx.doi.org/10.1016/j.renene.2020.05.089]

[135] Abdel-Rahman MA, Tashiro Y, Zendo T, Sakai K, Sonomoto K. Enterococcus faecium QU 50: a novel thermophilic lactic acid bacterium for high-yield l-lactic acid production from xylose. FEMS Microbiol Lett 2015; 362(2): 1-7.
[http://dx.doi.org/10.1093/femsle/fnu030] [PMID: 25670701]

[136] Hofvendahl K, Hahn-Hägerdal B. Factors affecting the fermentative lactic acid production from renewable resources1. Enzyme Microb Technol 2000; 26(2-4): 87-107.
[http://dx.doi.org/10.1016/S0141-0229(99)00155-6] [PMID: 10689064]

[137] Koutinas AA, Xu Y, Wang R, Webb C. Polyhydroxybutyrate production from a novel feedstock derived from a wheat-based biorefinery. Enzyme Microb Technol 2007; 40(5): 1035-44.
[http://dx.doi.org/10.1016/j.enzmictec.2006.08.002]

[138] John RP, Nampoothiri KM, Pandey A. Solid-state fermentation for l-lactic acid production from agro wastes using Lactobacillus delbrueckii. Process Biochem 2006; 41(4): 759-63.
[http://dx.doi.org/10.1016/j.procbio.2005.09.013]

[139] Tan J, Abdel-Rahman MA, Sonomoto K. Biorefinery-based lactic acid fermentation: microbial production of pure monomer product.Synthesis, Structure and Properties of Poly (lactic acid). Springer 2017; pp. 27-66.
[http://dx.doi.org/10.1007/12_2016_11]

[140] Wang L, Zhao B, Liu B, *et al.* Efficient production of l-lactic acid from corncob molasses, a waste by-product in xylitol production, by a newly isolated xylose utilizing Bacillus sp. strain. Bioresour Technol 2010; 101(20): 7908-15.
[http://dx.doi.org/10.1016/j.biortech.2010.05.031] [PMID: 20627714]

[141] Li Z, Lu J, Yang Z, Han L, Tan T. Utilization of white rice bran for production of l-lactic acid. Biomass Bioenergy 2012; 39: 53-8.
[http://dx.doi.org/10.1016/j.biombioe.2011.12.039]

[142] Nakano S, Ugwu CU, Tokiwa Y. Efficient production of d-(−)-lactic acid from broken rice by Lactobacillus delbrueckii using Ca(OH)2 as a neutralizing agent. Bioresour Technol 2012; 104: 791-4.
[http://dx.doi.org/10.1016/j.biortech.2011.10.017] [PMID: 22093977]

[143] Prückler M, Lorenz C, Endo A, *et al.* Comparison of homo- and heterofermentative lactic acid bacteria for implementation of fermented wheat bran in bread. Food Microbiol 2015; 49: 211-9.
[http://dx.doi.org/10.1016/j.fm.2015.02.014] [PMID: 25846933]

[144] Ding Z, Kumar V, Sar T, *et al.* Agro waste as a potential carbon feedstock for poly-3-hydroxy alkanoates production: Commercialization potential and technical hurdles. Bioresour Technol 2022; 364128058
[http://dx.doi.org/10.1016/j.biortech.2022.128058.] [PMID: 36191751]

[145] Ouyang J, Ma R, Zheng Z, Cai C, Zhang M, Jiang T. Open fermentative production of l-lactic acid by Bacillus sp. strain NL01 using lignocellulosic hydrolyzates as low-cost raw material. Bioresour Technol 2013; 135: 475-80.
[http://dx.doi.org/10.1016/j.biortech.2012.09.096] [PMID: 23127843]

[146] Ouyang S, Zou L, Qiao H, Shi J, Zheng Z, Ouyang J. One-pot process for lactic acid production from wheat straw by an adapted Bacillus coagulans and identification of genes related to hydrolysate-tolerance. Bioresour Technol 2020; 315123855
[http://dx.doi.org/10.1016/j.biortech.2020.123855.] [PMID: 32707506]

[147] Zhou Y, Kumar V, Harirchi S, *et al.* Recovery of value-added products from biowaste: A review. Bioresour Technol 2022; 360127565
[http://dx.doi.org/10.1016/j.biortech.2022.127565.] [PMID: 35788392]

[148] Taherzadeh M, Karimi K. Pretreatment of lignocellulosic wastes to improve ethanol and biogas production: a review. Int J Mol Sci 2008; 9(9): 1621-51.
[http://dx.doi.org/10.3390/ijms9091621] [PMID: 19325822]

[149] Duan Y, Tarafdar A, Kumar V, *et al.* Sustainable biorefinery approaches towards circular economy for conversion of biowaste to value added materials and future perspectives. Fuel 2022; 325124846
[http://dx.doi.org/10.1016/j.fuel.2022.124846.]

[150] Wischral D, Arias JM, Modesto LF, de França Passos D, Pereira N Jr. Lactic acid production from sugarcane bagasse hydrolysates by *Lactobacillus pentosus* : Integrating xylose and glucose fermentation. Biotechnol Prog 2019; 35(1)e2718
[http://dx.doi.org/10.1002/btpr.2718.] [PMID: 30295001]

[151] Wischral D. Statistical Optimization of Lactic Acid Production by Lactobacillus pentosus using Hemicellulosic Hydrolysate from Sugarcane Bagasse. Revista Ingeniería 2018; 29(1): 41-51.
[http://dx.doi.org/10.15517/ri.v29i1.33477]

[152] Rodrigues C, Vandenberghe L, Woiciechowski A, de Oliveira J, Letti L, Soccol C. Production and application of lactic Acid.Current Developments in Biotechnology and Bioengineering. Elsevier 2017; pp. 543-56.
[http://dx.doi.org/10.1016/B978-0-444-63662-1.00024-5]

[153] Salvachúa D, Smith H, St John PC, *et al.* Succinic acid production from lignocellulosic hydrolysate by Basfia succiniciproducens. Bioresour Technol 2016; 214: 558-66.
[http://dx.doi.org/10.1016/j.biortech.2016.05.018] [PMID: 27179951]

[154] Akhtar J, Idris A, Abd Aziz R. Recent advances in production of succinic acid from lignocellulosic biomass. Appl Microbiol Biotechnol 2014; 98(3): 987-1000.
[http://dx.doi.org/10.1007/s00253-013-5319-6] [PMID: 24292125]

[155] Cheng KK, Zhao XB, Zeng J, Zhang JA. Biotechnological production of succinic acid: current state and perspectives. Biofuels Bioprod Biorefin 2012; 6(3): 302-18.
[http://dx.doi.org/10.1002/bbb.1327]

[156] Borges ER, Pereira N Jr. Succinic acid production from sugarcane bagasse hemicellulose hydrolysate by Actinobacillus succinogenes. J Ind Microbiol Biotechnol 2011; 38(8): 1001-11.
[http://dx.doi.org/10.1007/s10295-010-0874-7] [PMID: 20882312]

[157] Li Q, Yang M, Wang D, *et al.* Efficient conversion of crop stalk wastes into succinic acid production by Actinobacillus succinogenes. Bioresour Technol 2010; 101(9): 3292-4.
[http://dx.doi.org/10.1016/j.biortech.2009.12.064] [PMID: 20061143]

[158] Lo E, Brabo-Catala L, Dogaris I, Ammar EM, Philippidis GP. Biochemical conversion of sweet sorghum bagasse to succinic acid. J Biosci Bioeng 2020; 129(1): 104-9.
[http://dx.doi.org/10.1016/j.jbiosc.2019.07.003] [PMID: 31400993]

[159] Corona-González RI, Varela-Almanza KM, Arriola-Guevara E, Martínez-Gómez ÁJ, Pelayo-Ortiz C, Toriz G. Bagasse hydrolyzates from Agave tequilana as substrates for succinic acid production by Actinobacillus succinogenes in batch and repeated batch reactor. Bioresour Technol 2016; 205: 15-23.
[http://dx.doi.org/10.1016/j.biortech.2015.12.081] [PMID: 26802183]

[160] Shi X, Chen Y, Ren H, *et al.* Economically enhanced succinic acid fermentation from cassava bagasse hydrolysate using Corynebacterium glutamicum immobilized in porous polyurethane filler. Bioresour Technol 2014; 174: 190-7.
[http://dx.doi.org/10.1016/j.biortech.2014.09.137] [PMID: 25463799]

[161] Zheng P, Dong JJ, Sun ZH, Ni Y, Fang L. Fermentative production of succinic acid from straw hydrolysate by Actinobacillus succinogenes. Bioresour Technol 2009; 100(8): 2425-9.
[http://dx.doi.org/10.1016/j.biortech.2008.11.043] [PMID: 19128958]

[162] Li Q, Siles JA, Thompson IP. Succinic acid production from orange peel and wheat straw by batch fermentations of Fibrobacter succinogenes S85. Appl Microbiol Biotechnol 2010; 88(3): 671-8.
[http://dx.doi.org/10.1007/s00253-010-2726-9] [PMID: 20645087]

[163] Lee PC, Lee SY, Chang HN. Succinic acid production by Anaerobiospirillum succiniciproducens ATCC 29305 growing on galactose, galactose/glucose, and galactose/lactose. J Microbiol Biotechnol

2008; 18(11): 1792-6.
[PMID: 19047823]

[164] Chen X, Jiang S, Zheng Z, Pan L, Luo S. Effects of culture redox potential on succinic acid production by Corynebacterium crenatum under anaerobic conditions. Process Biochem 2012; 47(8): 1250-5.
[http://dx.doi.org/10.1016/j.procbio.2012.04.026]

[165] Song H, Jang SH, Park JM, Lee SY. Modeling of batch fermentation kinetics for succinic acid production by Mannheimia succiniciproducens. Biochem Eng J 2008; 40(1): 107-15.
[http://dx.doi.org/10.1016/j.bej.2007.11.021]

[166] Lee SJ, Lee DY, Kim TY, Kim BH, Lee J, Lee SY. Metabolic engineering of Escherichia coli for enhanced production of succinic acid, based on genome comparison and in silico gene knockout simulation. Appl Environ Microbiol 2005; 71(12): 7880-7.
[http://dx.doi.org/10.1128/AEM.71.12.7880-7887.2005] [PMID: 16332763]

[167] Efe Ç, van der Wielen LAM, Straathof AJJ. Techno-economic analysis of succinic acid production using adsorption from fermentation medium. Biomass Bioenergy 2013; 56: 479-92.
[http://dx.doi.org/10.1016/j.biombioe.2013.06.002]

[168] Ur-Rehman S, Mushtaq Z, Zahoor T, Jamil A, Murtaza MA. Xylitol: a review on bioproduction, application, health benefits, and related safety issues. Crit Rev Food Sci Nutr 2015; 55(11): 1514-28.
[http://dx.doi.org/10.1080/10408398.2012.702288] [PMID: 24915309]

[169] Islam MS. Effects of xylitol as a sugar substitute on diabetes-related parameters in nondiabetic rats. J Med Food 2011; 14(5): 505-11.
[http://dx.doi.org/10.1089/jmf.2010.0015] [PMID: 21434778]

[170] C. Pierini and C. CNC, Xylitol: A sweet alternative, History, vol. 5, p. 6, 2001.

[171] Felipe Hernández-Pérez A, de Arruda PV, Sene L, da Silva SS, Kumar Chandel A, de Almeida Felipe MG. Xylitol bioproduction: state-of-the-art, industrial paradigm shift, and opportunities for integrated biorefineries. Crit Rev Biotechnol 2019; 39(7): 924-43.
[http://dx.doi.org/10.1080/07388551.2019.1640658] [PMID: 31311338]

[172] Azizah N. Biotransformation of Xylitol Production from Xylose of Lignocellulose Biomass Using Xylose Reductase Enzyme. Journal of Food and Life Sciences 2019; 3(2): 103-12.

[173] Rafiqul I, Sakinah AM. Bioproduction of xylitol by enzyme technology and future prospects. Int Food Res J 2012; 19(2): 405.

[174] Atzmüller D, Ullmann N, Zwirzitz A. Identification of genes involved in xylose metabolism of Meyerozyma guilliermondii and their genetic engineering for increased xylitol production. AMB Express 2020; 10(1): 78.
[http://dx.doi.org/10.1186/s13568-020-01012-8] [PMID: 32314068]

[175] Baptista SL, Carvalho LC, Romaní A, Domingues L. Development of a sustainable bioprocess based on green technologies for xylitol production from corn cob. Ind Crops Prod 2020; 156112867
[http://dx.doi.org/10.1016/j.indcrop.2020.112867.]

[176] Canilha L, Almeida e Silva JB, Felipe MGA, Carvalho W. Batch xylitol production from wheat straw hemicellulosic hydrolysate using Candida guilliermondii in a stirred tank reactor. Biotechnol Lett 2003; 25(21): 1811-4.
[http://dx.doi.org/10.1023/A:1026288705215] [PMID: 14677703]

[177] Li M, Meng X, Diao E, Du F. Xylitol production by Candida tropicalis from corn cob hemicellulose hydrolysate in a two-stage fed-batch fermentation process. J Chem Technol Biotechnol 2012; 87(3): 387-92.
[http://dx.doi.org/10.1002/jctb.2732]

[178] Bedő S, Antal B, Rozbach M, Fehér A, Fehér C. Optimised fractionation of wheat bran for arabinose biopurification and xylitol fermentation by Ogataea zsoltii within a biorefinery process. Ind Crops Prod 2019; 139111504

[http://dx.doi.org/10.1016/j.indcrop.2019.111504.]

[179] Kim TH, Ryu HJ, Oh KK. Low acid hydrothermal fractionation of Giant Miscanthus for production of xylose-rich hydrolysate and furfural. Bioresour Technol 2016; 218: 367-72.
[http://dx.doi.org/10.1016/j.biortech.2016.06.106] [PMID: 27380022]

[180] Kim SY, Kim JH, Oh DK. Improvement of xylitol production by controlling oxygen supply in Candida parapsilosis. J Ferment Bioeng 1997; 83(3): 267-70.
[http://dx.doi.org/10.1016/S0922-338X(97)80990-7]

[181] López F, Delgado OD, Martínez MA, Spencer JFT, Figueroa LIC. Characterization of a new xylitol-producer Candida tropicalis strain. Antonie van Leeuwenhoek 2004; 85(4): 281-6.
[http://dx.doi.org/10.1023/B:ANTO.0000020368.37876.1c] [PMID: 15031642]

[182] Mussatto SI, Roberto IC. Establishment of the optimum initial xylose concentration and nutritional supplementation of brewer's spent grain hydrolysate for xylitol production by Candida guilliermondii. Process Biochem 2008; 43(5): 540-6.
[http://dx.doi.org/10.1016/j.procbio.2008.01.013]

[183] Ramesh S, Muthuvelayudham R, Rajesh Kannan R, Viruthagiri T. Enhanced production of xylitol from corncob by Pachysolen tannophilus using response surface methodology," International Journal of Food Science, 2013; 2013.

[184] Yoshitake J. H. OHIwA, M. Shimamura, and T. Imai, "Production of polyalcohol by a Corynebacterium sp. Part I. Production of pentitol from aldopentose,". Agric Biol Chem 1971; 35(6): 905-11.

[185] Dahiya JS. Xylitol production by *Petromyces albertensis* grown on medium containing D -xylose. Can J Microbiol 1991; 37(1): 14-8.
[http://dx.doi.org/10.1139/m91-003]

[186] Antunes FAF, *et al.* Biotechnological production of xylitol from biomass.Production of Platform Chemicals from Sustainable Resources. Springer 2017; pp. 311-42.
[http://dx.doi.org/10.1007/978-981-10-4172-3_10]

[187] Delgado Arcaño Y, Valmaña García OD, Mandelli D, Carvalho WA, Magalhães Pontes LA. Xylitol: A review on the progress and challenges of its production by chemical route. Catal Today 2020; 344: 2-14.
[http://dx.doi.org/10.1016/j.cattod.2018.07.060]

[188] Rao RS, Jyothi CP, Prakasham RS, Sarma PN, Rao LV. Xylitol production from corn fiber and sugarcane bagasse hydrolysates by Candida tropicalis. Bioresour Technol 2006; 97(15): 1974-8.
[http://dx.doi.org/10.1016/j.biortech.2005.08.015] [PMID: 16242318]

[189] Li S, Zhang J, Xu H, Feng X. Improved xylitol production from D-Arabitol by enhancing the coenzyme regeneration efficiency of the pentose phosphate pathway in Gluconobacter oxydans. J Agric Food Chem 2016; 64(5): 1144-50.
[http://dx.doi.org/10.1021/acs.jafc.5b05509] [PMID: 26727541]

[190] Ko BS, Kim J, Kim JH. Production of xylitol from D-xylose by a xylitol dehydrogenase gene-disrupted mutant of Candida tropicalis. Appl Environ Microbiol 2006; 72(6): 4207-13.
[http://dx.doi.org/10.1128/AEM.02699-05] [PMID: 16751533]

[191] Kim H, Lee HS, Park H, *et al.* Enhanced production of xylitol from xylose by expression of Bacillus subtilis arabinose:H $^+$ symporter and Scheffersomyces stipitis xylose reductase in recombinant Saccharomyces cerevisiae. Enzyme Microb Technol 2017; 107: 7-14.
[http://dx.doi.org/10.1016/j.enzmictec.2017.07.014] [PMID: 28899489]

[192] Jeon WY, Yoon BH, Ko BS, Shim WY, Kim JH. Xylitol production is increased by expression of codon-optimized Neurospora crassa xylose reductase gene in Candida tropicalis. Bioprocess Biosyst Eng 2012; 35(1-2): 191-8.
[http://dx.doi.org/10.1007/s00449-011-0618-8] [PMID: 21922311]

[193] Dasgupta D, Junghare V, Nautiyal AK, Jana A, Hazra S, Ghosh D. Xylitol Production from Lignocellulosic Pentosans: A Rational Strain Engineering Approach toward a Multiproduct Biorefinery. J Agric Food Chem 2019; 67(4): 1173-86.
[http://dx.doi.org/10.1021/acs.jafc.8b05509] [PMID: 30618252]

[194] Arora R, Behera S, Kumar S. Bioprospecting thermophilic/thermotolerant microbes for production of lignocellulosic ethanol: A future perspective. Renew Sustain Energy Rev 2015; 51: 699-717.
[http://dx.doi.org/10.1016/j.rser.2015.06.050]

[195] Prior B, Kilian S, Du Preez J. Fermentation of smallcap˜ D-xylose by the yeasts Candida shehatae and Pichia stipitis. Process Biochem 1989; 24(1): 21-32.

[196] Mäki-Arvela P, Salmi T, Holmbom B, Willför S, Murzin DY. Synthesis of sugars by hydrolysis of hemicelluloses--a review. Chem Rev 2011; 111(9): 5638-66.
[http://dx.doi.org/10.1021/cr2000042] [PMID: 21682343]

[197] Delbecq F, Wang Y, Muralidhara A, El Ouardi K, Marlair G, Len C. Hydrolysis of hemicellulose and derivatives—A review of recent advances in the production of furfural. Front Chem 2018; 6: 146.
[http://dx.doi.org/10.3389/fchem.2018.00146] [PMID: 29868554]

[198] Lindroos M, Heikkila H, Nurmi J, Eroma O-P. Method for recovery of xylose from solutions. 2000.

[199] Zhang HJ, Fan XG, Qiu XL, *et al.* A novel cleaning process for industrial production of xylose in pilot scale from corncob by using screw-steam-explosive extruder. Bioprocess Biosyst Eng 2014; 37(12): 2425-36.
[http://dx.doi.org/10.1007/s00449-014-1219-0] [PMID: 24890135]

[200] Chen K, Luo G, Lei Z, Zhang Z, Zhang S, Chen J. Chromatographic separation of glucose, xylose and arabinose from lignocellulosic hydrolysates using cation exchange resin. Separ Purif Tech 2018; 195: 288-94.
[http://dx.doi.org/10.1016/j.seppur.2017.12.030]

[201] Krishania M, Kumar V, Sangwan RS. Integrated approach for extraction of xylose, cellulose, lignin and silica from rice straw. Bioresour Technol Rep 2018; 1: 89-93.
[http://dx.doi.org/10.1016/j.biteb.2018.01.001]

[202] Loow Y-L, *et al.* Improvement of xylose recovery from the stalks of oil palm fronds using inorganic salt and oxidative agent. Energy Convers Manage 2017; 138: 248-60.
[http://dx.doi.org/10.1016/j.enconman.2016.12.015]

[203] Unrean P. Optimized feeding schemes of simultaneous saccharification and fermentation process for high lactic acid titer from sugarcane bagasse. Ind Crops Prod 2018; 111: 660-6.
[http://dx.doi.org/10.1016/j.indcrop.2017.11.043]

[204] van der Pol EC, Eggink G, Weusthuis RA. Production of l(+)-lactic acid from acid pretreated sugarcane bagasse using *Bacillus coagulans* DSM2314 in a simultaneous saccharification and fermentation strategy. Biotechnol Biofuels 2016; 9(1): 248.
[http://dx.doi.org/10.1186/s13068-016-0646-3] [PMID: 27872661]

[205] Peng L, Xie N, Guo L, Wang L, Yu B, Ma Y. Efficient open fermentative production of polymer-grade L-lactate from sugarcane bagasse hydrolysate by thermotolerant Bacillus sp. strain P38. PLoS One 2014; 9(9)e107143
[http://dx.doi.org/10.1371/journal.pone.0107143.] [PMID: 25192451]

[206] Liu G, Sun J, Zhang J, Tu Y, Bao J. High titer l -lactic acid production from corn stover with minimum wastewater generation and techno-economic evaluation based on Aspen plus modeling. Bioresour Technol 2015; 198: 803-10.
[http://dx.doi.org/10.1016/j.biortech.2015.09.098] [PMID: 26454367]

[207] Moldes AB, Torrado A, Converti A, Domínguez JM. Complete bioconversion of hemicellulosic sugars from agricultural residues into lactic acid by Lactobacillus pentosus. Appl Biochem Biotechnol 2006; 135(3): 219-28.

[http://dx.doi.org/10.1385/ABAB:135:3:219] [PMID: 17299209]

[208] Vallejos ME, Chade M, Mereles EB, *et al.* Strategies of detoxification and fermentation for biotechnological production of xylitol from sugarcane bagasse. Ind Crops Prod 2016; 91: 161-9.
[http://dx.doi.org/10.1016/j.indcrop.2016.07.007]

[209] Dominguez JM, Gong CS, Tsao GT. Pretreatment of sugar cane bagasse hemicellulose hydrolysate for xylitol production by yeast Seventeenth Symposium on Biotechnology for Fuels and Chemicals. 49-56.
[http://dx.doi.org/10.1007/978-1-4612-0223-3_5]

[210] Huang CF, Jiang YF, Guo GL, Hwang WS. Development of a yeast strain for xylitol production without hydrolysate detoxification as part of the integration of co-product generation within the lignocellulosic ethanol process. Bioresour Technol 2011; 102(3): 3322-9.
[http://dx.doi.org/10.1016/j.biortech.2010.10.111] [PMID: 21095119]

[211] Cheng KK, Zhang J-A, Ling H-Z, *et al.* Optimization of pH and acetic acid concentration for bioconversion of hemicellulose from corncobs to xylitol by Candida tropicalis. Biochem Eng J 2009; 43(2): 203-7.
[http://dx.doi.org/10.1016/j.bej.2008.09.012]

[212] Li Q, Lei J, Zhang R, *et al.* Efficient decolorization and deproteinization using uniform polymer microspheres in the succinic acid biorefinery from bio-waste cotton (Gossypium hirsutum L.) stalks. Bioresour Technol 2013; 135: 604-9.
[http://dx.doi.org/10.1016/j.biortech.2012.06.101] [PMID: 22985822]

[213] Chen X, Jiang S, Li X, Pan L, Zheng Z, Luo S. Production of succinic acid and lactic acid by Corynebacterium crenatum under anaerobic conditions. Ann Microbiol 2013; 63(1): 39-44.
[http://dx.doi.org/10.1007/s13213-012-0441-8]

[214] Dorado MP, Lin SKC, Koutinas A, Du C, Wang R, Webb C. Cereal-based biorefinery development: Utilisation of wheat milling by-products for the production of succinic acid. J Biotechnol 2009; 143(1): 51-9.
[http://dx.doi.org/10.1016/j.jbiotec.2009.06.009] [PMID: 19539669]

[215] Paiva JE, Maldonade IR, Scamparini ARP. Xylose production from sugarcane bagasse by surface response methodology. Rev Bras Eng Agric Ambient 2009; 13(1): 75-80.
[http://dx.doi.org/10.1590/S1415-43662009000100011]

[216] Thamsee T, Cheirsilp B, Yamsaengsung R, Ruengpeerakul T, Choojit S, Sangwichien C. Efficient of acid hydrolysis of oil palm empty fruit bunch residues for xylose and highly digestible cellulose pulp productions. Waste Biomass Valoriz 2018; 9(11): 2041-51.
[http://dx.doi.org/10.1007/s12649-017-9965-2]

[217] Satari B, Karimi K, Kumar R. Cellulose solvent-based pretreatment for enhanced second-generation biofuel production: a review. Sustain Energy Fuels 2019; 3(1): 11-62.
[http://dx.doi.org/10.1039/C8SE00287H]

CHAPTER 2

Current Biotechnological Advancements in Lignin Valorization For Value-added Products

Muskan Pandey, **Richa Parashar**[1] and **Barkha Singhal**[1,*]

[1] *School of Biotechnology, Gautam Buddha University, Greater Noida (U.P.), India*

Abstract: Recent years have seen a tremendous demand in bioenergy. The technological advancements in the production of second-generation biofuels have opened a plethora of opportunities for the valorization of natural polymers. Lignin is one of the most abundant and recalcitrant materials available on earth. Advancements in genetic engineering, metabolic engineering and synthetic biology applications fueled tremendous interest in the valorization of lignin into fuels as well as platform and commodity chemicals. Though there is a growing continuum for biofuel advancements in recent years, at the same time, a rising upsurge has also been envisaged in the valorization of waste bioresources. Therefore, this chapter entails about various aspects and embodiments related to lignin bioconversion and their routes for obtaining various products. This chapter also highlights current biotechnological interventions for the improvement of the valorization process as well as the current challenges and future perspectives in this burgeoning area.

Keywords: Genetically encoded biosensors, Lignin valorization, Microbial fermentation, Metabolic engineering, Metagenomics, Synthetic biology, Value-added products.

INTRODUCTION

The overwhelming boost in attaining sustainability for energy requirements leads to rapid developments in exploring various natural and synthetic energy resources. The depletion of fossil fuels as well as rising environmental security urges the scientific community to develop and rely on biobased energy sources. Among this continuum, the development of bioethanol, biodiesel, and biohydrogen through renewable biomass imparts significant contributions to moving ahead in this direction. However, the utilization of biomass is a technologically demanding task that leads to the voluminous generation of lignin, one of the most recalcitrant and The current research estimates 300 billion tons of available lignin with an annual

* **Corresponding author Barkha Singhal:** School of Biotechnology, Gautam Buddha University, Greater Noida (U.P.), India ; E-mail: gupta.barkha@gmail.com,

Vinay Kumar, Sivarama Krishna Lakkaboyana & Neha Sharma (Eds.)

increment of approximately 20 billion tons complex polymers on the earth [1]. Lignin is released mainly through various industrial sectors including paper and pulp industries and second-generation biofuel plants [2]. Thus, it was estimated that 140 million tons of lignin have been simply burned per year despite having immense hidden resources that have been unraveled through synthetic retrospection. Therefore, the valorization of lignin is currently a thrust area of research for up-scaling bio-based economy. Currently, the potential of lignin valorization is not limited to the production of various commodity chemicals like alcohol, hydrocarbons, ketones, and acids but also bio-based value-added products like coumarins, flavonoids, stilbenoids, poly-hydroxy butyrate (PHA), *etc* [3]. The value of currently produced lignin is estimated to be 3.3 Billion dollars with an energy occupancy of 89% of the market [4]. The percentage of lignin available in different countries across the globe is depicted in Fig. (**1**).

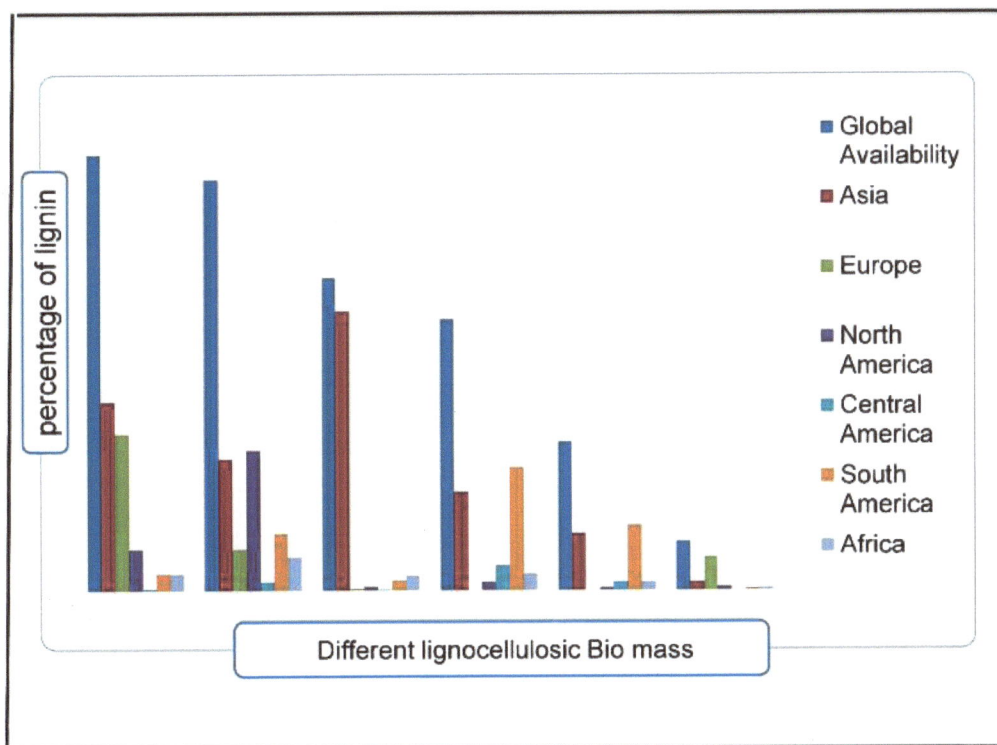

Fig. (1). Availability of lignin in various lignocellulose biomass at global scale.

Though nature embraced lignin with marvelous credentials of making an essential component of living systems as plants but still its utilization by humans for commercial use is an arduous task and a long road to cover till date [5]. Research studies speculated various physio-chemical and biological approaches for lignin valorization but the ecofriendly approach of bioconversion of lignin into value-added products is the most feasible approach utilized till date and there have been continuous advancements in this realm [6]. The current interventions of "omics" technology and the advent of synthetic biology lead to a paradigmatic shift in lignin depolymerization and modification. Therefore, this chapter advocates about recent biotechnological advancements in lignin valorization as well as emphasizes on the challenges and future prospects in this budding area.

Chemical Structure of Lignin

Lignin is considered the most heterogeneous polymer and a renewable resource for the production of aromatics. It is generally considered a side product in biorefineries but holds tremendous potential for harnessing sustainable bioproducts such as bioactive compounds, fuels and other useful industrial chemicals. The chemistry of lignin is quite complex as it's mainly comprised of three phenylpropanoid units: the monolignols coniferyl alcohol (G), sinapyl alcohol (S) and p-coumaroyl alcohol (H) [5]. Various oxidoreductases such as laccases and peroxidases enzymes present in plants are being used for assembly of these sub-units through reactive radical intermediates to form lignin polymers [7]. The linkage of these subunits is characterized by carbon-carbon and ether bonds and the most common linkage is β-O-4 ether linkage [8]. Apart from that, subunits are also connected through α-O-4 linkages, β-β linkages, 5-5 linkages, β-5 linkages, and biphenyl and diaryl ether structures resulting in the enhancement of complexity of a tridimensional framework of lignin [9]. The various linkages present in softwood and hardwood lignin are represented in Fig. (**2**). However, the lignin composition and its percentage vary with respect to plant species. Softwood contains the highest G-type lignin content and hardwood contains an equal proportion of G/S-type lignin [1]. Thus, for complete utilization of such an important renewable biomass, all three major subunits must be efficiently transfigured to value added compounds that lead to the fulfillment of sustainable and cost-effective bio refineries.

Biological Valorization of Lignin

Lignin has diverse structural heterogeneity resulting in various classes of products followed by the application of various chemical procedures for the breakdown of lignin. The well appreciated procedures that have been documented immensely in the literature for the modification of lignin are the application of heat, supercritical fluids, ionic fluids, and fractionation by ultrafiltration. Currently, there is a fascinating research frontier in the biological route for lignin depolymerization and fractionation. In nature, both bacteria and fungi have been

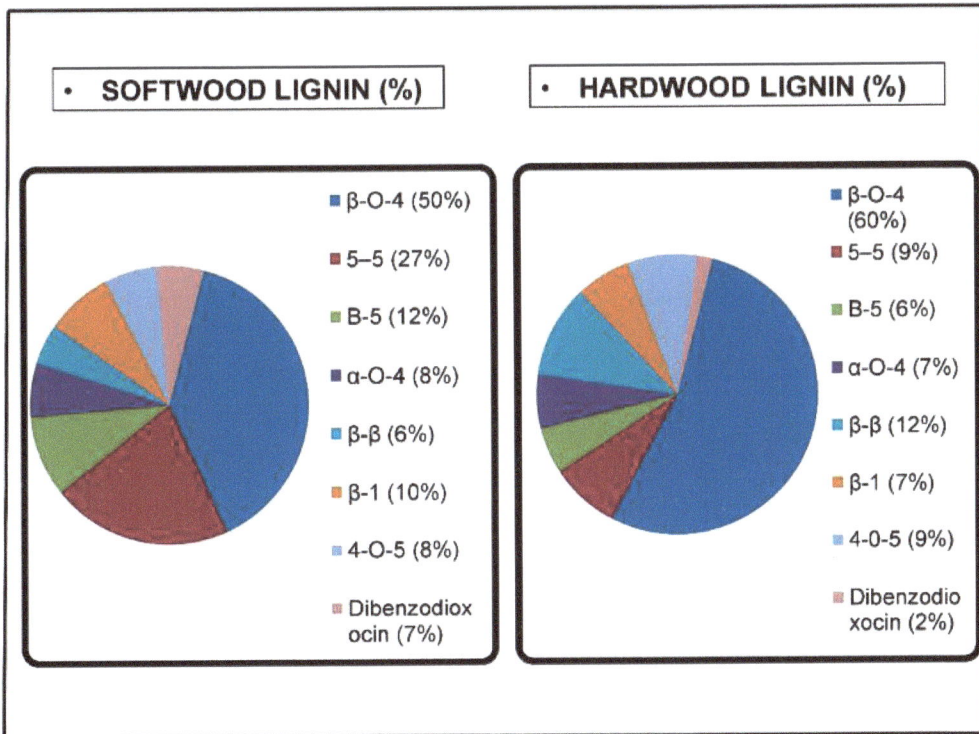

Fig. (2). Proportions of linkages present in softwood and hardwood lignin in Plants.

found capable of depolymerizing lignin with the help of multiple enzymes machinery on various natural biomass resources *in-vitro* and *in-vivo*. Earlier, fungi such as white and brown rot fungi such as *Phanerochaete chrysosporium Ceriporiopsis subvermispora, Echinodontium taxodii, Pleurotus ostreatus, Cyathus stercoeus,* and *Trametes versicolor* are considered important sources for metabolizing lignin [6]. These fungi have been explored due to their inherent enzymes such as laccases, lignin peroxidase (LiP), manganese peroxidase (MnP), versatile peroxidase (VP), aryl alcohol oxidase (AAO), H_2O_2-generating enzymes, and β-esterase boxes [10]. To date, fungi have been more widely explored for lignin degradation but there is still a dearth of studies related to the complete understanding of degradation pathways. Thus, various bacterial systems have been explored and a variety of species such as *Rhodococcus jostii*, Pseudomonas *putida, Rhodococcus opacus, Acinetobacter baylyi, Amycolatopsis sp* and *Sphingomonas* have been found to be the native degraders of lignin. But their capacity for lignin degradation has been found to be lower as compared with fungal systems. Thus, owing to easy genetic manipulation, they are the more amenable choice for genetic modification to enhance lignin degradation. The diverse enzymes catalyze a variety of reactions to form aromatic intermediates that further cleave other linkages through non-enzymatic reactions along with various oxygenated aromatic compounds.

Biological lignin valorization consists of three steps involving lignin depolymerization, aromatics degradation and target product biosynthesis. An overview of lignin valorization has been depicted in Fig. (**3**). The initial lignin breakdown results in lignin monomers such as *p*-coumarate, vanillin, cresols, guaiacol, ferulate and phenol. Furthermore, these heterogeneous aromatic derivatives have been converted into central intermediates, such as protocatechuate and catechol. The G-type and H-type lignin is converted into two major aromatic metabolites protocatechuate or catechol through various enzymatic machineries like acyl-CoA synthetases, acyl-CoA hydratases/lyases, decarboxylases, and dehydrogenases [11]. Further catechol and protocatechuate have been catabolized into various ring compounds by the β-ketoadipate pathway through ortho-cleavage using dioxygenases. Meta-cleaving dioxygenases independent of the β-ketoadipate pathway have also been reported in different microorganisms [12]. Similarly, S-type lignin is converted into other aromatic compounds through the non-interacting branch. In addition to that, anaerobic bacteria have also been explored for the catabolism of aromatics for lignin valorization. This class of bacteria generates benzoyl-CoA as a central intermediate in the central pathway that has been further processed by ring-reducing enzymes such as ATP-dependent or ATP-independent reductases [13]. Furthermore, the biosynthesis of value-added products has been channelized through "biological funnels" in which the depolymerized heterogeneous product of lignin is converted into various bioactive and valuable products. Although technological advances have been reported. But still various potential organisms and enzymes are unexplored. Genetic engineering of lignin-degrading pathways and bioprospecting of lignin degrading environments will strengthen lignin valorization.

Value Added Products from Lignin

The lignin holds intricate complexity thus challenges for its depolymerization. In recent years, there has been a continuous upsurge in bio-based products due to biodegradability and lesser side effects [14]. There are a range of compounds available which can be converted into high-value and can contribute significantly to the development of the bio-based economy. A variety of chemical platforms have been produced but currently, with the intervention of biological methods, a range of value-added compounds have been synthesized [15]. A detailed overview of various products from lignin valorization has been depicted in Fig. (**4**).

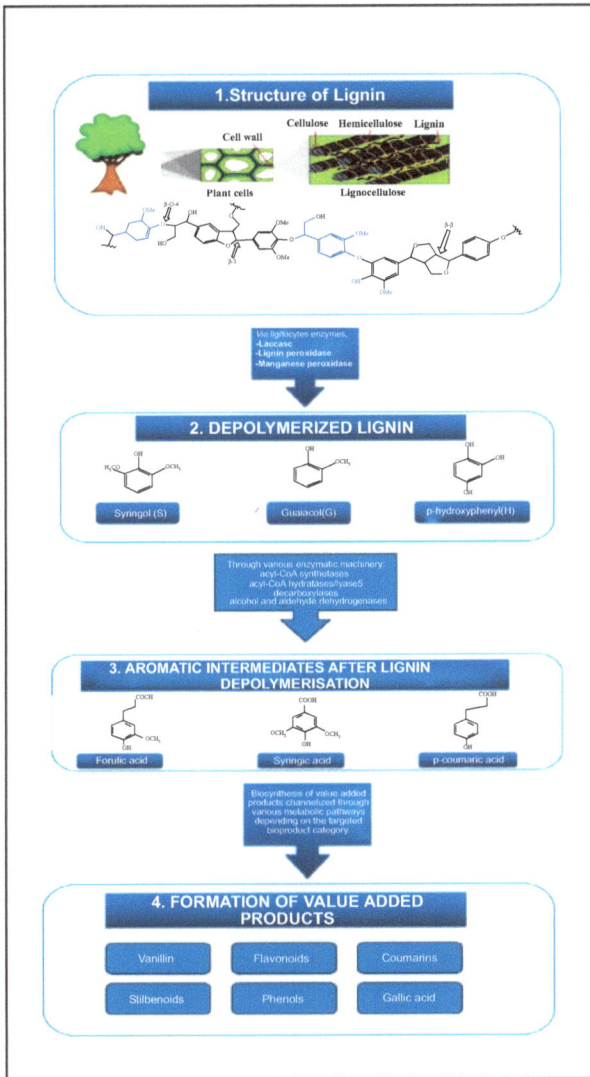

Fig. (3). An overview of the process of biological valorization for harnessing value-added products.

A variety of aromatic compounds like vanillin (food flavoring agent), p-hydroxybenzoic acid (an anti-microbial compound) and pyrogallol (used as bioactive compounds) can be produced from biological lignin valorization. In addition, cis-muconic acid (cis, cis-MA) captured more than $22 billion market and acts as a major feedstock for plastics, fibers, and adipic acid production [16]. Apart from that, lignin derivatives have been well-suited to produce PHAs.

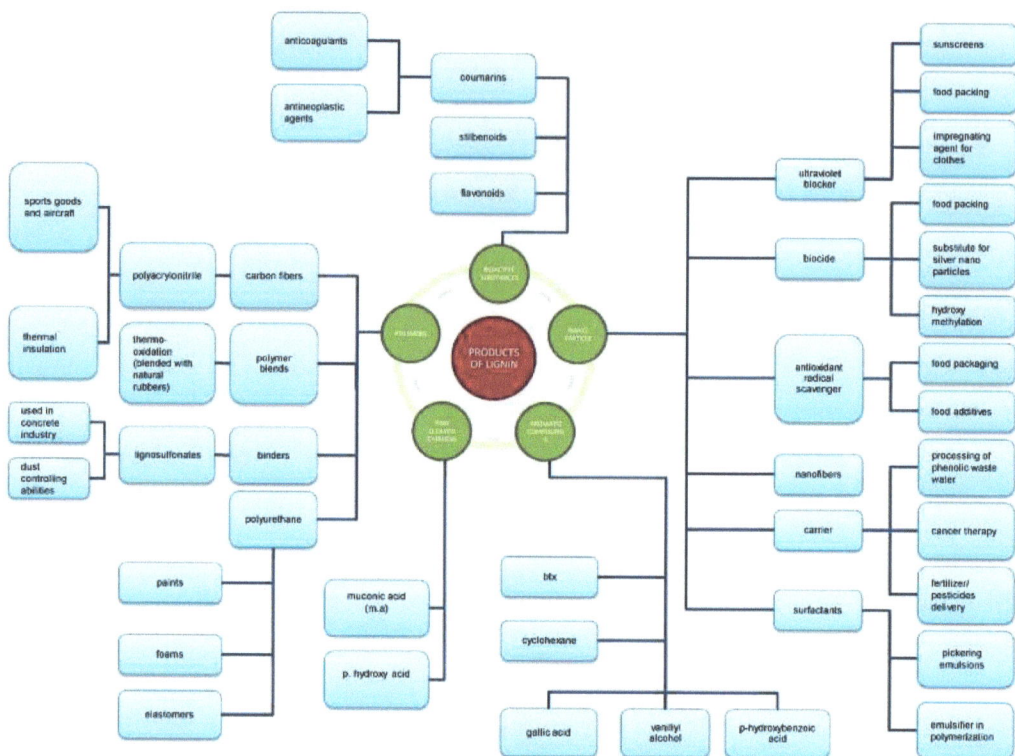

Fig. (4). An overview of value-added products from lignin valorization.

Role of Metagenomics for the advancements in Lignin Valorization

The past decade has witnessed tremendous development in exploring the culture-independent techniques for the identification of genetic tractability of various microorganisms present in environmental samples. It is estimated that to date, culture-dependent techniques enabled the identification of only 1-10% of microbial biodiversity, therefore, molecular level methods based on DNA/RNA have been explored for abrogating technical challenges associated with conventional cultivation techniques that are better known as metagenomics [17, 18]. The plethora of literature is associated with the analysis of genes encoding the cellulose/hemicellulose metabolism but the identification of strains for lignin biotransformation is still in the embryonic stage.

However, next-generation sequencing platforms (NGS) gave a comprehensive idea about various genes encoding for enzymatic lignin degradation. The available information can be used for useful applications. The genetic engineering of the desired genes can be pursued after the functional annotation of genes and their evolutionary relationship by establishing phylogenetic analysis. The whole metagenomic sequencing facilitates the comprehensive designing and understanding of metabolic interaction, metabolic reconstruction and flux balance analysis. Various databases can be used to complete the process. These include Kyoto Encyclopedia of Genes and Genomes (KEGG), NCBI, PATRIC, Pfam, ExPASy and UniProt [19 - 21]. The computational approaches applied to metagenomic studies for lignin valorization are summarized in Fig. (5).

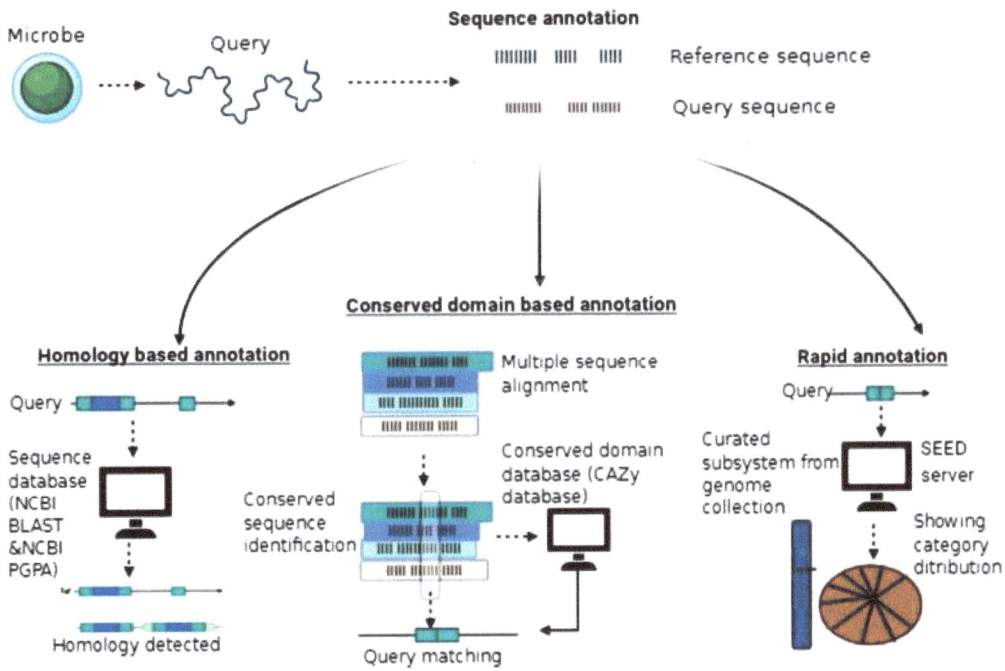

Fig. (5). Workflow for metagenomic studies in lignin valorization.

More recently, the eLignin database provides an avenue for gathering and processing information on microbial diversity, genes, metabolic pathways, enzymes, and metabolites for lignin valorization [19]. Similarly, enzyme genes have been analyzed through conserved domain-based annotation. The database (CAZy) on CAZymes (carbohydrate-active enzymes) responsible for cleavage and control of carbohydrates metabolism has been developed based on the similarity of a protein structure and gene sequences [22]. To date, six classes of the CAZy database have been published that include glycosyl hydrolases, carbohydrate esterases, glycosyltransferases, polysaccharide lyases, carbohydrate-binding modules and redox type enzymes (AAs) [23 - 26]. Currently, the database depicted more richness on fungal species but recently bacterial populations such as Rhodococcus jostii RHA1, Sphingobium sp. SYK-6 Streptomyces viridosporus

T7A, Pseudomonas putida mt-2, Rhizobium, Arthrobacter and Thauera have been used for the lignin depolymerization process [27, 28].

Recently, more pseudomonas strains have been used for lignin valorization. The metagenomic analysis of Pseudomonas sp. Q18 revealed various genes and enzymes such as DyP-peroxidase, vanillate O-demethylase, β-etherase, laccase, feruloyl esterase, carboxylesterase cytochrome P450 and chloroperoxidase for degradation of lignin and its derived compounds [29, 30]. More recently, targeted metagenomic sequencing based on 16S rRNA has been performed with a sugarcane soil sample to establish a lignin-degrading consortium (LigMet) for the identification of novel bacterial biocatalysts and metabolic pathways for bio-transformation of lignin and derived compounds [31]. The taxonomic analysis revealed the abundance of genus *Actinobacteria*, *Proteobacteria* and *Firmicutes* members in the *Micrococcaceae* and *Alcaligenaceae* families [31].

The studies reported a novel strain of Paenarthrobacter carrying eight gene clusters for aromatic degradation of lignin. Furthermore, the vanillin synthesis has been performed with novel identified genes encoding for enzymes feruloyl-CoA synthetase and enoyl-coa hydratase/aldolase [32]. In addition to that, genetically-encoded biosensors based on EmrR aTF and its cognate promoter have been reported for the high-throughput metagenomic screening of bacterial DNA samples retrieved from a coal bed in *E. coli* [33]. The studies identified a multicopper oxidase (CopA) enzyme that has been successfully used for the degradation of mono-aromatic compounds derived from lignin breakdown [34]. Thus, the functional exploration of microbial diversity and their associated pathways through metagenomics are in the budding stage of development. But a highly bright future can be foreseen in lignin valorization.

Role of Genetically-Encoded Biosensors for the Advancements in Lignin Valorization

In nature, an organism possesses a sophisticated gene regulatory framework to sense various environmental cues that involve sensing of targeted metabolites. Thus, this system is mimicked to construct genetically encoded biosensors for measurement and regulation of metabolite productivity. In lignin valorization, metabolic engineering and synthetic biology techniques can be used for novel pathway construction but the optimization of these pathways involves screening a number of genetic variants that are generated by either random mutagenesis or directed evolution [35 - 37].

Therefore, genetically-encoded biosensors based on various molecular systems like RNA riboswitches, ligand-based fluorescent proteins, receptors (allosteric transcription factors (aTFs)), and histidine kinases (HK) play a key role in lignin valorization [38 - 40]. These sensors have been well-reported for high-throughput screening of large mutant libraries. The applications of genetically-encoded biosensors in various domains of lignin valorization have been represented in Fig. (**6**).

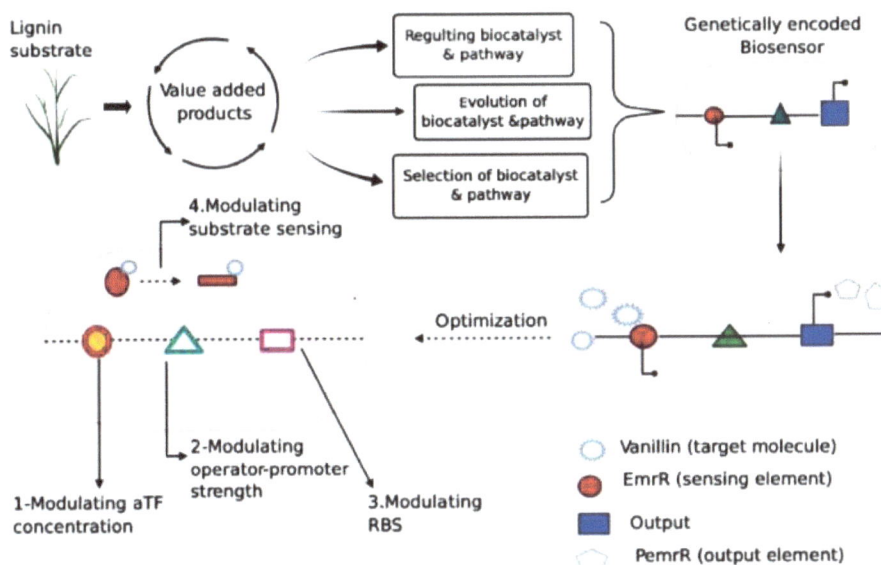

Fig. (6). Applications of Genetically Encoded Biosensors in various domains of Lignin Valorization.

Recent research speculated various successful efforts for the phenotypic screening and categorizing high-yielding enzymes and phenotypes for lignin-degrading microbes using directed evolution. These biosensors along with microfluidics and fluorescence-activated cell sorting (FACS) can screen up to 10^7-10^8 clones per day [36, 41, 42]. Recently Lee *et al.* [43], reported the high-throughput screening of improved lignin-degrading enzymes. It produced mutants up to 6×10^6 based on the tryptophan-indole lyase library. More recently, this sensing technology has been well utilized for analyzing aromatic monomers, sugars and other relevant intermediate compounds produced through the process of lignin valorization. Most of the sensors reported were constituting *E. coli* as host based on aromatic-responsive transcription factors. The sensors were used either in the selection of lignin degrading enzymes, or in the evolution of more adaptive strains for lignin valorization. Sugars (xylose, arabinose, rhamnose, mannose and maltose) [44 - 48] and aromatic compounds (vanillin, vanillic acid, protocatechuic acid, syringaldehyde, benzaldehyde and benzoic acid derivatives) [49, 33, 50 - 53] have been analyzed using genetically encoded biosensors. Recently, aromatic-responsive transcription factors and their promoters/operators derived from

Acinetobacter spp., Pseudomonas spp., Sphingobium spp, and Rhodococcus jostii RHA1 were used for biosensors' development [54 - 56].

A genetically-encoded biosensor based on the FerC repressor was reported for feruloyl esterase enzyme (CE1) activity [57]. Similarly, a PCA-responsive biosensor was reported for the optimized selection of dehydroshikimate dehydratase enzymes (AsbF) based on the pcaU transcription factor derived from Acinetobacter sp. ADP1 [58]. The FRET-based sugar-responsive biosensors have also been developed through random mutagenesis. In addition, metabolic engineering applications of biosensors have also been reported for enhanced productivity. Recently, Lo *et al.* [59] reported a two-layered gene circuit that was based on a *p*-coumaric acid sensing module. This methodology reduced the metabolic burden to two-fold by the selective uptake of target substrates on the depletion of nutrients. It also enhanced lignin breakdown efficiency.

In a similar line, a biosensor based on ADH7 vanillin-inducible promoter [60] was reported that can regulate the pathway for catechol biosynthesis. It only requires vanillin, without any use of exogenous inducers to enhance productivity. An example of enhanced vanillin production has been demonstrated through genetically-encoded biosensors in Fig. (7). It was observed that genetically improved biosensors can be used for useful applications in lignin valorization. Protein and computational engineering techniques can be used to produce value-added products from the lignin.

Role of Metabolic Engineering and System Biology in the Advancements of Lignin Valorization

In recent years, heterologous expressions can be seen as a technique to produce value-added products. The system metabolic engineering represents an example of in strain development for industries by understanding the metabolic and regulatory processes and response to various environmental signals.

The introduction of novel metabolic pathways for the catabolism of aromatic compounds released during lignin depolymerization in the non-native host is quite a daunting task. The metabolic engineering is an essentially required genetic tool for abrogating the complexity of lignin and its conversion through various pathways. A variety of genetic elements like synthetic promoters, ribosomal binding sites, riboswitches, cistronic elements, operators and genome editing tools like CRISPR-Cas [61], TALENS, ZFNs enabled the high-through put screening [62]. The system metabolic engineering has been applied for priming various hosts for lignin valorization.

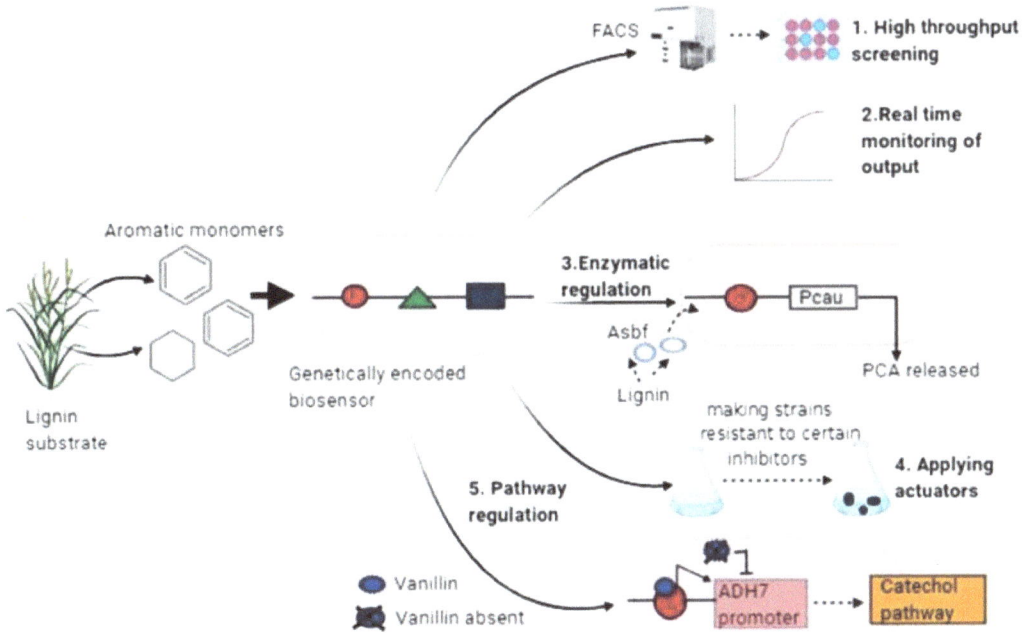

Fig. (7). Application of Genetically encoded biosensor for enhanced vanillin production.

E.coli is one of the widely used bacterial strains for the production of lignin degradation products. Recently, the heterologous expression of nine genes from *P. putida* KT2440 in *E.coli* enabled the degradation of 4-hydroxybenzoate and protocatechuate by 4-hydroxybenzoate 3-monooxygenase enzyme [63]. In addition, the production of catechol from vanillin has been reported by the heterologous expression of ligV (vanillin dehydrogenase) and ligM (vanillate--demethylase) genes isolated from *Sphingomonas paucimobilis* SKY-6 and aroY (protocatechuate decarboxylase) derived from *Klebsiella pneumoniae* under the control of the vanillin-inducible ADH7 promoter in *E. coli* [11]. Furthermore, structure-guided protein engineering was performed through crystallographic studies and molecular dynamic simulation (MD) in which phenyl alanine amnio acid was mutated for the modification of enzyme GcoA-F169 activity [64]. This modification leads to the alteration of specificity of one of the aromatic compounds including syringol released during the depolymerization of S-type lignin. In addition, an engineered phenol-inducible promoter was constructed by the replacement of a spacer region of an endogenous promoter PemrR to create promoter variants Pvtac, Pvtrc and Pvtic [65]. Thus, a variety of chemicals like vanillin, muconic acid, fatty acids, medium chain-length polyhydroxyalkanoates (mcl-PHB) and pyridine dicarboxylic acids can be used as platform chemicals to produce high-value compounds. The above discussion revealed the fact that metabolic engineering needs an hour for lignin valorization.

Role of Synthetic Biology for the Advancements in Lignin Valorization

The intrinsic heterogeneity and recalcitrant nature of lignin pose a significant challenge. Although several pretreatment methodologies have been reported for lignin breakdown. But microbial degradation has played a key role in their mineralization. The rising availability of whole genome sequencing data facilitates the molecular-level analysis of microbial populations, their enzymes and metabolic pathways for lignin degradation. The current advancements in synthetic biology have led to the design, construction, and optimization of novel metabolic pathways. It can be used to rebuild natural pathways for efficient bio-conversion of lignin into value-added products. Recent research highlighted various applications of synthetic biological tools to harness a variety of value-added products including aromatic compounds and bioactive substances [66]. The role of synthetic biology to produce value-added products from lignin valorization has been represented in Fig. (**8**).

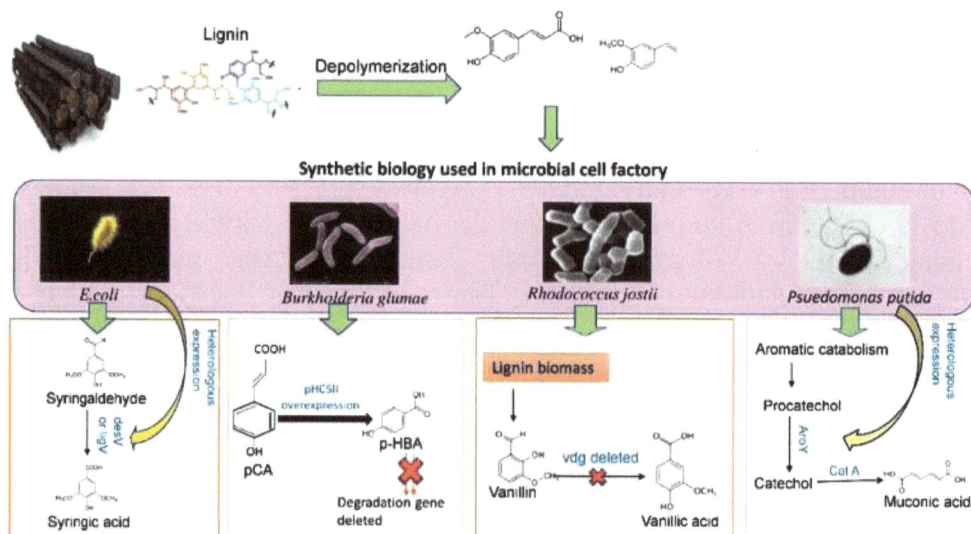

Fig. (8). Application of synthetic biology for improving microbial cell factories for lignin valorization.

Among aromatic compounds, vanillin, gallic acid and *p*-hydroxybenzoic acid (pHBA) are the major value-added products from lignin valorization [66]. For pHBA production, the metabolic pathway was reconstructed by the deletion of two genes for pHBA degradation and the overexpression of phcs II encoding p-hydroxy cinnmaoyl-CoA synthetase II gene for pHBA production from

Burkholderia glumae strain BGR [67]. Similarly, bio-phenol was produced from pHBA through the heterologous expression of pHBA decarboxylase gene in *E. coli* [68].

Also, vanillin has been produced through wheat straw lignocellulosic biomass by the deletion of vanillin dehydrogenase genes from *Rhodococcus jostii* RHA1 (R. jostii RHA1) [69]. In addition, vanillin can be produced through the catalysis of ferulic acid by the heterologous expression of Fcs and Ech genes in *E. coli* isolated from thermophilic actinomycete *A. thermoflava* N1165 in temperature-controlled reaction [70]. The expression of these two genes has been limited by the continuous requirement of ATP and CoA that can be alleviated by the introduction of Pad and Ado genes in *E.coli* that lead to the development of coenzymes [71]. Moreover, the heterologous expression of VpVAN gene derived from *Vanilla planifolia* into S. cerevisiae has also been reported for vanillin production [72]. More recently, vanillic acid, a deaminated product of vanillin possessing excellent pharmacological properties has been produced in *E.coli* through the heterologous expression of CV2025 *w*-transaminase (*w*-Tam) gene isolated from *Chromobacterium violaceum* DSM3019 [73]. Gallic acid is produced either through the catalysis of syringic acid or pHBA. Syringic acid has been produced in *E. coli* through the heterologous expression of desV or ligV genes followed by the conversion to gallic acid through the expression of demethylase (desA and ligM) enzymes [74, 75]. Apart from that, gallic acid can be produced from pHBA in *E.coli* through the heterologous expression of the mutant gene of pHBA hydroxylase (PobA) derived from *Pseudomonas aeruginosa* [76].

Moreover, the metabolic construction of pathways for the synthesis of important ring-cleaved aromatic compounds such as muconic acid (MA) and PHA has been reported through synthetic biology approaches. Muconic acid has been produced by ortho-cleavage of catechol through the heterologous expression of catA and aroY genes (encoding dioxygenase and PCA decarboxylase respectively) in *P. putida* KT2440 and from phenol, MA has been produced in same organism by the integration of dmpKLMNOP gene encoding phenol monooxygenase in same strain [77].

Furthermore, MA synthesis can also be performed by the development of a double knock-out mutant by the deletion of two muconate cyclo-isomerases from the mutant *Amycolatopsis sp*. ATCC 39116 MA-2 [78]. In addition, the heterologous expressions of ligV, ligM, and aroY genes in *E. coli* enabled the production of MA from vanillin [60]. Apart from that, aromatic carboxylic acids such as Pyridine 2,4-dicarboxylic acid and pyridine 2,5-dicarboxylic acid have also been generated through lignin valorization. Recently, the introduction of genes for PCA

4,5-dioxygenase or PCA 2,3-dioxygenase enzymes into *R. jostii RHA1* has been reported to produce these acids from wheat straw biomass [79]. More recently, an important application has been envisaged to produce biodegradable medium chain-length polyhydroxyalkanoates (mcl-PHAs) from *P. putida* [80]. The bacterial enzymatic machinery with the dyp peroxidase expression was enhanced by modification and optimization of various central and peripheral pathways by synthetic biology approaches [79]. Furthermore, bioactive compounds such as coumarins, stilbenoids, flavonoids, resveratrol, p-coumaric acid and ferulic acid have been produced during lignin valorization. The widely adopted *E. coli* or *S. cerevisiae* have been utilized for the above purpose. The *denovo* pathway has been built in *E. coli* by introducing various genes from other organisms such as *Nocardia* carboxylic acid reductase, *Rhodiola* glycosyltransferase UGT73B6 and endogenous alcohol dehydrogenases [81]. Similarly, the glycosylated form of gallic acid can be used to produce β-Glucogallin through the heterologous expression of glucosyl-transferase gene VvGT2 in *E. coli* derived from *Vitis vinifera* [82]. Similarly, the co-expression of two genes VpScVAN and AtUGT72E2 in S. cerevisiae has been reported to produce vanillin glucoside [83]. Therefore, based on the above discussion, synthetic biology tools provide an excellent platform for the production of diversified value-added products from lignin valorization, however, more comprehensive studies must be required to elucidate the complex degradation pathways of microbial cell factories.

Current Challenges and Future Prospects

The lignin valorization and subsequent set-up of lignin based biorefinery have been considered as one of the sustainable alternatives for biobased energy resources. The technology has shown promising potentials in saving the rich reserves of our earth in the form of fossil fuels that need millions of years to replenish in the environment. In spite of showing marvelous credentials, the technology is far behind in commercialization and gross availability. In 2011, National Regional Energy Laboratory (NREL) proposed various scenarios for lignin utilization. Scenario 1 proposed that 23.9% of energy after the combustion of lignin-rich wastes was used for heat or steam, and 23.6% of energy was used for the grid purpose in revenue generation. In addition, 52.5% was used for electricity consumption. In scenario 2, 23.6% of the lignin has been broken down for the conversion of various products while the rest of the lignin biomass has been utilized for heat, steam, and electricity purposes. In scenario 3, 76.1% of lignin biomass converted into value-added products and the rest 23.9% was used for heat and steam generation [84]. Thus, in the current scenario, the landscape of lignin valorization is still skeptical.

The recalcitrant nature of lignin itself imposes a considerable technical challenge. Though genetic engineering has been advanced, but developing single efficient microbial cell factory for lignin valorization is still a formidable task. The intermediate aromatic compounds released during lignin depolymerization act as growth inhibitors effects. Therefore manipulation of cells toward product inhibition is also a significant challenge. The optimization of pentose sugar utilization pathway in microbial cell factories is also a current challenge, however, a lot of progress has been envisaged with respect to cellulosic and hemi-cellulosic biomass [85]. Heterogeneity in lignin composition, also confers a substantial difference in biomass pretreatment, and separation of value-added products. In the past decade, a breakthrough has been seen in the biological valorization of lignin. The research studies have been advanced in identifying novel ligninolytic enzymes but studies related to their secretion pathway as well as molecular mechanism need to be elucidated more comprehensively for expediting protein engineering tools. Moreover, there is a pressing need to enhance their catalytic efficiency in normal culture. There are scanty studies on characterized enzymes from bacterial cell factories that lead to the extracellular conversion of HMW to LMW lignin compounds [86]. The research could be further directed to develop a mechanism for consolidated bioprocessing in which lignin depolymerization and value-added products' production could take place simultaneously. Apart from that, various depolymerized compounds after lignin breakdown do not possess the properties required for further conversion.

Thus, more thorough studies accompanying synthetic biology are required. The studies should be more focused on flux balance analysis and redox balance to up-scale value-added products from lignin. The analysis of system biology metabolic pathway of various lignolytic organisms should be more precisely done for the biosynthesis of targeted value-added products [87]. Various resources have been used to produce value-added products [87 - 93]. In addition, nanotechnology approaches can be used to produce valuable products [94 - 98]. The futuristic studies will be done to find novel strains and enzymes based on culture-independent approaches. Moreover, the recently discovered genome editing tools like Crispr-Cas, TALENS should also be applied to achieve desired characteristics. Simultaneously, computational databases should be strengthened for the utilization of data related to lignocellulosic biomass utilization. In addition to that, the economic analysis of lignin-based biorefinery will be done in detail for evaluating the cost drivers and economic impact at social level. Thus, focus on these techno-economic parameters will help develop sustainable industries with judicious utilization of lignin-based bioresources.

CONCLUSION

It is clearly contemplated that renewability and abundance of waste lignin paved the way for the replacement of chemical-based non-renewable and hazardous chemical entities in various manufacturing industries. The valorization of lignin into value-added products holds tremendous potential in closing the door to the overwhelming use of our fossil fuels and leads to the development of sustainable economy. However, prodigious research must be needed to upgrade the conversion of various aromatic compounds into more beneficial compounds for societal relevance. Microbial cell factories hold great prospects for lignin valorization but still have many technical challenges that need to be overcome. Therefore, the application of advanced genome editing tools, "omics" technologies and synthetic biology will definably give new dimensions to this research. Thus, by recognizing the numerous value-added products from lignin, it a pressing need to set up a complete biorefinery process that will impart a sustainable and profitable economy in the future.

ACKNOWLEDGEMENTS

The authors are highly grateful to the Gautam Buddha University for providing technical support for writing this chapter.

REFERENCES

[1]   Lourenço A, Pereira H. Compositional Variability of Lignin in Biomass. Intechopen 2018; pp. 1-34.

[2]   Hodásová L, Jablonsky M, Skulcova AB, Haz A. Lignin, potential products and their market value. Wood Res 2015; 60(6): 973-86.

[3]   Rabinovitch-Deere CA, Oliver JWK, Rodriguez GM, Atsumi S. Synthetic biology and metabolic engineering approaches to produce biofuels. Chem Rev 2013; 113(7): 4611-32.
        [http://dx.doi.org/10.1021/cr300361t] [PMID: 23488968]

[4]   Hämäläinen V, Grönroos T, Suonpää T, *et al.* Enzymatic Processes to Unlock the Lignin Value Front Bioeng Biotechnol 2018; 6: 1-10.

[5]   Amthor JS. Efficiency of lignin biosynthesis: a quantitative analysis. Ann Bot (Lond) 2003; 91(6): 673-95.
        [http://dx.doi.org/10.1093/aob/mcg073] [PMID: 12714366]

[6]   Wan C, Li Y. Solid-State Biological Pretreatment of Lignocellulosic Biomass.Green Biomass Pretreatment for Biofuels Production SpringerBriefs in Molecular Science. Dordrecht: Springer 2013; pp. 67-86.
        [http://dx.doi.org/10.1007/978-94-007-6052-3_3]

[7]   Heinzkill M, Bech L, Halkier T, Schneider P, Anke T. Characterization of laccases and peroxidases from wood-rotting fungi (family Coprinaceae). Appl Environ Microbiol 1998; 64(5): 1601-6.
        [http://dx.doi.org/10.1128/AEM.64.5.1601-1606.1998] [PMID: 9572923]

[8]   Lu Y, Lu YC, Hu HQ, Xie FJ, Wei XY, Fan X. Structural Characterization of Lignin and Its Degradation Products with Spectroscopic Methods. J Spectrosc 2017; 2017: 1-15.
        [http://dx.doi.org/10.1155/2017/8951658]

[9]   Abdelaziz OY, Brink DP, Prothmann J, *et al.* Biological valorization of low molecular weight lignin. Biotechnol Adv 2016; 34(8): 1318-46.
[http://dx.doi.org/10.1016/j.biotechadv.2016.10.001] [PMID: 27720980]

[10]   Hatakka A, Hammel KE. Fungal Biodegradation of Lignocelluloses Mycota 2019; 10: 319-40.

[11]   Becker J, Wittmann C. A field of dreams: Lignin valorization into chemicals, materials, fuels, and health-care products. Biotechnol Adv 2019; 37(6)107360.
[http://dx.doi.org/10.1016/j.biotechadv.2019.02.016] [PMID: 30959173]

[12]   Stainer RY, Ornston LN. The β-Ketoadipate Pathway. Adv Microb Physiol 1973; 9(0): 89-151.
[http://dx.doi.org/10.1016/S0065-2911(08)60377-X] [PMID: 4599397]

[13]   Durante G, Álvarez HG, Blázquez B, *et al.* Anaerobic pathways for the catabolism of aromatic compounds.Lignin Valorization: Emerging Approaches. The Royal Society of Chemistry 2018; pp. 330-90.
[http://dx.doi.org/10.1039/9781788010351-00333]

[14]   Cannatelli MD, Ragauskas AJ. Conversion of lignin into value-added materials and chemicals via laccase-assisted copolymerization. Appl Microbiol Biotechnol 2016; 100(20): 8685-91.
[http://dx.doi.org/10.1007/s00253-016-7820-1] [PMID: 27645296]

[15]   Arevalo-Gallegos A, Ahmad Z, Asgher M, Parra-Saldivar R, Iqbal HMN. Lignocellulose: A sustainable material to produce value-added products with a zero waste approach—A review. Int J Biol Macromol 2017; 99: 308-18.
[http://dx.doi.org/10.1016/j.ijbiomac.2017.02.097] [PMID: 28254573]

[16]   Wang YY, Meng X, Pu Y, Ragauskas AJ. Recent Advances in the Application of Functionalized Lignin in Value-Added Polymeric Materials. Polymers (Basel) 2020; 12(10): 1-25.
[http://dx.doi.org/10.3390/polym12102277] [PMID: 33023014]

[17]   Zhu W, Westman G, Theliander H. Investigation and characterization of lignin precipitation in the lignoboost process. J Wood Chem Technol 2014; 34(2): 77-97.
[http://dx.doi.org/10.1080/02773813.2013.838267]

[18]   Vitorino L, Bessa L. Microbial diversity: The gap between the estimated and the known. Diversity (Basel) 2018; 10(2): 46.
[http://dx.doi.org/10.3390/d10020046]

[19]   Brink DP, Ravi K, Lidén G, Gorwa-Grauslund MF. Mapping the diversity of microbial lignin catabolism: experiences from the eLignin database. Appl Microbiol Biotechnol 2019; 103(10): 3979-4002.
[http://dx.doi.org/10.1007/s00253-019-09692-4] [PMID: 30963208]

[20]   Artimo P, Jonnalagedda M, Arnold K, *et al.* ExPASy: SIB bioinformatics resource portal. Nucleic Acids Res 2012; 40(Web Server issue)W597-603.
[PMID: 22661580]

[21]   Kanehisa M, Goto S, Sato Y, Kawashima M, Furumichi M, Tanabe M. Data, information, knowledge and principle: back to metabolism in KEGG. Nucleic Acids Res 2014; 42(D1): D199-205.
[http://dx.doi.org/10.1093/nar/gkt1076] [PMID: 24214961]

[22]   Huang L, Zhang H, Wu P, *et al.* dbCAN-seq: a database of carbohydrate-active enzyme (CAZyme) sequence and annotation. Nucleic Acids Res 2018; 46(D1): D516-21.
[http://dx.doi.org/10.1093/nar/gkx894] [PMID: 30053267]

[23]   Cantarel BL, Coutinho PM, Rancurel C, Bernard T, Lombard V, Henrissat B. The Carbohydrate-Active EnZymes database (CAZy): an expert resource for Glycogenomics. Nucleic Acids Res 2009; 37(Database): D233-8.
[http://dx.doi.org/10.1093/nar/gkn663] [PMID: 18838391]

[24]   Quinlan RJ, Sweeney MD, Lo Leggio L, *et al.* Insights into the oxidative degradation of cellulose by a

copper metalloenzyme that exploits biomass components. Proc Natl Acad Sci USA 2011; 108(37): 15079-84.
[http://dx.doi.org/10.1073/pnas.1105776108] [PMID: 21876164]

[25] Li X, Beeson WT IV, Phillips CM, Marletta MA, Cate JHD. Structural basis for substrate targeting and catalysis by fungal polysaccharide monooxygenases. Structure 2012; 20(6): 1051-61.
[http://dx.doi.org/10.1016/j.str.2012.04.002] [PMID: 22578542]

[26] Bey M, Zhou S, Poidevin L, *et al.* Cello-oligosaccharide oxidation reveals differences between two lytic polysaccharide monooxygenases (family GH61) from Podospora anserina. Appl Environ Microbiol 2013; 79(2): 488-96.
[http://dx.doi.org/10.1128/AEM.02942-12] [PMID: 23124232]

[27] Bugg TDH, Ahmad M, Hardiman EM, Singh R. The emerging role for bacteria in lignin degradation and bio-product formation. Curr Opin Biotechnol 2011; 22(3): 394-400.
[http://dx.doi.org/10.1016/j.copbio.2010.10.009] [PMID: 21071202]

[28] Wang L, Nie Y, Tang YQ, *et al.* Diverse bacteria with lignin degrading potentials isolated from two ranks of coal. Front Microbiol 2016; 7: 1428.
[http://dx.doi.org/10.3389/fmicb.2016.01428] [PMID: 27667989]

[29] Yang C, Yue F, Cui Y, *et al.* Biodegradation of lignin by *Pseudomonas* sp. Q18 and the characterization of a novel bacterial DyP-type peroxidase. J Ind Microbiol Biotechnol 2018; 45(10): 913-27.
[http://dx.doi.org/10.1007/s10295-018-2064-y] [PMID: 30051274]

[30] Prabhakaran M, Couger MB, Jackson CA, Weirick T, Fathepure BZ. Genome sequences of the lignin-degrading Pseudomonas Sp. strain YS-1p and Rhizobium Sp. strain YS-1r isolated from decaying wood. Genome Announc 2015; 3(2)e00019-15.
[http://dx.doi.org/10.1128/genomeA.00019-15.] [PMID: 25744986]

[31] Moraes EC, Alvarez TM, Persinoti GF, *et al.* Lignolytic-consortium omics analyses reveal novel genomes and pathways involved in lignin modification and valorization. Biotechnol Biofuels 2018; 11(1): 75.
[http://dx.doi.org/10.1186/s13068-018-1073-4] [PMID: 29588660]

[32] Gasson MJ, Kitamura Y, McLauchlan WR, *et al.* Metabolism of ferulic acid to vanillin. A bacterial gene of the enoyl-SCoA hydratase/isomerase superfamily encodes an enzyme for the hydration and cleavage of a hydroxycinnamic acid SCoA thioester. J Biol Chem 1998; 273(7): 4163-70.
[http://dx.doi.org/10.1074/jbc.273.7.4163] [PMID: 9461612]

[33] Strachan CR, Singh R, VanInsberghe D, *et al.* Metagenomic scaffolds enable combinatorial lignin transformation. Proc Natl Acad Sci USA 2014; 111(28): 10143-8.
[http://dx.doi.org/10.1073/pnas.1401631111] [PMID: 24982175]

[34] Granja-Travez RS, Bugg TDH. Characterization of multicopper oxidase CopA from Pseudomonas putida KT2440 and Pseudomonas fluorescens Pf-5: Involvement in bacterial lignin oxidation. Arch Biochem Biophys 2018; 660(1): 97-107.
[http://dx.doi.org/10.1016/j.abb.2018.10.012] [PMID: 30347180]

[35] Nielsen J, Keasling JD. Engineering cellular metabolism. Cell 2016; 164(6): 1185-97.
[http://dx.doi.org/10.1016/j.cell.2016.02.004] [PMID: 26967285]

[36] Rogers JK, Taylor ND, Church GM. Biosensor-based engineering of biosynthetic pathways. Curr Opin Biotechnol 2016; 42: 84-91.
[http://dx.doi.org/10.1016/j.copbio.2016.03.005] [PMID: 26998575]

[37] Holtz WJ, Keasling JD. Engineering static and dynamic control of synthetic pathways. Cell 2010; 140(1): 19-23.
[http://dx.doi.org/10.1016/j.cell.2009.12.029] [PMID: 20085699]

[38] Paepe B De, Peters G, Coussement P, et al. Tailor-made transcriptional biosensors for optimizing

microbial cell factories. J Ind Microbiol Biotechnol 201; 44(4-5): 623-45

[39] Shi S, Ang EL, Zhao H. *In vivo* biosensors: mechanisms, development, and applications. J Ind Microbiol Biotechnol 2018; 45(7): 491-516.
[http://dx.doi.org/10.1007/s10295-018-2004-x] [PMID: 29380152]

[40] Ravikumar S, Baylon MG, Park SJ, Choi J. Engineered microbial biosensors based on bacterial two-component systems as synthetic biotechnology platforms in bioremediation and biorefinery. Microb Cell Fact 2017; 16(1): 62.
[http://dx.doi.org/10.1186/s12934-017-0675-z] [PMID: 28410609]

[41] Rogers JK, Guzman CD, Taylor ND, Raman S, Anderson K, Church GM. Synthetic biosensors for precise gene control and real-time monitoring of metabolites. Nucleic Acids Res 2015; 43(15): 7648-60.
[http://dx.doi.org/10.1093/nar/gkv616] [PMID: 26152303]

[42] Siedler S, Khatri NK, Zsohár A, *et al.* Development of a bacterial biosensor for rapid screening of yeast p-coumaric acid production. ACS Synth Biol 2017; 6(10): 1860-9.
[http://dx.doi.org/10.1021/acssynbio.7b00009] [PMID: 28532147]

[43] Kwon KK, Lee DH, Kim SJ, *et al.* Evolution of enzymes with new specificity by high-throughput screening using DmpR-based genetic circuits and multiple flow cytometry rounds. Sci Rep 2018; 8(1): 2659.
[http://dx.doi.org/10.1038/s41598-018-20943-8] [PMID: 29422524]

[44] Ribeiro LF, Tullman J, Nicholes N, *et al.* A xylose-stimulated xylanase–xylose binding protein chimera created by random nonhomologous recombination. Biotechnol Biofuels 2016; 9(1): 119.
[http://dx.doi.org/10.1186/s13068-016-0529-7] [PMID: 27274356]

[45] Teo WS, Chang MW. Bacterial XylRs and synthetic promoters function as genetically encoded xylose biosensors in *Saccharomyces cerevisiae*. Biotechnol J 2015; 10(2): 315-22.
[http://dx.doi.org/10.1002/biot.201400159] [PMID: 24975936]

[46] Ribeiro LF, Bressan F, Furtado GP, Meireles F, Ward RJ. d-Xylose detection in Escherichia coli by a xylose binding protein-dependent response. J Biotechnol 2013; 168(4): 440-5.
[http://dx.doi.org/10.1016/j.jbiotec.2013.10.019] [PMID: 24161920]

[47] Kaper T, Lager I, Looger LL, Chermak D, Frommer WB. Fluorescence resonance energy transfer sensors for quantitative monitoring of pentose and disaccharide accumulation in bacteria. Biotechnol Biofuels 2008; 1(1): 11.
[http://dx.doi.org/10.1186/1754-6834-1-11] [PMID: 18522753]

[48] Kelly CL, Liu Z, Yoshihara A, *et al.* Synthetic chemical inducers and genetic decoupling enable orthogonal control of the rhaBAD promoter. ACS Synth Biol 2016; 5(10): 1136-45.
[http://dx.doi.org/10.1021/acssynbio.6b00030] [PMID: 27247275]

[49] Kunjapur AM, Prather KLJ. Development of a vanillate biosensor for the vanillin biosynthesis pathway in E. coli. ACS Synth Biol 2019; 8(9): 1958-67.
[http://dx.doi.org/10.1021/acssynbio.9b00071] [PMID: 31461264]

[50] F M Machado L, Currin A, Dixon N. Directed evolution of the PcaV allosteric transcription factor to generate a biosensor for aromatic aldehydes. J Biol Eng 2019; 13(1): 91.
[http://dx.doi.org/10.1186/s13036-019-0214-z] [PMID: 31798685]

[51] Ho JCH, Pawar SV, Hallam SJ, Yadav VG. An improved whole-cell biosensorfor the discovery of lignin-transforming enzymes in functional metagenomic screens. ACS Synth Biol 2018; 7(2): 392-8.
[http://dx.doi.org/10.1021/acssynbio.7b00412] [PMID: 29182267]

[52] Fiorentino G, Ronca R, Bartolucci S. A novel E. coli biosensor for detecting aromatic aldehydes based on a responsive inducible archaeal promoter fused to the green fluorescent protein. Appl Microbiol Biotechnol 2009; 82(1): 67-77.
[http://dx.doi.org/10.1007/s00253-008-1771-0] [PMID: 18998120]

[53] Xue H, Shi H, Yu Z, *et al.* Design, construction, and characterization of a set of biosensors for aromatic compounds. ACS Synth Biol 2014; 3(12): 1011-4.
[http://dx.doi.org/10.1021/sb500023f] [PMID: 25524112]

[54] Kamimura N, Takahashi K, Mori K, *et al.* Bacterial catabolism of lignin-derived aromatics: New findings in a recent decade: Update on bacterial lignin catabolism. Environ Microbiol Rep 2017; 9(6): 679-705.
[http://dx.doi.org/10.1111/1758-2229.12597] [PMID: 29052962]

[55] Ahmad M, Roberts JN, Hardiman EM, Singh R, Eltis LD, Bugg TDH. Identification of DypB from Rhodococcus jostii RHA1 as a lignin peroxidase. Biochemistry 2011; 50(23): 5096-107.
[http://dx.doi.org/10.1021/bi101892z] [PMID: 21534568]

[56] Roberts JN, Singh R, Grigg JC, Murphy MEP, Bugg TDH, Eltis LD. Characterization of dye-decolorizing peroxidases from Rhodococcus jostii RHA1. Biochemistry 2011; 50(23): 5108-19.
[http://dx.doi.org/10.1021/bi200427h] [PMID: 21534572]

[57] Machado LFM, Dixon N. Development and substrate specificity screening of an *in vivo* biosensor for the detection of biomass derived aromatic chemical building blocks. Chem Commun (Camb) 2016; 52(76): 11402-5.
[http://dx.doi.org/10.1039/C6CC04559F] [PMID: 27722239]

[58] Jha RK, Kern TL, Fox DT, M Strauss CE. Engineering an Acinetobacter regulon for biosensing and high-throughput enzyme screening in E. coli via flow cytometry. Nucleic Acids Res 2014; 42(12): 8150-60.
[http://dx.doi.org/10.1093/nar/gku444] [PMID: 24861620]

[59] Lo TM, Chng SH, Teo WS, Cho HS, Chang MW. A two-layer gene circuit for decoupling cell growth from metabolite production. Cell Syst 2016; 3(2): 133-43.
[http://dx.doi.org/10.1016/j.cels.2016.07.012] [PMID: 27559924]

[60] Wu W, Liu F, Singh S. Toward engineering *E. coli* with an autoregulatory system for lignin valorization. Proc Natl Acad Sci USA 2018; 115(12): 2970-5.
[http://dx.doi.org/10.1073/pnas.1720129115] [PMID: 29500185]

[61] Zhou Y, Lin L, Wang H, Zhang Z, Zhou J, Jiao N. Development of a CRISPR/Cas9n-based tool for metabolic engineering of Pseudomonas putida for ferulic acid-to-polyhydroxyalkanoate bioconversion. Commun Biol 2020; 3(1): 98.
[http://dx.doi.org/10.1038/s42003-020-0824-5] [PMID: 32139868]

[62] Gaj T, Gersbach CA, Barbas CF III. ZFN, TALEN, and CRISPR/Cas-based methods for genome engineering. Trends Biotechnol 2013; 31(7): 397-405.
[http://dx.doi.org/10.1016/j.tibtech.2013.04.004] [PMID: 23664777]

[63] Clarkson SM, Giannone RJ, Kridelbaugh DM, Elkins JG, Guss AM, Michener JK. Construction and Optimization of a Heterologous Pathway for Protocatechuate Catabolism in Escherichia coli Enables Bioconversion of Model Aromatic Compounds. Appl Environ Microbiol 2017; 83(18)e01313-17.
[http://dx.doi.org/10.1128/AEM.01313-17.] [PMID: 28733280]

[64] Machovina MM, Mallinson SJB, Knott BC, *et al.* Enabling microbial syringol conversion through structure-guided protein engineering. Proc Natl Acad Sci USA 2019; 116(28): 13970-6.
[http://dx.doi.org/10.1073/pnas.1820001116] [PMID: 31235604]

[65] Varman AM, Follenfant R, Liu F, Davis RW, Lin YK, Singh S. Hybrid phenolic-inducible promoters towards construction of self-inducible systems for microbial lignin valorization. Biotechnol Biofuels 2018; 11(1): 182.
[http://dx.doi.org/10.1186/s13068-018-1179-8] [PMID: 29988329]

[66] Zhang R, Zhao CH, Chang HC, Chai MZ, Li BZ, Yuan YJ. Lignin valorization meets synthetic biology. Eng Life Sci 2019; 19(6): 463-70.
[http://dx.doi.org/10.1002/elsc.201800133] [PMID: 32625023]

[67]   Jung DH, Kim EJ, Jung E, Kazlauskas RJ, Choi KY, Kim BG. Production of *p*- hydroxybenzoic acid from *p*- coumaric acid by *Burkholderia glumae* BGR1. Biotechnol Bioeng 2016; 113(7): 1493-503.
[http://dx.doi.org/10.1002/bit.25908] [PMID: 26693833]

[68]   Miao L, Li Q, Diao A, Zhang X, Ma Y. Construction of a novel phenol synthetic pathway in Escherichia coli through 4-hydroxybenzoate decarboxylation. Appl Microbiol Biotechnol 2015; 99(12): 5163-73.
[http://dx.doi.org/10.1007/s00253-015-6497-1] [PMID: 25758959]

[69]   Sainsbury PD, Hardiman EM, Ahmad M, *et al.* Breaking down lignin to high-value chemicals: the conversion of lignocellulose to vanillin in a gene deletion mutant of Rhodococcus jostii RHA1. ACS Chem Biol 2013; 8(10): 2151-6.
[http://dx.doi.org/10.1021/cb400505a] [PMID: 23898824]

[70]   Ni J, Gao YY, Tao F, Liu HY, Xu P. Temperature-directed biocatalysis for the sustainable production of aromatic aldehydes or alcohols. Angew Chem Int Ed 2018; 57(5): 1214-7.
[http://dx.doi.org/10.1002/anie.201710793] [PMID: 29178412]

[71]   Ni J, Wu YT, Tao F, Peng Y, Xu P. A Coenzyme-free biocatalyst for the value-added utilization of lignin-derived aromatics. J Am Chem Soc 2018; 140(47): 16001-5.
[http://dx.doi.org/10.1021/jacs.8b08177] [PMID: 30376327]

[72]   Hansen EH, Møller BL, Kock GR, *et al.* De novo biosynthesis of vanillin in fission yeast (Schizosaccharomyces pombe) and baker's yeast (Saccharomyces cerevisiae). Appl Environ Microbiol 2009; 75(9): 2765-74.
[http://dx.doi.org/10.1128/AEM.02681-08] [PMID: 19286778]

[73]   Du CJ, Rios-Solis L, Ward JM, Dalby PA, Lye GJ. Evaluation of CV2025 ω-transaminase for the bioconversion of lignin breakdown products into value-added chemicals: synthesis of vanillylamine from vanillin. Biocatal Biotransform 2014; 32(5-6): 302-13.
[http://dx.doi.org/10.3109/10242422.2014.976632]

[74]   Wu W, Dutta T, Varman AM, *et al.* Lignin valorization: two hybrid biochemical routes for the conversion of polymeric lignin into value-added chemicals. Sci Rep 2017; 7(1): 8420.
[http://dx.doi.org/10.1038/s41598-017-07895-1] [PMID: 28827602]

[75]   Kamimura N, Goto T, Takahashi K, *et al.* A bacterial aromatic aldehyde dehydrogenase critical for the efficient catabolism of syringaldehyde. Sci Rep 2017; 7(1): 44422.
[http://dx.doi.org/10.1038/srep44422] [PMID: 28294121]

[76]   Chen Z, Shen X, Wang J, Wang J, Yuan Q, Yan Y. Rational engineering of *p* -hydroxybenzoate hydroxylase to enable efficient gallic acid synthesis via a novel artificial biosynthetic pathway. Biotechnol Bioeng 2017; 114(11): 2571-80.
[http://dx.doi.org/10.1002/bit.26364] [PMID: 28650068]

[77]   Vardon DR, Franden MA, Johnson CW, *et al.* Adipic acid production from lignin. Energy Environ Sci 2015; 8(2): 617-28.
[http://dx.doi.org/10.1039/C4EE03230F]

[78]   Barton N, Horbal L, Starck S, Kohlstedt M, Luzhetskyy A, Wittmann C. Enabling the valorization of guaiacol-based lignin: Integrated chemical and biochemical production of cis,cis-muconic acid using metabolically engineered Amycolatopsis sp ATCC 39116. Metab Eng 2018; 45: 200-10.
[http://dx.doi.org/10.1016/j.ymben.2017.12.001] [PMID: 29246517]

[79]   Lee S, Kang M, Bae JH, Sohn JH, Sung BH. Bacterial Valorization of Lignin: Strains, Enzymes, Conversion Pathways, Biosensors, and Perspectives. Front Bioeng Biotechnol 2019; 7: 209.
[http://dx.doi.org/10.3389/fbioe.2019.00209] [PMID: 31552235]

[80]   Linger JG, Vardon DR, Guarnieri MT, *et al.* Lignin valorization through integrated biological funneling and chemical catalysis. Proc Natl Acad Sci USA 2014; 111(33): 12013-8.
[http://dx.doi.org/10.1073/pnas.1410657111] [PMID: 25092344]

[81] Bai Y, Yin H, Bi H, Zhuang Y, Liu T, Ma Y. De novo biosynthesis of Gastrodin in Escherichia coli. Metab Eng 2016; 35: 138-47.
[http://dx.doi.org/10.1016/j.ymben.2016.01.002] [PMID: 26804288]

[82] De Bruyn F, De Paepe B, Maertens J, *et al.* Development of an in vivo glucosylation platform by coupling production to growth: Production of phenolic glucosides by a glycosyltransferase of *Vitis vinifera.* Biotechnol Bioeng 2015; 112(8): 1594-603.
[http://dx.doi.org/10.1002/bit.25570] [PMID: 25728421]

[83] Gallage NJ, Hansen EH, Kannangara R, *et al.* Vanillin formation from ferulic acid in Vanilla planifolia is catalysed by a single enzyme. Nat Commun 2014; 5(1): 4037.
[http://dx.doi.org/10.1038/ncomms5037] [PMID: 24941968]

[84] Shen R, Tao L, Yang B. Techno-economic analysis of jet-fuel production from biorefinery waste lignin. Biofuels Bioprod Biorefin 2018; 13(10): 486-501.

[85] Francois JM, Alkim C, Morin N. Engineering microbial pathways for production of bio-based chemicals from lignocellulosic sugars: current status and perspectives. Biotechnol Biofuels 2020; 13(1): 118.
[http://dx.doi.org/10.1186/s13068-020-01744-6] [PMID: 32670405]

[86] Gonçalves CC, Bruce T, Silva COG, *et al.* Bioprospecting Microbial Diversity for Lignin Valorization: Dry and Wet Screening Methods. Front Microbiol 2020; 11: 1081.
[http://dx.doi.org/10.3389/fmicb.2020.01081] [PMID: 32582068]

[87] Xu Z, Lei P, Zhai R, Wen Z, Jin M. Recent advances in lignin valorization with bacterial cultures: microorganisms, metabolic pathways, and bio-products. Biotechnol Biofuels 2019; 12(1): 32.
[http://dx.doi.org/10.1186/s13068-019-1376-0] [PMID: 30815030]

[88] Awasthi MK, Kumar V, Yadav V, *et al.* Current state of the art biotechnological strategies for conversion of watermelon wastes residues to biopolymers production: A review. Chemosphere 2022; 290133310.
[http://dx.doi.org/10.1016/j.chemosphere.2021.133310.] [PMID: 34919909]

[89] Liu H, Kumar V, Jia L, *et al.* Biopolymer poly-hydroxyalkanoates (PHA) production from apple industrial waste residues: A review. Chemosphere 2021; 284131427
[http://dx.doi.org/10.1016/j.chemosphere.2021.131427] [PMID: 34323796]

[90] Awasthi SK, Kumar M, Kumar V, *et al.* A comprehensive review on recent advancements in biodegradation and sustainable management of biopolymers. Environ Pollut 2022; 307119600
[http://dx.doi.org/10.1016/j.envpol.2022.119600.] [PMID: 35691442]

[91] Duan Y, Tarafdar A, Kumar V, *et al.* Sustainable biorefinery approaches towards circular economy for conversion of biowaste to value added materials and future perspectives. Fuel 2022; 325124846.
[http://dx.doi.org/10.1016/j.fuel.2022.124846.]

[92] Kumar V, Sharma N, Umesh M, *et al.* Emerging challenges for the agro-industrial food waste utilization: A review on food waste biorefinery. Bioresour Technol 2022; 362127790.
[http://dx.doi.org/10.1016/j.biortech.2022.127790.] [PMID: 35973569]

[93] Awasthi SK, Sarsaiya S, Kumar V, *et al.* Processing of municipal solid waste resources for a circular economy in China: An overview. Fuel 2022; 317123478.
[http://dx.doi.org/10.1016/j.fuel.2022.123478.]

[94] Kumar V, Sharma N, Maitra SS. *In vitro* and *in vivo* toxicity assessment of nanoparticles. Int Nano Lett 2017; 7(4): 243-56.
[http://dx.doi.org/10.1007/s40089-017-0221-3]

[95] K. Vinay, S. Neha, and S. Maitra, Protein and Peptide Nanoparticles: Preparation and Surface Modification, in Functionalized Nanomaterials I: CRC Press, 2020, pp. 191-204.

[96] S. Vallinayagam et al., Recent developments in magnetic nanoparticles and nano-composites for

wastewater treatment, Journal of Environmental Chemical Engineering, vol. 9, no. 6, p. 106553, 2021.

[97]   V. Kumar, N. Sharma, S. K. Lakkaboyana, and S. S. Maitra, Silver nanoparticles in poultry health: Applications and toxicokinetic effects, in Silver Nanomaterials for Agri-Food Applications: Elsevier, 2021, pp. 685-704.

[98]   T. C. Egbosiuba, Biochar and bio-oil fuel properties from nickel nanoparticles assisted pyrolysis of cassava peel, Heliyon, vol. 8, no. 8, p. e10114, 2022/08/01/ 2022

CHAPTER 3

Food Waste Bioconversion To High-value Products

Anjali Khajuria[1], Abhinay Thakur[2] and Rahul Datta[3,*]

[1] *Department of Zoology, Central University of Jammu, Rahya-Suchani (Bagla), District Samba, J&K-181143, India*

[2] *Assistant Professor, PG Department of Zoology, DAV College Jalandhar (Punjab)-144008, India*

[3] *Centre for Agricultural Research and Innovation, Guru Nanak Dev University, Amritsar, Punjab, 143005, India*

Abstract: During the last few decades, food remains a primary concern throughout the world as it is depleting day by day. On the other side, its residual waste is accumulating over time. Around one-third of food produced for human consumption is wasted which escalates the environmental issues and ecological burden. Management of waste food by current methods is cost-ineffective with adverse impacts on the environment. Therefore, attempts have been made to convert food waste into high-value by-products. Being a rich source of carbohydrates, proteins, sugars, and fats, it acts as a potential source for high-value products. The organic nature of food makes it a raw material for industries related to biofuel, bioactive compounds, prebiotics, livestock food, and biodegradable plastics. Bioconversion of food waste into valuable products not only provides economic advantage but reduces stress on landfills. The valorization of low-cost, abundantly available food waste into biofuel can decrease the demand for fossil fuels and economic loss for their manufacturing. Minimum food wastage and re-utilization of wasted food can be a sustainable approach to combating this problem. In this chapter, various techniques used for bioconversion and the valuable products produced by waste food processing have been discussed with their prospects.

Keywords: Bioconversion, Food waste, Sustainable, Value-added products, Valorization.

INTRODUCTION

Food waste (FW) has gained attention in the last few years due to several environmental, social and economic concerns, as well as climate change and scarcity of fossil fuel resources [1, 2]. Around 1.3 billion tonnes of food are wasted each year throughout the world which cost $750 billion, causing huge economic losses [3, 4]. In Asian countries, FW has seen a continuous increase from 278 to 416 million tonnes from 2005-2025 [5]. An alarming rise in the

---
[*] **Corresponding author Rahul Datta:** Centre for Agricultural Research and Innovation, Guru Nanak Dev University Amritsar, Punjab, India; Tel:9149688353; E-mail: rahuldutta1709@gmail.com

Vinay Kumar, Sivarama Krishna Lakkaboyana & Neha Sharma (Eds.)

human population leads to an increase in food requirements and consequently in food waste. Efforts have been made to convert food waste into high value products. According to FAO (Food and Agricultural Organization), around one-third of food produced globally for human consumption is lost along the food supply chain. About less than 30% of municipal solid waste comprises FW in all countries except highly populated countries like India and China where it ranges from 30-60% [6]. The lack of effective waste management, disposal and treatment strategies results in environmental problems [7]. Waste collection, storage and proper segregation are major concerns in suitable waste conversion. However, inappropriate management of waste resulted in several environmental issues and health hazards [8]. Being high in nutritional content, purefaction of FW occurs rapidly which creates a breeding ground for several disease-causing organisms [7]. Food waste is rich in several molecules *viz.* carbohydrates (starch, cellulose and hemicelluloses), protein, lipids, lignin, and organic acids, [9, 7]. In order to produce heat or energy from FW, it is incinerated which leads to air pollution [10]. Management of FW can be done by the conversion of food waste into value-added products *viz.* ethanol, enzymes, organic acid, biopolymers and bioplastic [11 - 21]. Several types of innovative strategies are being utilized for waste valo-rization such as the conversion of FW into biofuel and animal feed. Various methods such as biological, chemical as well as thermal are used to recover nutrients and high-value products from FW which are an important source of energy [22 - 27].

Food waste (FW) is divided into two types. These include pre-consumption food wastes (PrCFWs) and post-consumption food wastes (PCFWs) [28]. PrCFWs include vegetables, fruits, and other peeling wastes. PCFWs include 40-60% starchy waste (meats and meat trimmings, cheese whey and coffee filters), 5-10% protein (fish processing wastes and eggshells) and 10-40% various other fatty or oily contents [29 - 32]. PrCFWs waste is easy to decompose whereas decom-posing of PCFWs management is challenging because of separation issues and the huge amount of oil contents [28].

Diminishing natural resources, such as petroleum, rising fuel prices and increasing environmental concerns have enforced us to look for alternative sources of energy [2, 33]. Several types of food wastes are generated worldwide in huge quantities which are rich in important constituents that may serve as a starting point for the production of various types of valuable products, through several bioconversion pathways [33].The food industry is responsible for one of the highest consumptions of natural resources [34]. Food processing by-products also account for the huge amount of leftover resources that could be valued for the recovery of value-added products [35].

CURRENT SCENARIO

About $1/3^{rd}$ of food is wasted globally which could be used to feed millions of people around the globe [36]. Food waste occurs at different levels *viz:* prematurely harvesting of crops by farmers, lack of processing technologies, inefficient storage, market system, and sales conditions, overproduction than the requirement and many more [37]. Displayed high standards of fresh products in supermarkets or retail stores make them unsalable, contributing to countless food waste [38].

FW has the potential to transmute into economically valued products [39]. The most widely used approaches for food valorization are composting, using the animal feed, landfilling, and incineration [40]. Composting is one of the most important approaches for bioconversion. It is an eco-friendly and highly acceptable practice because it reduces stress on landfills and provides fertilizers. Thereby it also helps the farmers to reduce or eliminate the need to rely on chemical fertilizers [41, 42]. One of the important advantages of composting is that it avoids the emission of methane [42]. The most cost-effective method for food supply chain waste is animal feed unless there are regulatory issues as well as the nature of the co-product generated in the process [2]. A large amount of food waste ends up in landfills. Dumping of a huge amount of food waste in landfills is very costly and it also poses serious environmental concerns *i.e.* by the production of greenhouse gases (Methane and Co_2) directly or indirectly [42]. To exploit value-added products, advanced conversion, and extraction technologies should be implemented on the basis of green chemistry. The diversity of food composition reflects its potential which affects food valorization to be converted into an economic value-added product [43].

BIOLOGICAL AGENTS USED FOR BIOCONVERSION

Insects

Food waste bioconversion using insects involves the breakdown of food waste into smaller biomass [44, 45]. It is one of the most economically viable methods for turning large quantities of food waste into valuable materials such as feed for animals, biofuel, lubricants, pharmaceuticals, dyes, *etc* [45]. On the other hand, it has the advantage of reducing the load on the environment [46]. Commercial rearing of insects can efficiently turn several tonnes of FW feedstock into valuable products [47]. Few species of insects have been used so far for insect-based bioconversion of food waste. Some of the important species are *Hermetia illucens* (L.) commonly known as Soldier fly larvae, *Musca domestica* (Housefly), *Cydia pomonella* (Codling moth), *Teleogryllus testaceus* (Cambodian field crickets) and *Tenebrio molitor* (Yellow Mealworm) [48, 52] (Table **1**). The process can be performed in controlled conditions that favor their growth and bioconversion. However, valuable products are produced at multiple steps during bioconversion [44]. Table **1** presents different types of insects species which can be used for food waste bioconversion.

Table 1. Insect species used for food waste bioconversion into valuable products.

S.No.	Insect species	Food waste	Product	References
1.	Black soldier fly (*Hermetia illucens*)	Rice straw, restaurant waste	Biofuel	[53]
		Rice straw	Biomass	[54]
		Coffee husk	Biomass, Fertilizers	[55]
		Waste from pears, banana, and cucumber	Biomass	[56]
		Corn stover	Biofuel, soil Amendment	[46]
		Sorghum	Biofuel	[57]
2.	Housefly (*Musca domestica*)	Restaurant waste corn silage, Sawdust	Biomass, biofuel, fertilizer	[58]
3.	Codling moth (*Cydia pomonella*)	Starch and cheese wastewater sludge	Biomass	[50]

Fungus

It has been observed that fungal biomass has the capacity to convert food waste into several lipid-rich raw materials that can be used for the production of feed and biodiesel. Various types of fungal species have been identified such as *Candida tropicalis, Aspergillus oryzae, Fusarium oxysporum, Mucor circinelloides, Lipomyces starkeyi, Saccharomyces cerevisiae,* and *Yarrowia lipolytica* which grow on different food wastes and produce high-added value metabolites, such as ethanol SCOs, mannitol and citric acid [58, 60]. Various extracellular enzymes such as amylases and proteases have been secreted by different species of fungi such as *Rhizopus, Aspergillus*, and *Monascus* which are capable of digesting starch and proteins in the food waste. The bioconversion produces amino acids and fermentable sugars which can be used for their growth

[61]. It has been reported that immobilized β-galactosidase from *Aspergillus oryzae* can be used for ice-cream formulations, replacing 25% of sucrose with differences in sensory quality [62]. On the other hand, the conversion of glucose into fructose by immobilizing enzyme glucose isomerase from *Actinoplanes missouriensis* aids in the replacement of sucrose. Similarly, Lorenzen and his coworkers [63] studied the addition of enzymatically treated lactose from skim milk, sweet whey, and acid whey to yogurt. Conversion of lactose into galacto-oligosaccharides, glucose, lactulose, and galactose by using enzymes β-galactosidase (*Kluyveromyces lactis*) and glucose isomerase (*Streptomyces rubiginous*) in yogurt enhanced its sweetness. Immobilization of calcium alginate and gelatine spheres cross-linked with glutaraldehyde and lectin concanavalin A further enabled more than 70% of lactose hydrolysis [64]. Immobilization of enzymes on the substrate further enhanced the potential of β-galactosidase for lactose valorization over multiple cycles of reuse.

Bacteria

Cheese whey can be utilized to produce polymer-grade lactic acid. The lactose content of cheese whey can be effectively fermented to lactic acid using a microbial culture of *Lactobacillus helveticus*. Alternately, a culture of *Lactobacillus bulgaricus* may also be employed [65]. In another report, organic d-Lactic acid is produced from cotton seed by *Sporolactobacillus inulinus* YBS1-5 which reduces the raw material cost of fermentation. Cotton seed is a good source of nitrogen. For efficient utilization, enzymatic hydrolysis of the cotton seed meal was carried out and concurrently used for d-lactic acid fermentation. Also, corncob residues rich in lignocellulosic content were subjected to hydrolysis by *Sporolactobacillus inulinus* for d-lactic acid production [66]. Several bacterial species have been identified that produce succinic acid. These bacteria include *Actinobacillus succinogenes*, *Anaerobiospirillum*, *succiniciproducens*, *Bacteroides fragilis* and *Mannheimia succiniciproducens*. Fermentation of bakery waste (cakes and pastries) by *Actinobacillus succinogenes* resulted in a 28-35% yield of succinic acid [67, 68].

Enzyme Immobilization

Bioconversion of FW into high-value bioproducts at the industrial level frequently demands high investment costs for infrastructure and equipment. Often preliminary bench-scale experimental trials are required for large-scale bioconversion processes. To achieve it, an immobilized enzyme system has been introduced for FW bioconversion instead of a chemical catalytic system for long stability and reusability throughout the food waste processing [69]. Enzyme immobilization activates biocatalysts to act efficiently in non-ideal conditions.

The immobilized enzyme retained activity over 5 cycles, with ~70% conversion efficiency achieved after each cycle [70]. Enzyme immobilization methods have been seen as a well-engineered tool to facilitate the biocatalysts to catalyze the bioconversion process even under non-ideal conditions. Recently, Feng *et al.* [71] reported the production of biodiesel from soya sauce residue oil by using Macroporous acrylic resin-lipase. During bioconversion, *Aspergillus niger* lipase (ANL) has been widely used for its immobilization over various species.

BIOMASS BIOCONVERSION AND THEIR PRODUCTS

FW has been used as the sole microbial feedstock for the development of various value-added bioproducts which are economically beneficial. These include methane, hydrogen, ethanol, enzymes, organic acid, biopolymers, bioplastics, *etc.*

Carbohydrates Bioconversion and their Products

Food waste carbohydrates consisting of crude fibers and free sugars establish the principal source of carbon which can be used as a raw material in several processes resulting in the production of economically essential products such as biofuels, hydrogen, bioethanol, prebiotics, oils, and sweeteners [72, 2]. Food waste from restaurants, cafeterias, and hotels can be hydrolysed by the immobilization of enzymes to break down into sugar monomers. Simple forms of carbohydrates are more efficient and amenable to enzymatic valorization of different value-added products [9]. According to the United States Department of Energy (US DOE), there are various sugar-derived building blocks identified which can be used as a raw material for valuable products or carbon sources for fermentation (Table. **2**). The value-added products may include 1,4-diacids, 2,5-furan dicarboxylic acid, 3-hydroxy propionic acid, aspartic acid, glucaric acid, glutamic acid, itaconic acid, levulinic acid, 3-10 hydroxybutyrolactone, glycerol, sorbitol, and xylitol [73]. These constituents' sugars can be potential alternatives for petroleum-based materials [33].

Table 2. Products obtained from various chemicals after the bioconversion of food waste.

S.No.	Chemicals from Food waste	Valuable Products
1.	1,4-diacids	Platform chemical for specialized polyester, raw material for nylon precursor, solvents, fibers, water-soluble polymers, acidulants and taste enhancers for food and beverages.
2.	2,5-furan dicarboxylic acid	PET analogs used in bottles, films, and containers
3.	3-hydroxy propionic acid,	Carpet fibers, contact lenses, polymers for Diapers
4.	Aspartic acid	Sweeteners, chelating salts
5.	Glucaric acid	Solvents, hyperbranched polyesters, nylons

(Table 2) cont.....

S.No.	Chemicals from Food waste	Valuable Products
6.	Glutamic acid	Polyesters and polyamides precursor
7.	Itaconic acid	Monomer for the production of resins, plastics.
8.	Levulinic acid	Solvents, fuel oxygenates, replace bisphenol A in polycarbonate synthesis
9.	3-10 hydroxybutyrolactone	A chiral building block chemical for pharmaceutical products.
10.	Glycerol	Food and beverage, pharmaceuticals, Polyether polyols, personal and oral care.
11.	Sorbitol	Food sweeteners, emulsion stabilizers, antifreeze, PET-like polymers.
12.	Xylitol	Antifreeze, unsaturated polyester resins, non-nutritive sweeteners.
13.	Ethanol	Biofuel, solvent.
14.	Lactic acid	Solvent, acidulant, flavor enhancer, shelf-life extender, biodegradable plastic.
15.	Isoprene	Rubber for tires and coatings, adhesives, fuel Additive.

Depending upon the complexity and the type of carbohydrate sugars present in food waste, different products can be produced by the valorization process.

Sweeteners

Food waste from dairy industries such as whey protein, whey permeates and acid is a rich source of lactose. Lactose is less sweet (16%) as compared to sucrose (100%). The less sweetening effect decreases its commercial value with respect to the increasing demand of the population in the United State [62]. However, enzymatic-valorization by β-galactosidase increases the sweetening effect of food waste. In a study conducted by Lorenzen *et al.* [63], it was revealed that the immobilization of lactose by β-galactosidase and glucose isomerase results in the production of lactulose, glucose, and galactose. Polysaccharides containing food waste can be converted into valuable sweeteners and biofuels after enzymatic valorization. Cellulose present in food waste (dairy whey) can be subjected to hydrolysis and enzyme immobilization to monosaccharides, and can be used as a sweetener or converted into rare sugar such as tagatose [74]. Food waste such as vegetables and fruits containing pectic substances such as pectic acid and pectin can be used as thickeners, texturizers, fillers, and glazes after valorization. Also, lignocellulosic materials containing (lignin, cellulose, and hemicellulose) food waste can be used as clarification agents in fruit juices [75].

Prebiotics

These are substances that aid in the growth and enhancement of microorganisms' activity. In addition to the sweetening effects of lactose, transglycosylation of lactose converts it into prebiotics such as lactosucrose, lactulose, and galacto-oligosaccharides. Isomerization of lactose through transgalactosylation using fructose as a co-substrate for β-galactosidase results in a disaccharide called lactulose [76]. Some of the most profitable, well-established processes in food and agricultural systems begin with a carbohydrate substrate; namely, the production of high fructose corn syrup, a $1.7 billion industry in the United States in 2016 [77].

Bioethanol

Bioethanol produced from starch and sucrose juice (carbohydrate) is in priority nowadays due to the increasing fuel demand. On the contrary, non-renewal sources are depleting at a speedy rate. Recently, mixed food waste, kitchen waste, wheat-rye, bread mashed, banana peel, and potato peel are exploited for bioethanol production. Bioethanol can be a good alternative for fuel consumption. FW is subjected to enzymatic hydrolysis or saccharification after pre-treatment. Generally, a mixture of α-amylase, β- amylase and glucoamylase is used for the hydrolysis followed by fermentation. Finally, distillation results in the production of pure ethanol. In a study conducted by Dourou *et al.* [78], wastewater from the olive mill enriched with starch, carbohydrates, and lipids, was used as a raw material for ethanol production by using *Candida tropicalis* and *S. cerevisiae* microorganisms. Food waste produced during harvesting of crops such as sugarcane bagasse, wheat and rice straws is composed of lignin, cellulose, hemicellulose, pectin, polysaccharides, proteins, ash, and minerals [79, 80]. The complexity of the lignocellulose structure makes the process more expensive during valorization. This is due to the linkage of cellulose components with glucose chains through β-1,4 linkages [81]. On the other hand, hemicellulose is composed of polymerized monosaccharides such as glucose, galactose, mannose, xylose, arabinose, 4-O-methyl glucuronic acid, and galacturonic acid residues. Lignin, forming the other major proportion of lignocellulose, acts as a binder between plant cells, which is highly resistant to biological degradation [82].

Protein Bioconversion and their Products

Proteins as a building block form an integral part of human meals. Most protein-rich foods get wasted during their processing. About 3.9-21.9% of food waste consists of protein. Poultry industries are among the largest producers of protein-rich food items, out of which most of the food is wasted in the form of blood, feathers, dead remains, soft meat, keratinized nails, *etc* [83]. Along with this, other food industries like dairy, soybeans, and oilseeds contribute a major portion of protein found in FW. Proteolytic enzymes can be used to valorize the protein content in FW and produce value-added products such as bioactive peptides. Immobilized enzymes can be used for whey protein and convert it into an antioxidative peptide [2]. A commercially immobilized enzyme presents significant reusability, as even after repeated use for ten cycles, 50% of its activity remains preserved. It has been reported that the immobilized enzyme derivative possesses a higher affinity for α-lactalbumin protein hydrolysis than that for β-lactoglobulin, indicating that immobilization can change the selectivity and cleavage affinity of the biocatalyst [84].

Lipids Bioconversion and their Products

Biodiesel Production

Biodiesel is a promising non-fossil fuel used in the US, Europe and many more countries in the world. It is rich in saturated and unsaturated fatty acid methyl esters present in oils [85]. FW is rich in natural fat contents such as palm oil cake, soyabean oil, and sunflower oil and restaurant waste can consist of raw materials with the added benefit of reducing the FW problem while generating sustainable biodiesel [86]. Biodiesel production is cost-effective when supplied with a low-cost raw material for oil production. But production from edible oils is raising its production cost. Food waste and waste cooking oil can be alternatives for producing biodiesel as it is of low-cost and non-competitive with respect to edible pieces of stuff [9, 85]. Also, vegetable oils can be replaced by renewal microbial oils using various food waste materials such as orange peels and Ricotta cheese whey (RCW), valorized by various species of fungi *viz:Cunninghamella echinulata, Mucor circinelloides, Mucor miehei, Aspergillus tubingensis, Mortierellaisabellina* and *Mucor racemosus* [87]. Fungal hydrolysis of food waste involves the use of enzyme catalysis containing *Aspergillus awamori* and *Aspergillus oryzae* or other species of fungi to separate the lipid content from carbohydrates, amino acids, and phosphate content [88]. The lipid content obtained is further subjected to extraction and purification for use as biodiesel directly.

Biogas

Biogas is one of the highly efficient and important commodities, produced by the valorization of food waste in an anaerobic reactor, as a sustainable alternative fuel to petroleum gas. It is mainly composed of 50-75% of methane, and 25-50% of carbon dioxide which can replace fossil fuels in various applications [69]. It is produced via anaerobic digestion (AD), in which biologically active microorganisms valorize the food waste material transforming fatty acids into biogas in methanogenesis. AD comprises hydrolysis, acidogenesis, and acetogenesis followed by methanogenesis [89, 90]. The initial step involves the liquefaction process, the breakdown of insoluble complex molecules into smaller and soluble organic substrates. The second step can be divided into acidogenesis or acetogenesis involving conversion into volatile fatty acid (VFA) and further its digestion into CH_3COOH, H_2, and CO_2. The third step involves methanogenesis which facilitates the further breakdown of acetic acid and the reduction of carbon dioxide producing methane and carbon dioxide [91]. According to Fisgativa & Tremier [92], food waste with high water-soluble organic matter content and low carbohydrate content enhances methane production.

Bioplastic

Plastics are usually synthesized from petrochemicals through irreversible processes [93]. Petroleum-derived polymers pose a serious threat to the environment as their degradation is very slow thus bioplastic comes into play and serves as an alternative to petroleum-based plastics [94]. The lactic acid synthesized from cheese whey is of polymer grade which can be used as a raw material to produce bioplastic, polylactic acid (PLA). Being of biological nature, it is eco-friendly and easily degraded by soil microbes. PLA bioplastic fitted the best substitute for popular synthetic plastics such as polystyrene and polypropylene [95]. Starch, cellulose, chitin, and caprolactone waste are being used to convert into polyhydroxyalkanoates (PHA) and PHA-based products [93]. There are reports that indicate the conversion of food waste into bioplastics and energy with the most emphasis on bioplastic production [96 - 101]. However, very few studies focus on the production of PHA from FW such as cooking oil and cheese whey were reported [102, 103]. PHA is the best substitute to other plastic materials being biodegradable in nature [93]. In addition, other resources have been used to do recovery and production of value-added products [104 - 110]. Moreover, nanoparticle approaches have also been used [111 - 115]. PHA has been used for several purposes *viz*. packaging, medical applications, energy, and fine chemicals [116 - 119].

CONCLUSION AND FUTURE PERSPECTIVES

Food wastes led to noteworthy environmental impacts, such as the emission of greenhouse gases and secondary pollution with severe loss of land, water, and energy resources. The biological conversion of food wastes into value-added products compensates for the use of non-renewal resources. The production of potential products from FW serves as an alternate for residual-free disposal and added economic status by excluding investments in non-renewal products. FW valorization converts an organic fraction of waste variety of chemicals and raw materials for bio-based products. Food wastes act as economical raw materials for the commercial manufacturing of various products by microbial fermentation. Further research is required for optimizing the reaction parameters, elucidating, and maneuvering the underlying mechanisms, and developing greener and more sustainable approaches.

REFERENCES

[1] Papargyropoulou E, Lozano R, Steinberger JK, Wright N. bin Ujang Z. The food waste hierarchy as a framework for the management of food surplus and food waste. J Clean Prod 2014; 76: 106-15.
[http://dx.doi.org/10.1016/j.jclepro.2014.04.020]

[2] Bilal M, Iqbal HMN. Sustainable bioconversion of food waste into high-value products by immobilized enzymes to meet bio-economy challenges and opportunities – A review. Food Res Int 2019; 123: 226-40.
[http://dx.doi.org/10.1016/j.foodres.2019.04.066] [PMID: 31284972]

[3] Paritosh K, Kushwaha SK, Yadav M, Pareek N, Chawade A, Vivekanand V. Food waste to energy: an overview of sustainable approaches for food waste management and nutrient recycling. BioMed Res Int 2017; 2017: 1-19.
[http://dx.doi.org/10.1155/2017/2370927] [PMID: 28293629]

[4] Gustavsson J, Cederberg C, Sonesson U, Van Otterdijk R, Meybeck A. Global food losses and food waste 2011.

[5] Melikoglu M, Lin CS, Webb C. Analysing global food waste problem: pinpointing the facts and estimating the energy content. Central European Journal of Engineering 2013; 3(2): 157-64.

[6] Zhou Y, Engler N, Nelles M. Symbiotic relationship between hydrothermal carbonization technology and anaerobic digestion for food waste in China. Bioresour Technol 2018; 260: 404-12.
[http://dx.doi.org/10.1016/j.biortech.2018.03.102] [PMID: 29657110]

[7] Ravindran R, Jaiswal AK. Exploitation of food industry waste for high-value products. Trends Biotechnol 2016; 34(1): 58-69.
[http://dx.doi.org/10.1016/j.tibtech.2015.10.008] [PMID: 26645658]

[8] Sindhu R, Gnansounou E, Rebello S, *et al.* Conversion of food and kitchen waste to value-added products. J Environ Manage 2019; 241: 619-30.
[http://dx.doi.org/10.1016/j.jenvman.2019.02.053] [PMID: 30885564]

[9] Uçkun Kiran E, Trzcinski AP, Ng WJ, Liu Y. Bioconversion of food waste to energy: A review. Fuel 2014; 134: 389-99.
[http://dx.doi.org/10.1016/j.fuel.2014.05.074]

[10] Ma H, Wang Q, Qian D, Gong L, Zhang W. The utilization of acid-tolerant bacteria on ethanol production from kitchen garbage. Renew Energy 2009; 34(6): 1466-70.
[http://dx.doi.org/10.1016/j.renene.2008.10.020]

[11] Sanders J, Scott E, Weusthuis R, Mooibroek H. Bio-refinery as the bio-inspired process to bulk chemicals. Macromol Biosci 2007; 7(2): 105-17.
[http://dx.doi.org/10.1002/mabi.200600223] [PMID: 17295397]

[12] Han S, Shin HS. Biohydrogen production by anaerobic fermentation of food waste. Int J Hydrogen Energy 2004; 29(6): 569-77.
[http://dx.doi.org/10.1016/j.ijhydene.2003.09.001]

[13] Ohkouchi Y, Inoue Y. Impact of chemical components of organic wastes on l(+)-lactic acid production. Bioresour Technol 2007; 98(3): 546-53.
[http://dx.doi.org/10.1016/j.biortech.2006.02.005] [PMID: 16546378]

[14]   Sakai K, Ezaki Y. Open L-lactic acid fermentation of food refuse using thermophilic Bacillus coagulans and fluorescence *in situ* hybridization analysis of microflora. J Biosci Bioeng 2006; 101(6): 457-63.
[http://dx.doi.org/10.1263/jbb.101.457] [PMID: 16935246]

[15]   Wang Q, Wang X, Wang X, Ma H, Ren N. Bioconversion of kitchen garbage to lactic acid by two wild strains of Lactobacillus species. J Environ Sci Health Part A Tox Hazard Subst Environ Eng 2005; 40(10): 1951-62.
[http://dx.doi.org/10.1080/10934520500184624] [PMID: 16194915]

[16]   Yang SY, Ji KS, Baik YH, Kwak WS, McCaskey TA. Lactic acid fermentation of food waste for swine feed. Bioresour Technol 2006; 97(15): 1858-64.
[http://dx.doi.org/10.1016/j.biortech.2005.08.020] [PMID: 16257200]

[17]   Zhang C, Xiao G, Peng L, Su H, Tan T. The anaerobic co-digestion of food waste and cattle manure. Bioresour Technol 2013; 129: 170-6.
[http://dx.doi.org/10.1016/j.biortech.2012.10.138] [PMID: 23246757]

[18]   Koike Y, An MZ, Tang YQ, *et al.* Production of fuel ethanol and methane from garbage by high-efficiency two-stage fermentation process. J Biosci Bioeng 2009; 108(6): 508-12.
[http://dx.doi.org/10.1016/j.jbiosc.2009.06.007] [PMID: 19914584]

[19]   He Y, Bagley DM, Leung KT, Liss SN, Liao BQ. Recent advances in membrane technologies for biorefining and bioenergy production. Biotechnol Adv 2012; 30(4): 817-58.
[http://dx.doi.org/10.1016/j.biotechadv.2012.01.015] [PMID: 22306168]

[20]   Pan J, Zhang R, Elmashad H, Sun H, Ying Y. Effect of food to microorganism ratio on biohydrogen production from food waste via anaerobic fermentation. Int J Hydrogen Energy 2008; 33(23): 6968-75.
[http://dx.doi.org/10.1016/j.ijhydene.2008.07.130]

[21]   Rao M, Singh SP. Bioenergy conversion studies of organic fraction of MSW: kinetic studies and gas yield?organic loading relationships for process optimisation. Bioresour Technol 2004; 95(2): 173-85.
[http://dx.doi.org/10.1016/j.biortech.2004.02.013] [PMID: 15246442]

[22]   Lin CSK, Koutinas AA, Stamatelatou K, *et al.* Current and future trends in food waste valorization for the production of chemicals, materials and fuels: a global perspective. Biofuels Bioprod Biorefin 2014; 8(5): 686-715.
[http://dx.doi.org/10.1002/bbb.1506]

[23]   Varelas V. Food wastes as a potential new source for edible insect mass production for food and feed: A review. Fermentation (Basel) 2019; 5(3): 81.
[http://dx.doi.org/10.3390/fermentation5030081]

[24]   di Bitonto L, Antonopoulou G, Braguglia C, *et al.* Lewis-Brønsted acid catalysed ethanolysis of the organic fraction of municipal solid waste for efficient production of biofuels. Bioresour Technol 2018; 266: 297-305.
[http://dx.doi.org/10.1016/j.biortech.2018.06.110] [PMID: 29982051]

[25]   Saqib NU, Sharma HB, Baroutian S, Dubey B, Sarmah AK. Valorisation of food waste via hydrothermal carbonisation and techno-economic feasibility assessment. Sci Total Environ 2019; 690: 261-76.
[http://dx.doi.org/10.1016/j.scitotenv.2019.06.484] [PMID: 31288117]

[26]   Papanikola K, Papadopoulou K, Tsiliyannis C, *et al.* Food residue biomass product as an alternative fuel for the cement industry. Environ Sci Pollut Res Int 2019; 26(35): 35555-64.
[http://dx.doi.org/10.1007/s11356-019-05318-4] [PMID: 31069656]

[27]   Antonopoulou G, Alexandropoulou M, Ntaikou I, Lyberatos G. From waste to fuel: Energy recovery from household food waste via its bioconversion to energy carriers based on microbiological processes. Sci Total Environ 2020; 732139230.

[http://dx.doi.org/10.1016/j.scitotenv.2020.139230.] [PMID: 32438165]

[28] Awasthi MK, Selvam A, Chan MT, Wong JWC. Bio-degradation of oily food waste employing thermophilic bacterial strains. Bioresour Technol 2018; 248(Pt A): 141-7.
[http://dx.doi.org/10.1016/j.biortech.2017.06.115] [PMID: 28684181]

[29] Aluyor EO, Obahiagbon KO, Ori-Jesu M. Biodegradation of vegetable oils: A review. Sci Res Essays 2009; 4(6): 543-8.

[30] Mohanan S, Maruthamuthu S, Muthukumar N, Rajasekar A, Palaniswamy N. Biodegradation of palmarosa oil (green oil) by Serratia marcescens. Int J Environ Sci Technol 2007; 4(2): 279-83.
[http://dx.doi.org/10.1007/BF03326285]

[31] Pleissner D, Lin CS. Valorisation of food waste in biotechnological processes 2013.
[http://dx.doi.org/10.1186/2043-7129-1-21]

[32] Demichelis F, Fiore S, Pleissner D, Venus J. Technical and economic assessment of food waste valorization through a biorefinery chain. Renew Sustain Energy Rev 2018; 94: 38-48.
[http://dx.doi.org/10.1016/j.rser.2018.05.064]

[33] Cho EJ, Trinh LTP, Song Y, Lee YG, Bae HJ. Bioconversion of biomass waste into high value chemicals. Bioresour Technol 2020; 298122386.
[http://dx.doi.org/10.1016/j.biortech.2019.122386.] [PMID: 31740245]

[34] García Herrero MI, Margallo Blanco M, Laso Cortabitarte J, *et al.* Towards a sustainable agri-food system by an energetic and environmental efficiency assessment

[35] Manara P, Vamvuka D, Sfakiotakis S, Vanderghem C, Richel A, Zabaniotou A. Mediterranean agri-food processing wastes pyrolysis after pre-treatment and recovery of precursor materials: A TGA-based kinetic modeling study. Food Res Int 2015; 73: 44-51.
[http://dx.doi.org/10.1016/j.foodres.2014.11.033]

[36] IBA, An International Forum on Industrial Bioprocesses, International Bioprocessing Association, 2018.

[37] Xiong X, Yu IKM, Tsang DCW, *et al.* Value-added chemicals from food supply chain wastes: State-of-the-art review and future prospects. Chem Eng J 2019; 375121983
[http://dx.doi.org/10.1016/j.cej.2019.121983.]

[38] Sarkar O, Butti SK, Mohan SV. Acidogenic biorefinery: food waste valorization to biogas and platform chemicals 2018.
[http://dx.doi.org/10.1016/B978-0-444-63992-9.00006-9]

[39] Lin CSK, Pfaltzgraff LA, Herrero-Davila L, *et al.* Food waste as a valuable resource for the production of chemicals, materials and fuels. Current situation and global perspective. Energy Environ Sci 2013; 6(2): 426-64.
[http://dx.doi.org/10.1039/c2ee23440h]

[40] Cortés A, Oliveira LFS, Ferrari V, Taffarel SR, Feijoo G, Moreira MT. Environmental assessment of viticulture waste valorisation through composting as a biofertilisation strategy for cereal and fruit crops. Environ Pollut 2020; 264114794.
[http://dx.doi.org/10.1016/j.envpol.2020.114794.] [PMID: 32428819]

[41] World Economic Forum, Driving Sustainable Consumption, Value Chain Waste, 2010, http://www.members.wef orum.org/pdf/sustainableconsumptionaccessed 2 July 2012.

[42] Liguori R, Faraco V. Biological processes for advancing lignocellulosic waste biorefinery by advocating circular economy. Bioresour Technol 2016; 215: 13-20.
[http://dx.doi.org/10.1016/j.biortech.2016.04.054] [PMID: 27131870]

[43] Barry T. Evaluation of the economic, social, and biological feasibility of bioconverting food wastes with the black soldier fly (Hermetia illucens). University of North Texas 2004.

[44] Fowles TM, Nansen C. Insect-based bioconversion: value from food waste InFood waste management.

Cham: Palgrave Macmillan 2020; pp. 321-46.

[45] Mutafela RN. High value organic waste treatment via black soldier fly bioconversion: Onsite pilot study 2010.

[46] Rosales E, Rodríguez Couto S, Sanromán A. New uses of food waste: application to laccase production by Trametes hirsuta. Biotechnol Lett 2002; 24(9): 701-4.
[http://dx.doi.org/10.1023/A:1015234100459]

[47] Ortiz JC, Ruiz AT, Morales-Ramos JA, Thomas M, Rojas MG, Tomberlin JK. Insect mass production technologies.Insects as sustainable food ingredients. Amsterdam: Elsevier 2019; pp. 153-201.

[48] Wang YS, Shelomi M. Review of black soldier fly (Hermetia illucens) as animal feed and human food. Foods 2017; 6(10): 91.
[http://dx.doi.org/10.3390/foods6100091] [PMID: 29057841]

[49] Niu Y, Zheng D, Yao B, *et al.* A novel bioconversion for value-added products from food waste using Musca domestica. Waste Manag 2017; 61: 455-60.
[http://dx.doi.org/10.1016/j.wasman.2016.10.054] [PMID: 28017550]

[50] Brar SK, Verma M, Tyagi RD, Valéro JR, Surampalli RY. Wastewater sludges as novel growth substrates for rearing codling moth larvae. World J Microbiol Biotechnol 2008; 24(12): 2849-57.
[http://dx.doi.org/10.1007/s11274-008-9818-z]

[51] Miech P, Berggren Å, Lindberg JE, Chhay T, Khieu B, Jansson A. Growth and survival of reared Cambodian field crickets (*Teleogryllus testaceus*) fed weeds, agricultural and food industry by-products. J Insects Food Feed 2016; 2(4): 285-92.
[http://dx.doi.org/10.3920/JIFF2016.0028]

[52] Wang H, Rehman K, Liu X, *et al.* Insect biorefinery: a green approach for conversion of crop residues into biodiesel and protein. Biotechnol Biofuels 2017; 10(1): 304.
[http://dx.doi.org/10.1186/s13068-017-0986-7]

[53] Zheng L, Hou Y, Li W, Yang S, Li Q, Yu Z. Biodiesel production from rice straw and restaurant waste employing black soldier fly assisted by microbes. Energy 2012; 47(1): 225-9.
[http://dx.doi.org/10.1016/j.energy.2012.09.006]

[54] Manurung R, Supriatna A, Esyanthi RR, Putra RE. Bioconversion of rice straw waste by black soldier fly larvae (Hermetia illucens L.): optimal feed rate for biomass production. J Entomol Zool Stud 2016; 4(4): 1036-41.

[55] Suantika G, Putra RE, Hutami R, Rosmiati M. Application of compost produced by bioconversion of coffee husk by black soldier fly larvae (Hermetia illucens) as solid fertilizer to lettuce (Lactuca sativa Var. Crispa). InProceedings of the International Conference on Green Technology 2017 Nov 1 (Vol. 8, No. 1, pp. 20-26).

[56] Tinder AC, Puckett RT, Turner ND, Cammack JA, Tomberlin JK. Bioconversion of sorghum and cowpea by black soldier fly (*Hermetia illucens* (L.)) larvae for alternative protein production. J Insects Food Feed 2017; 3(2): 121-30.
[http://dx.doi.org/10.3920/JIFF2016.0048]

[57] Niu Y, Zheng D, Yao B, *et al.* A novel bioconversion for value-added products from food waste using Musca domestica. Waste Manag 2017; 61: 455-60.
[http://dx.doi.org/10.1016/j.wasman.2016.10.054] [PMID: 28017550]

[58] Dourou M, Kancelista A, Juszczyk P, *et al.* Bioconversion of olive mill wastewater into high-added value products. J Clean Prod 2016; 139: 957-69.
[http://dx.doi.org/10.1016/j.jclepro.2016.08.133]

[59] Tzimorotas D, Afseth NK, Lindberg D, Kjørlaug O, Axelsson L, Shapaval V. Pretreatment of different food rest materials for bioconversion into fungal lipid-rich biomass. Bioprocess Biosyst Eng 2018; 41(7): 1039-49.
[http://dx.doi.org/10.1007/s00449-018-1933-0] [PMID: 29654357]

[60] Prasoulas G, Gentikis A, Konti A, Kalantzi S, Kekos D, Mamma D. Bioethanol production from food waste applying the multienzyme system produced on-site by Fusarium oxysporum F3 and mixed microbial cultures. Fermentation (Basel) 2020; 6(2): 39.
[http://dx.doi.org/10.3390/fermentation6020039]

[61] Pleissner D, Lin CS. Valorisation of food waste in biotechnological processes 2013.
[http://dx.doi.org/10.1186/2043-7129-1-21]

[62] Andler SM, Goddard JM. Transforming food waste: how immobilized enzymes can valorize waste streams into revenue streams. npj Science of Food. 2018 Oct 29;2(1):1-1.

[63] Lorenzen PC, Breiter J, Clawin-Rädecker I, Dau A. A novel bi-enzymatic system for lactose conversion. Int J Food Sci Technol 2013; 48(7): 1396-403.
[http://dx.doi.org/10.1111/ijfs.12101]

[64] Mörschbächer AP, Volpato G, Souza CFV. Kluyveromyces lactis β-galactosidase immobilization in calcium alginate spheres and gelatin for hydrolysis of cheese whey lactose. Cienc Rural 2016; 46(5): 921-6.
[http://dx.doi.org/10.1590/0103-8478cr20150833]

[65] Plessas S, Bosnea L, Psarianos C, Koutinas AA, Marchant R, Banat IM. Lactic acid production by mixed cultures of Kluyveromyces marxianus, Lactobacillus delbrueckii ssp. bulgaricus and Lactobacillus helveticus. Bioresour Technol 2008; 99(13): 5951-5.
[http://dx.doi.org/10.1016/j.biortech.2007.10.039] [PMID: 18155517]

[66] Bai Z, Gao Z, Sun J, Wu B, He B. d-Lactic acid production by Sporolactobacillus inulinus YBS1-5 with simultaneous utilization of cottonseed meal and corncob residue. Bioresour Technol 2016; 207: 346-52.
[http://dx.doi.org/10.1016/j.biortech.2016.02.007] [PMID: 26897413]

[67] Zhang C, Xiao G, Peng L, Su H, Tan T. The anaerobic co-digestion of food waste and cattle manure. Bioresour Technol 2013; 129: 170-6.
[http://dx.doi.org/10.1016/j.biortech.2012.10.138] [PMID: 23246757]

[68] Pleissner D, Qi Q, Gao C, et al. Valorization of organic residues for the production of added value chemicals: A contribution to the bio-based economy. Biochem Eng J 2016; 116: 3-16.
[http://dx.doi.org/10.1016/j.bej.2015.12.016]

[69] Ng HS, Kee PE, Yim HS, Chen PT, Wei YH, Chi-Wei Lan J. Recent advances on the sustainable approaches for conversion and reutilization of food wastes to valuable bioproducts. Bioresour Technol 2020; 302122889.
[http://dx.doi.org/10.1016/j.biortech.2020.122889.] [PMID: 32033841]

[70] Itoh H, Sato T, Takeuchi T, Khan AR, Izumori K. Preparation of d-sorbose from d-tagatose by immobilized d-tagatose 3-epimerase. J Ferment Bioeng 1995; 79(2): 184-5.
[http://dx.doi.org/10.1016/0922-338X(95)94091-5]

[71] Feng K, Huang Z, Peng B, et al. Immobilization of Aspergillus niger lipase onto a novel macroporous acrylic resin: Stable and recyclable biocatalysis for deacidification of high-acid soy sauce residue oil. Bioresour Technol 2020; 298122553.
[http://dx.doi.org/10.1016/j.biortech.2019.122553.] [PMID: 31846852]

[72] Carmona-Cabello M, Garcia IL, Leiva-Candia D, Dorado MP. Valorization of food waste based on its composition through the concept of biorefinery. Curr Opin Green Sustain Chem 2018; 14: 67-79.
[http://dx.doi.org/10.1016/j.cogsc.2018.06.011]

[73] Dessie W, Luo X, Wang M, et al. Current advances on waste biomass transformation into value-added products. Appl Microbiol Biotechnol 2020; 104(11): 4757-70.
[http://dx.doi.org/10.1007/s00253-020-10567-2] [PMID: 32291487]

[74] Periyasamy K, Santhalembi L, Mortha G, Aurousseau M, Subramanian S. Carrier-free co-immobilization of xylanase, cellulase and β-1,3-glucanase as combined cross-linked enzyme

aggregates (combi-CLEAs) for one-pot saccharification of sugarcane bagasse. RSC Advances 2016; 6(39): 32849-57.
[http://dx.doi.org/10.1039/C6RA00929H]

[75]   Baldassarre S, Babbar N, Van Roy S, *et al.* Continuous production of pectic oligosaccharides from onion skins with an enzyme membrane reactor. Food Chem 2018; 267: 101-10.
[http://dx.doi.org/10.1016/j.foodchem.2017.10.055] [PMID: 29934143]

[76]   Hua X, Yang R, Zhang W, Fei Y, Jin Z, Jiang B. Dual-enzymatic synthesis of lactulose in organic-aqueous two-phase media. Food Res Int 2010; 43(3): 716-22.
[http://dx.doi.org/10.1016/j.foodres.2009.11.008]

[77]   Coherent Market Insights. Global High Fructose Corn Syrup Market Report by Type (HFCS 42, HFCS 55, HFCS 65 and HFCS 90), by End-Use Industry (Food Industry, Beverage Industry, Pharmaceuticals and Others) and by Geography - Trends and Forecast to 2025. (2018).

[78]   Dourou M, Kancelista A, Juszczyk P, *et al.* Bioconversion of olive mill wastewater into high-added value products. J Clean Prod 2016; 139: 957-69.
[http://dx.doi.org/10.1016/j.jclepro.2016.08.133]

[79]   Van Dyk JS, Pletschke BI. A review of lignocellulose bioconversion using enzymatic hydrolysis and synergistic cooperation between enzymes—Factors affecting enzymes, conversion and synergy. Biotechnol Adv 2012; 30(6): 1458-80.
[http://dx.doi.org/10.1016/j.biotechadv.2012.03.002] [PMID: 22445788]

[80]   Kumar R, Tabatabaei M, Karimi K, Sárvári Horváth I. Recent updates on lignocellulosic biomass derived ethanol - A review. Biofuel Research Journal 2016; 3(1): 347-56.
[http://dx.doi.org/10.18331/BRJ2016.3.1.4]

[81]   Hamad WY, Hu TQ. Structure–process–yield interrelations in nanocrystalline cellulose extraction. Can J Chem Eng 2010; 88(3): 392-402.

[82]   Lin SY, Lebo SE. Lignin.John Wiley and Sons I. Chem. Technol 2001; pp. 1-32.

[83]   Brandelli A, Sala L, Kalil SJ. Microbial enzymes for bioconversion of poultry waste into added-value products. Food Res Int 2015; 73: 3-12.
[http://dx.doi.org/10.1016/j.foodres.2015.01.015]

[84]   Rocha GF, Kise F, Rosso AM, Parisi MG. Potential antioxidant peptides produced from whey hydrolysis with an immobilized aspartic protease from Salpichroa originifolia fruits. Food Chem 2017; 237: 350-5.
[http://dx.doi.org/10.1016/j.foodchem.2017.05.112] [PMID: 28764006]

[85]   Karmee SK. Liquid biofuels from food waste: Current trends, prospect and limitation. Renew Sustain Energy Rev 2016; 53: 945-53.
[http://dx.doi.org/10.1016/j.rser.2015.09.041]

[86]   Fowles TM, Nansen C. Insect-based bioconversion: value from food waste InFood waste management. Cham: Palgrave Macmillan 2020; pp. 321-46.

[87]   Carota E, Crognale S, D'Annibale A, Petruccioli M. Bioconversion of agro-industrial waste into microbial oils by filamentous fungi. Process Saf Environ Prot 2018; 117: 143-51.
[http://dx.doi.org/10.1016/j.psep.2018.04.022]

[88]   Pleissner D, Lam WC, Sun Z, Lin CSK. Food waste as nutrient source in heterotrophic microalgae cultivation. Bioresour Technol 2013; 137: 139-46.
[http://dx.doi.org/10.1016/j.biortech.2013.03.088] [PMID: 23587816]

[89]   Capson-Tojo G, Rouez M, Crest M, Steyer JP, Delgenès JP, Escudié R. Food waste valorization via anaerobic processes: a review. Rev Environ Sci Biotechnol 2016; 15(3): 499-547.
[http://dx.doi.org/10.1007/s11157-016-9405-y]

[90]   Villano M, Aulenta F, Majone M. Perspectives of biofuels production from renewable resources with

bioelectrochemical systems. Asia-Pac J Chem Eng 2012; 7: S263-74.
[http://dx.doi.org/10.1002/apj.1643]

[91]   Kumar P, Chandrasekhar K, Kumari A, Sathiyamoorthi E, Kim B. Electro-fermentation in aid of bioenergy and biopolymers. Energies 2018; 11(2): 343.
[http://dx.doi.org/10.3390/en11020343]

[92]   Fisgativa H, Tremier A. Influence of food waste characteristics variations on treatability through anaerobic digestion.

[93]   Tsang YF, Kumar V, Samadar P, *et al.* Production of bioplastic through food waste valorization. Environ Int 2019; 127: 625-44.
[http://dx.doi.org/10.1016/j.envint.2019.03.076] [PMID: 30991219]

[94]   Sharma P, Gaur VK, Kim SH, Pandey A. Microbial strategies for bio-transforming food waste into resources. Bioresour Technol 2020; 299122580.
[http://dx.doi.org/10.1016/j.biortech.2019.122580.] [PMID: 31877479]

[95]   Narayanan CM, Das S, Pandey A. Food waste utilization: green technologies for manufacture of valuable products from food wastes and agricultural wastes 2017.
[http://dx.doi.org/10.1016/B978-0-12-811413-1.00001-2]

[96]   Khanna S, Srivastava AK. Recent advances in microbial polyhydroxyalkanoates. Process Biochem 2005; 40(2): 607-19.
[http://dx.doi.org/10.1016/j.procbio.2004.01.053]

[97]   Chee JY, Yoga SS, Lau NS, Ling SC, Abed RM, Sudesh K. Bacterially produced polyhydroxyalkanoate (PHA): converting renewable resources into bioplastics. Current research, technology and education topics in Applied Microbiology and Microbial Biotechnology. 2010 Jan 1;2:1395-404

[98]   Chen GQ, Patel MK. Plastics derived from biological sources: present and future: a technical and environmental review. Chem Rev 2012; 112(4): 2082-99.
[http://dx.doi.org/10.1021/cr200162d] [PMID: 22188473]

[99]   Bugnicourt E, Cinelli P, Lazzeri A, Alvarez VA. Polyhydroxyalkanoate (PHA): Review of synthesis, characteristics, processing and potential applications in packaging.

[100]   Koutinas AA, Vlysidis A, Pleissner D, *et al.* Valorization of industrial waste and by-product streams via fermentation for the production of chemicals and biopolymers. Chem Soc Rev 2014; 43(8): 2587-627.
[http://dx.doi.org/10.1039/c3cs60293a] [PMID: 24424298]

[101]   Salgaonkar B, Bragança J. Utilization of sugarcane bagasse by Halogeometricum borinquense strain E3 for biosynthesis of poly (3-hydroxybutyrate-co-3-hydroxyvalerate). Bioengineering (Basel) 2017; 4(2): 50.
[http://dx.doi.org/10.3390/bioengineering4020050] [PMID: 28952529]

[102]   Desroches M, Escouvois M, Auvergne R, Caillol S, Boutevin B. From vegetable oils to polyurethanes: synthetic routes to polyols and main industrial products. Polym Rev (Phila Pa) 2012; 52(1): 38-79.
[http://dx.doi.org/10.1080/15583724.2011.640443]

[103]   Valentino F, Riccardi C, Campanari S, Pomata D, Majone M. Fate of β-hexachlorocyclohexane in the mixed microbial cultures (MMCs) three-stage polyhydroxyalkanoates (PHA) production process from cheese whey. Bioresour Technol 2015; 192: 304-11.
[http://dx.doi.org/10.1016/j.biortech.2015.05.083] [PMID: 26048084]

[104]   Awasthi MK, Kumar V, Yadav V, *et al.* Current state of the art biotechnological strategies for conversion of watermelon wastes residues to biopolymers production: A review. Chemosphere 2022; 290133310.
[http://dx.doi.org/10.1016/j.chemosphere.2021.133310.] [PMID: 34919909]

[105]   Liu H, Kumar V, Jia L, *et al.* Biopolymer poly-hydroxyalkanoates (PHA) production from apple

industrial waste residues: A review. Chemosphere 2021; 284131427.
[http://dx.doi.org/10.1016/j.chemosphere.2021.131427.] [PMID: 34323796]

[106] Awasthi SK, Kumar M, Kumar V, *et al.* A comprehensive review on recent advancements in biodegradation and sustainable management of biopolymers. Environ Pollut 2022; 307119600.
[http://dx.doi.org/10.1016/j.envpol.2022.119600] [PMID: 35691442]

[107] Duan Y, Tarafdar A, Kumar V, *et al.* Sustainable biorefinery approaches towards circular economy for conversion of biowaste to value added materials and future perspectives. Fuel 2022; 325124846.
[http://dx.doi.org/10.1016/j.fuel.2022.124846]

[108] Kumar V, Sharma N, Umesh M, *et al.* Emerging challenges for the agro-industrial food waste utilization: A review on food waste biorefinery. Bioresour Technol 2022; 362127790.
[http://dx.doi.org/10.1016/j.biortech.2022.127790] [PMID: 35973569]

[109] Awasthi SK, Sarsaiya S, Kumar V, *et al.* Processing of municipal solid waste resources for a circular economy in China: An overview. Fuel 2022; 317123478.
[http://dx.doi.org/10.1016/j.fuel.2022.123478]

[110] Kumar V, Sharma N, Maitra SS. *In vitro* and in vivo toxicity assessment of nanoparticles. Int Nano Lett 2017; 7(4): 243-56.
[http://dx.doi.org/10.1007/s40089-017-0221-3]

[111] Vinay K, Neha S, Maitra S. "Protein and Peptide Nanoparticles: Preparation and Surface Modification," in Functionalized Nanomaterials I. CRC Press 2020; pp. 191-204.

[112] Vallinayagam S, *et al.* Recent developments in magnetic nanoparticles and nano-composites for wastewater treatment. J Environ Chem Eng 2021; 9(6)106553.
[http://dx.doi.org/10.1016/j.jece.2021.106553]

[113] Kumar V, Sharma N, Lakkaboyana SK, Maitra SS. "Silver nanoparticles in poultry health: Applications and toxicokinetic effects," in Silver Nanomaterials for Agri-Food Applications. Elsevier 2021; pp. 685-704.
[http://dx.doi.org/10.1016/B978-0-12-823528-7.00005-6]

[114] Egbosiuba T C. Biochar and bio-oil fuel properties from nickel nanoparticles assisted pyrolysis of cassava peel 2022.
[http://dx.doi.org/10.1016/j.heliyon.2022.e10114]

[115] Chen GQ, Wu Q. The application of polyhydroxyalkanoates as tissue engineering materials. Biomaterials 2005; 26(33): 6565-78.
[http://dx.doi.org/10.1016/j.biomaterials.2005.04.036] [PMID: 15946738]

[116] Chen GQ. A microbial polyhydroxyalkanoates (PHA) based bio- and materials industry. Chem Soc Rev 2009; 38(8): 2434-46.
[http://dx.doi.org/10.1039/b812677c] [PMID: 19623359]

[117] Chen GQ, Patel MK. Plastics derived from biological sources: present and future: a technical and environmental review. Chem Rev 2012; 112(4): 2082-99.
[http://dx.doi.org/10.1021/cr200162d] [PMID: 22188473]

[118] Liang Q, Qi Q. From a co-production design to an integrated single-cell biorefinery. Biotechnol Adv 2014; 32(7): 1328-35.
[http://dx.doi.org/10.1016/j.biotechadv.2014.08.004] [PMID: 25172032]

[119] Koch D, Mihalyi B. Assessing the change in environmental impact categories when replacing conventional plastic with bioplastic in chosen application fields. Chem Eng Trans 2018; 70: 853-8.

CHAPTER 4

Olive Oil Wastes Valorization for High Value Compounds Production

Pritha Chakraborty[1,*]

[1] School of Allied Healthcare and Sciences, Jain (Deemed to be) University, Bengaluru, India.

Abstract: The consumption of olive oil is deeply rooted in human history and the production of olive oil contributes greatly to the economy of Mediterranean countries. Olive oil is generally extracted following three different methods; the traditional pressing method, two-phase decantation system and three-phase decantation system. These extraction processes generate mainly two different types of waste which are olive mill solid waste (OMSW) and olive mill wastewater (OMWW). Olive mill by-products are considered a major environmental hazard in Mediterranean regions as they are high in phenol, lipid and organic acid content. To eliminate this problem, valorization of these waste products is the need of the hour. Phytochemical compounds like phenols, and flavonoids are important and useful for pharmaceutical industries. Other than the recovery of these value-added compounds, olive waste can be used as animal feed and a source of clean energy. Biological treatment of these wastes reduces the percentage of phenols and organic acids and then it can be used in agricultural applications. The valorization strategies of olive mill wastes depend on factors like socio-economic conditions, and agricultural and industrial environments. In this chapter, the olive oil production process, phytochemical characteristics of generated waste and their environmental impact are discussed. This discussion also emphasized the available valorization techniques of olive oil by-products, their advantages, and disadvantages.

Keywords: Anaerobic treatment, Animal feed, Biological oxygen demand, Biofuel, Chemical oxygen demand, High value added compound, Microbiological treatment, Olive wastes, Olive mill waste water, Olive mill solid waste, Pressing method, Phyto-toxicity, Physical treatment, Phytochemical characteristics, Phenolic acids, *Phanerochaete chrysoporium*, *Pleurotus ostreatus*, Two phase decantation system, Three phase decantation system, Valorization techniques.

---

* **Corresponding author Pritha Chakraborty:** School of Allied Healthcare and Sciences, Jain (Deemed to be) University, Bengaluru, India; E-mail: prithachakraborty7@gmail.com

Vinay Kumar, Sivarama Krishna Lakkaboyana & Neha Sharma (Eds.)

INTRODUCTION

Olive (*Olea europaea* L.) trees are evergreen and commercial crop with major economic importance in the Mediterranean region. The cultivation and production of olive oil is an important agricultural sector in Europe [1]. Olives are consumed either as table olives or as olive oil. As a rich source of essential fatty acids and antioxidants, olive oil is widely consumed all over the world and deeply rooted in the diet of the Mediterranean world. In recent times, olive oil is on high demand and cultivation of olive trees has increased in Greece, Italy, Spain and other countries. In 2013, the global production of olives reached 20,000,000 tons per year and total production of table olives reached 2,900,000 tons. In 2018, the estimated consumption of olive oil worldwide exceeded 3,300,000 tons per year [2]. According to a study by Khdair and Abu-Rumman [3], total 11 million hectare of land was used for olive tree cultivation in 2015 and almost 50% of the total land was covered by European Union countries. Approximately 72% of total annual olive oil produced comes from Europe. Other than Mediterranean countries, Asia, Africa and America are also producing 15%, 12%, and 2% of global olive production.

Olive oil production practices can be dated back to 6500 years ago [2]. The recent increase in olive oil consumption can be explained by exploring its health benefits. Slow aging, decline in age-related cognitive issues, improvement in thrombosis and gastric issues, and reduction in lipoproteins and cholecystokinin bile secretion are among several health benefits that have been linked to the incorporation of olive oil in everyday diet [4]. These health benefits can be ascribed to oleic acid (55-83%) [5] (Miranda *et al.,* 2019) and phenolic compounds present in olive oil [6]. Phenolic compounds are known for their antioxidant, anti-inflammatory, anti-proliferative, anti-atherogenic, antimicrobial and anticancer properties [7 - 13]. The Health benefits of olive oil are explained in detail in Table **1**.

Table 1. Health benefits of olive oil according to Ciriminna *et al.,* [14].

	HEALTH BENEFITS
Joint	Decreases swelling and maintains bone joint health, reduces joint inflammation and pain, increases joint flexibility and improves mobility.
Skin	Helps skin conditions caused by auto immune diseases, improves skin moisture retention, reduces premature aging skin, supports healthy and radiant skin, and reduces damage from sun exposure.
Other	Reduces cardiovascular issues, helps in the repair of cartilage, and reduces fatigue.

Increased production of olive and unregulated disposal of olive mill waste into the immediate environment have raised serious environmental concerns in olive-producing countries [15]. Industrial production of olive oil generates mainly two types of waste which include olive mill wastewater (OMWW) and olive mill solid waste (OMSW). High concentrations of different phenolic compounds and fatty acids are associated with the phytotoxicity of olive mill waste. These bioactive compounds are reported to inhibit plant and bacterial growth [2]. Disposal of both solid and liquid waste into agricultural soil affects the chemical and physical properties of soil like porosity, acidity, salinity and heavy metal content [1, 16]. Oxidation and further polymerization of tannins result in the discolouration of water and are difficult to remove from water. The lipid content of liquid waste forms a thin layer of film that blocks the penetration of sunlight and oxygen, inhibiting microbial growth. High phosphorus content leads to eutrophication and fatty acid content produces a pungent odor during dry warm weather [17].

Olive processing by-products are a richand abundant source of macromolecules (proteins, sugars, fatty acids, plant enzymes and pigments) and bioactive compounds like polyphenols, vitamins, and many other aromatic and aliphatic compounds. These compounds have great importance in pharmaceutical, cosmetics and food industries and can be recovered by valorization of the waste products. Waste valorization is the most recent approach involving different modern technologies for recycling or reuse of waste materials to convert them into high-value products instead of dumping them into the environment [18, 19]. Vandermeersch *et al.,* [20] have explained a detailed hierarchy of waste management such as prevention, use for human nutrition, conversion for human nutrition, use for animal feed, use as raw materials in industry (a biobased economy), process into fertilizer by anaerobic digestion or composting, and use as renewable energy, incineration, and landfill [21]. In this chapter, we have discussed olive oil production process and generated waste material and the available valorization methods.

OLIVE OIL PRODUCTION PROCESS AND GENERATED WASTE MATERIALS

Olive oil is extracted and separated from olive fruits by both traditional and industrial processes. The quality of olive oil depends on different factors like the quality of olives, the time of harvest and extraction process [1]. The production of olive oil involves picking the fruits, removal of leaves and washing, crushing, mixing, pressing the fruits and separating the oil [2]. There are two main olive oil production processes available. These include discontinuous and continuous processes. The discontinuous process involves a traditional pressing procedure and the continuous process involves the centrifugation process [6].

Three main extraction processes are in use in recent days: i) The traditional pressing process, ii) The two-phase decantation process and iii) The three-phase decantation process [22] (Fig. **1**).

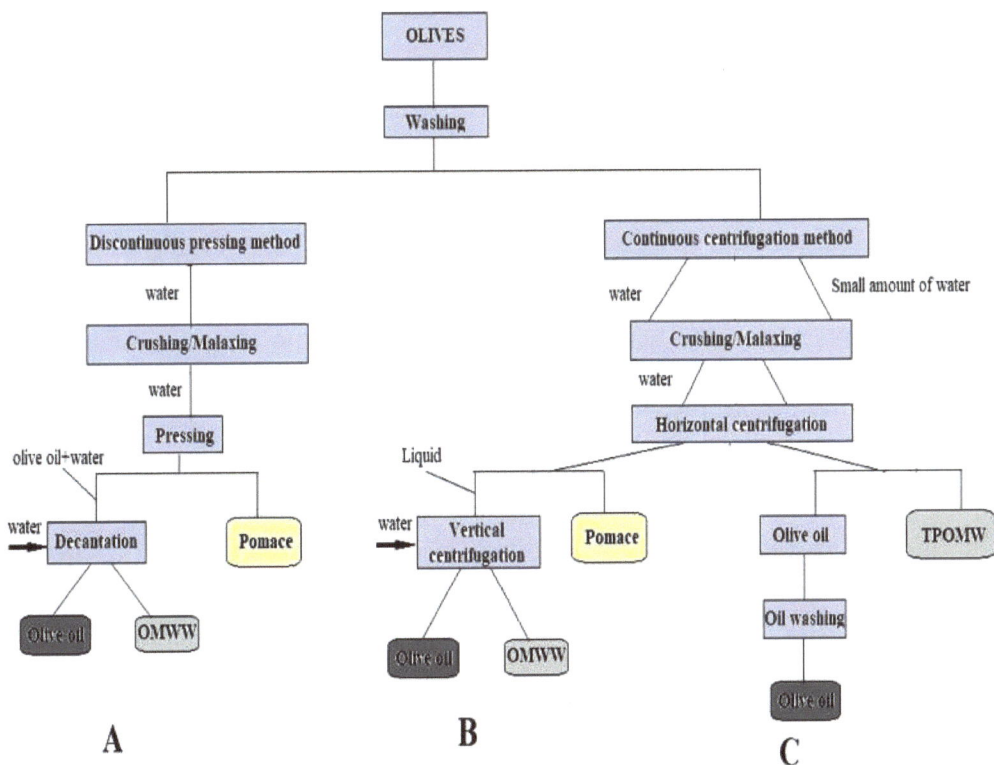

Fig. (1). Flow chart of different olive oil extraction processes; **A**. Traditional pressing method, **B**. Three-phase extraction process, **C**. Two phase extraction process [6].

In general, most olive oil is extracted from the fruit using water. In the next step, the resulting paste is mixed and malaxed at 27 °C for 20-30 mins to agglomerate small oil droplets increasing the percentage of available oil and thus separating the oil phase from water. The oil produced at a lower temperature is labeled as cold pressed which is preferred by customers [23].

The traditional pressing process is an obsolete and oldest olive oil processing method. Before the invention of the hydraulic press, traditional pressing was the only method for olive oil production but now it is restricted only to a few family-run small olive mills. In this process, olives are washed, crushed, and mixed with warm water. Before pressing, the olive paste is placed over fiber disks and stacked over each other. The traditional disks were made of coconut fiber and hemp but now synthetic fiber is used for easier maintenance. The liquid phase (oil and water) is drained by applying pressure on the solid phase of the olive paste. The resulting solid mixture is called olive pomace and the olive oil is separated from the remaining liquid by decantation [6]. Till 1970, almost all olive oil was produced by this method because the equipment and technique are cheap and simple [17, 24]. Though the main disadvantage of this method is that it produces less volume of oil (50-60 L per 100 kg fruits) as very little water is used in this process. The chemical oxygen demand (COD) of the produced liquid wastewater is much higher than the other processes [25]. These disadvantages eventually made the traditional pressing method less popular for industrial-scale olive oil production and it was out ruled by other methods.

The continuous processes are used for industrial-scale production and oil is separated by the centrifugation process. The density and composition of olive paste are different from the discontinuous process. In the three phase decantation process, almost 1.5-1.75 times more water is added producing a larger volume of oil (80-120 L per 100 kg of fruits) [26]. Horizontal centrifugation at the second step produces olive pomace and liquid phase. Another round of vertical centrifugation of the liquid phase is done to separate olive oil leaving olive mill wastewater as a by-product. As this method produces three different fractions which are olive pomace (30%), olive oil (20%) and wastewater (50%), it is called a three-phase decantation process [27, 28]. A large number of mills in Italy, Portugal and 70% of all the functioning mills in Greece use this three-phase decantation process for olive oil production. They produce a considerable amount of olive mill wastewater and impose environmental pollution [17, 29].

In 1990, to reduce water consumption in the three-phase decantation process, a two-phase decantation process was introduced. In the two-phase decantation process, pressing and centrifugation produce only two fractions; olive pomace and olive oil. The semi-solid olive pomace is called Two-Phase Olive Mill Waste (TPOMW) which is a combination of olive husk and olive wastewater (OWW). The pomace is further processed to extract the oil-producing two types of oil, i) Oil extracted using solvents, and ii) Oil obtained by the second centrifugation. Almost 90% of all the functioning mills in Spain produce olive oil using this technology while the other countries are yet to adapt [22, 16, 30].

In Tables **2, 3** and **4**, different aspects of these three olive oil extraction techniques such as advantages, disadvantages, usage and production of different fractions are enlisted in detail.

Table 2. Main advantages and disadvantages of olive oil extraction methods [6].

Methods	Advantages	Disadvantages
Traditional pressing method	i) Cheap equipment ii) Simple technology, iii) Relatively less water consumption and iv) Small volume of OMWW production	i) Discontinuous process, ii) High manpower and iii) Higher COD of OMWW
Two-phase decantation method	i) Less water consumption	i) High concentration of pollution load, ii) High moisture content, iii) Higher energy consumption
Three phase decantation method	ii) Larger volume of water consumption iii) Larger quantity of OMWW production iv) Expensive installation	iv) Complete automation v) Better oil quality vi) Smaller area requirement

Table 3. Olive oil extraction technologies used in different European Union countries [22].

Countries	Approximate percentage (%)		
	Traditional	Two phase decantation	Three phase decantation
Spain	NIL	98	2
Italy	42%	2	56
Greece	18%	NIL	82
Portugal	85%	3	12
Cyprus	12%	4	84%
Croatia	42%	56	2
Malta	40%	NIL	60

Table 4. The quantity and composition of generated waste vary with processing techniques, variety and maturity of olive and are enlisted [3, 31, 32].

Methods	Input	Quantity	Output	Quantity
Traditional pressing	Olive	1000kg	Oil	230.4 kg
			Pomace	500 kg
	Washing water	100-200L	Wastewater	650 L
Two-phase decantation	Olive	1000 kg	Oil	256 kg
	Washing water	100-200L	Pomace	581 kg
	Hot water	700-1000L	Wastewater	1200 L

(Table 4) cont.....

			Oil	247 kg
Three phase decantation	Olive	1000kg	Pomace	Wastewater735 kg
	Washing water	100-200L	Waste water	200-300 L

PHYTOCHEMICAL COMPOSITION OF OLIVE WASTES AND THEIR ENVIRONMENTAL IMPACT

Olive oil production processes generated mainly two fractions of waste; semi-solid and liquid. Semi-solid fraction of waste is called olive pomace or olive cake which is made up of olive peel, pulp, pits, and seeds. In 2017-2018, approximately 13 million tonnes of wet pomace were discarded around the world [5]. The olive pulp consists of 70-90% of the total weight of the pomace [6]. The high content of phenols, tannins and suspended solids with mildly acidic pH (approximately 5.2) makes it difficult for any conventional treatment [33 - 37]. The moisture content of olive pomace differs with the processing methods; traditional pressing (20-25%), two-phase decantation method (55-75%), three-phase decantation process (40-50%) [38 - 40].

The liquid fraction is called olive mill wastewater (OMWW). OMWW is reddish black in color and has a strong unpleasant odor [3]. OMWW consists of water (83-94%), organic compounds (4-16%) and mineral salt (0.4-2.5%). The organic matter consists of oil, carbohydrates, lipids, pectin, organic acids, mucilage, phenols, and lignin. The phenolic content (caffeic acid, tyrosol, tannins and anthocyanin) ranges from 2 to 15% of the total available organic content. Due to higher organic content, OMWW has higher biological and chemical oxygen demand (BOD and COD) [41 - 48]. A study by Niaounakis and Halvadakis [17], has reported that the concentration of organic matter is 400 times higher in OMWW in compared to typical household wastewater. It has an acidic pH (4-5) with electrical conductivity of 5.5 -12.0 dS/m.

The thick sludge-like waste, generated from two-phase extraction systems is called two-phase olive mill waste (TPOMW). It has different characteristics from olive pomace and OMWW and contains olive stone and pulp from fruit and vegetation water. The moisture content is higher (65-75%) than olive pomace. Organic matter and dry ash range from 1.70-4% and 1.42-4% respectively [6, 26,49] (Table **5**).

Table 5. Main properties and phytochemical composition of different olive mill wastewater (OMWW) and olive mill solid waste (OMSW) [1, 3, 45, 50].

Parameters	OMWW	OMSW
pH	5-6.7	4.8- 5.8
Water (%)	82	71
Dry matter (%)	6-7	50-72
Redox potential (mv)	-80 to-330	NA
Bod (g/l)	35-110	-
Cod (g/l)	40-120	34-112
Alkalinity (g caco$_3$/l)	NA	0.7-0.3
Density (g/l)	1-1.2	NA
Turbidity (ntu)	42000-62000	
Suspended solids (g/l)	1-10	35-105
Total volatile fatty acids (TVFA) (g acetic acid/ l)		0.76-2.20
Electrical conductivity (ds/m)	5-13	1-4
Organic matter	46-62	845-980
C/n	52-54	28-71
Phenols (%)	1-11	0.5-2.4
Fats (g/kg)	NA	120
Proteins (g/kg)	NA	71
K (g/kg)	2-9	7-30
Na (g/kg)	0.1-0.4	0.5-1.5
Ca (g/kg)	0.2-0.6	1.5-9
Mg (g/kg)	0.04-0.21	0.7-3.9
Fe (mg/kg)	18-129	80-1468
Cu (mg/kg)	1.5-5.5	12-29
Mn (mg/kg)	1-11	5-38
Zn (mg/kg)	2.4-11.9	10-36

Phenolic compounds found in OMWW are different from olive pomace Table **6**. Phenolic compounds are organic molecules consisting of the aromatic ring, hydroxyl group and functional side chain. Most phenolic acids found in nature are combined with sugar, lipids, organic acids, and other phenols [51]. The main classes of phenolic acids found in nature are tannin, lignin, aldehydes, flavonoids and lignans [52]. They impart a strong color in fruits and vegetables and also protect plants from pathogens and UV radiation [6]. The major phenolic acids present in OMWW are phenyl alcohol, flavonoids, secoiridoid derivatives, oleuropein, hydroxytyrosol and lignans [53, 54]. The presence of several other phenolic acids such as 4-methylcatechol, 4-hydroxybenzoic acid, protocatechuic acid, vanillic acid, 3,4-dihydroxyphenylglycol, homovanillic alcohol, 4- hydroxy-3,5-dimethoxybenzoic acid, 2-(4-hydroxy-3-methoxy) phenylethanol, and 2-(3,--dihydroxyphenyl)-1,2-ethandiol is also detected [55 - 57]. Solid waste from the olive mill is also rich and contains different classes of phenolic acids than liquid waste. The main phenolic acids found in solid wastes are salidroside, nuezhenide (only in seed), nuezhenide-oleoside (present in olive stone) and verbascoside (present in seed and pulp) [58]. Tyrosol, hydroxytyrosol, and decarboxymethyl oleuropein are some other phenolic acids found in seed, stone, and pulp. In TPOMW, tyrosol, and hydroxytyrosol are found combined with ρ-coumaric acid and vanillic acid [15, 59, 60]. Verbascoside, rutin, caffeoylquinic acid, luteo-li--4-glucoside, 11-methyloleoside, hydroxytyrosol-10-b-glucoside, luteo-lin-7-rutinoside and oleoside are some minor phenolic acids found in TPOMW [15].

Table 6. Main phenolic acids present in olive mill wastewater (OMWW) and olive mill solid waste (OMSW).

Phenolic acids	OMWW (%)	OMSW (%)
Cinnamic acid	+	+
P-coumaric acid	+	+
Caffeic acid	+	+
Ferulic acid	+	+
Vanillic acid	+	+
Gallic acid	+	+
Oleuropein	+	-
Demethyloleuropein	+	+
Verbascoside	+	+
Tyrosol	+	+
Hydroxytyrosol	+	+
Luteolin	+	+
Rutin	+	+
Quercetin	+	+
Apigenin	-	+

Different carbohydrates other major compounds found in olive mill waste. Dietary fiber consists of polysaccharides and lignin, which are present in olive fruits [61]. Carbohydrate content varies and changes with different ripening stages of olive and differs in OMWW and pomace [62]. Pectin (Arabinans, homogalacturonans, rhamnogalacturonans), cellulose, hemicellulose (xylan, arabinoxylan, xyloglucan), and lignin are some of the main polysaccharides present in olive waste residues [63]. Olive pulp is rich in arabinan pectins, though most of the water-soluble pectins in olive fruits end up in vegetation water [64]. Table 7 presents the carbohydrate composition of waste associated with olive mill.

Table 7. Carbohydrate composition in olive mill wastewater (OMWW) and olive mill solid waste (OMSW) [6].

Carbohydrate	Origin	Yield
Soluble fraction (%)	OMWW	2.32-3.84
	OMSW	23.3-28.7
Insoluble fraction (%)	OMWW	0.13-0.16
	OMSW	51.7-61
Pectin (%)	OMWW	2.32-3.83
	OMSW	1.2-8.9
Hemicellulose (g/kg)	OMSW	351
Cellulose (g/kg)	OMSW	194
Lignin (g/kg)	OMSW	426.3

Direct discharge of olive mill waste into soil leads to soil pollution and phytotoxicity due to their high phenolic acid content [65]. The lipids present in OMWW increase hydrophobicity in the soil thus water retention of soil decreases [66, 67]. Hajjouji *et al.,* [67], reported that diluted OMWW (10%) had shown a genotoxic effect on *Vicia faba* due to the presence of oleuropein and gallic acid. According to Mechri *et al.,* [68], application of OMWW in the soil also disturbs the function of arbuscular mycorrhizas. The discharge of OMWW in water bodies is not recommended as it can disrupt the balance of the entire ecosystem as it leads to microbial respiration and eutrophication. The dark color and high lipid content of OMWW block sunlight and oxygen and inhibit plant growth [16, 17]. Spandre *et al.,* [69], and Danellakis *et al.,* [70], reported that phenolic acids can enter the groundwater level and pathological changes occur in marine organisms when untreated OMWW is dumped in rivers or sea. Storing OMWW in open tanks and discharging of OMWW to open soil lead to fermentation and release methane and hydrogen sulfide producing a strong pungent smell [17].

VALORIZATION OF OLIVE MILL WASTES

The demand for agricultural products is expected to rise 1.1% more in 2050 than in 2010. In recent years, there is a growing interest in functional food which offers high nutrition to consumers. The increasing need for functional foods can threaten food security and result in the scarcity of natural resources; together with which can jeopardize the sustainability of the food system. There is a need to establish the sustainable use of food waste and by-products with a practical strategy encompassing bioeconomy, circular economy and sustainable resource policies [71].

Berbel and Posadillo [71], reported a concept of bioeconomy related to the biomass value pyramid. The biomass value pyramid explains the importance of each value-added product that can be recovered from the waste. Recovering pharmaceuticals and fine chemicals from waste biomass is given the highest priority followed by the production of food and animal feed. The production of compost for agricultural use and biofuels as energy resources is the next task in the priority list.

The different types of valorization techniques of olive mill solid waste and liquid waste and their fate are discussed in detail under this section.

Physical Treatment

Evaporation

Evaporation of OMWW in ponds or lagoons is the most economic and simple technique which is favorable for the warm climate of Mediterranean counties, Southern Europe and Northern Africa. Though it is simple and cheaper, due to several environmental disadvantages, the evaporation of OMWW is not widely used [72]. It consumes larger areas and takes longer treatment time. The decomposition and fermentation of organic materials leave an unpleasant odor and attract insect and rodents. This process also bears chances of groundwater contamination [22, 73].

Direct Application To Soil

The practice of direct application of OMWW to soil also has both positive and negative impacts. OMWW acts as an organic fertilizer increasing the nutrient content of soil. Increased salt content, and polyphenol content, and low pH cause phytotoxicity and reduce mobility of some organic compounds [74]. It is reported to leach herbicides like metamitron and clopyralid and also has antimicrobial activity against *Rhizoctonia solani*; a plant pathogen [29].

Physico-chemical Treatment

In the physico-chemical treatment of OMWW, electro-coagulation and precipitation with chemicals (calcium hydroxide, aluminium sulphate, and lime pre-treatment) are followed to reduce the pollution load [75]. Electro-osmosis dewatering is another effective technique where water is extracted from OMWW with colloidal material (sodium chloride) placed between two electrodes (iron aluminium electrode) which reduces COD and TSS by 47% and 82% respectively [76, 77]. The pre-treatment of OMWW with chitosan and alum coagulants reported a reduction in COD and phenols by 57% and 63% [78].

Recovery Of High Value-added Components From Omww And Omsw

Olive fruit is rich in phenolic acids with beneficial chemical and nutritional properties. Polyphenols inhibit lipid oxidation, and deterioration of food and improve health-promoting properties when added externally. The mechanical process of olive oil production leaves 98% phenolic compounds in the olive mill waste [79, 80]. Although several studies [32, 81 - 83] have confirmed their antioxidant, antimicrobial and molluscicidal activities, Mulinacci *et al.,* [47], have extensively studied several extraction methods used to recover phenolics from olive waste. Verbacoside, tyrosol and caffeic acid can be recovered by conventional Soxhlet extraction with ethanol, though longer extraction time is one of the disadvantages of this process [84]. Unconventional methods include Microwave assisted extraction (MAE), Pulsed electric fields (PEF), High voltage electrical discharge (HVED), Sub and supercritical fluid extraction (SFT) and Pressurized liquid extraction (PLE). These extraction processes are equipped with improved techniques, are relatively cheap, consume less energy and provide a faster mode of action and are environment-friendly [85].

In the MAE extraction process, the internal pressure in the cell is created to rupture the cell wall and release the bioactive compound into the solvent. Later the solvent is heated to separate the target compound [86 - 88]. Higher extraction efficiency, minimum deterioration of target compound, less energy consumption and short extraction time are some of the advantages of this process [89, 90].

Pulsed electric fields are a non-thermal, sustainable and economic method that is used to extract selective bioactive compounds from waste [91 - 93]. This process helps to soften and disrupt the cell membrane thus releasing the intracellular compounds in the solvent [94, 95].

High voltage electrical discharge is an electro-technology that provides high mechanical pressure to disintegrate the cell wall. This technique is used to extract soluble high-added value compounds, enhance the kinetic and quantity of extracted molecules and also inhibit microbial growth [96, 97].

Supercritical fluid technology is one of the green technologies which restrict the use of organic solvents due to their negative impact on health and the environment. Instead, it generates supercritical fluids by applying high pressure and temperature above the critical point. The unique characteristics of these fluids are their gas-like diffusivity and liquid-like power in addition to low cost, no toxicity and flammability that make them ideal solvents for the extraction of valuable components. Lower viscosity, high molecular diffusivity, and low surface tension of supercritical CO_2 (SC- CO_2) improve mass transfer and its separation [98, 99].

Pressurized liquid extraction is an advanced extraction technique that operates at higher pressure and temperature during the extraction process [100]. These conditions to increase the solubility of analytes, decrease water viscosity, improve the penetration and mass transfer rate and finally extraction yield [101]. Shorter extraction time, lower solvent and energy consumption and higher selectivity are the advantages of this process [102].

Production of Absorbents

Activated carbon is used in environmental applications and waste water treatment processes. Though the typical raw materials are coal, peat, and lignite, activated carbon produced from olive mill waste can be activated by physical and chemical means [1]. Physical activation is achieved with steam, CO_2, N_2 or a combination of all three at 750-800°C. Dehydrating agents like H_3PO_4, $ZnCl_2$, KOH, $CaCl_2$, and K_2CO_3 are used in chemical activation and temperature ranges from 400 to 650°C [103, 104]. Activation time for both physical and chemical processes ranges from 10-6400 mins with an optimum time of 60-120 mins [105].

Activated carbon prepared from olive mill waste is used mainly in the adsorption of heavy metals, volatile organic compounds and xenobiotic compounds like dyes and pesticides. After the second extraction of oil, the olive cake is incinerated at 600°C. The produced ash with a surface area of 146.4m²/g and at pH 7 is used for copper and dye removal [106]. Though Banat *et al.,* [107], have reported successful adsorption of methylene blue and no adsorption of methyl orange. Raw olive cakes are also reported to adsorb copper ions. The chemical affinity of hydrophilic extraction of olive cake is like humic acid with carboxylic acid groups as active sites. As the removal of chromium ions are highly pH sensitive, the ideal pH range for Cr^{3+} removal is 4-6, but the maximum removal of Cr^{6+} is achieved only at lower pH (pH 2 or below) [108 - 110].

Production of Animal feed and food

Treated olive wastes are used as animal feed as they are high in dietary fiber content. The dietary fiber present in the olive stone, cake, seed husks and whole stone is a good source of sugars and some saccharides. They can be recovered by various methods like acid hydrolysis, enzymatic hydrolysis, and steam explosion [111 - 115]. Dietary fiber plays an important role in several diseases prevention and reduction in the chances of different types of cancer [116]. Due to their positive effect on health, dietary fibers have the high demand as food supplements and gelling agents [117 - 119]. But the high concentration of phenolic acids acts as inhibitor and imposes a problem for valorization. To achieve the highest activity of pectic polysaccharides, cations and phenols are eliminated first by ultrafiltration with 25-100 kDa membranes [61, 120]. Fiber content binds to the released water and thus improves the water-holding capacity of food and can be used as an additive.

Aerobic Biotechnological Treatments

Microbiological Treatments

Due to the antimicrobial activity and high concentration of phytochemicals, OMWW cannot be used directly in the soil. Removal of phenolic acids and bioconversion of organic matter and plant nutrients transform OMWW into a soil conditioner and fertilizer [121, 122]. Filamentous fungus (*Phanerochaete chrysoporium, Lentinula edodes, Pleurotus ostreatus* and *Aspergillus* sp.) and yeasts are reported to produce extracellular enzymes like laccase, manganese peroxidase and lignin peroxidase which degrade phenolic acids, aromatic compounds and lignin and result in reduced toxicity and COD [123-131]. A study by Assas *et al.*, [132], reported that fungi *Geotrichum candidum* release peroxidase enzymes to degrade phenolic compounds and reduce COD by 75%, and colour by 65%. *Candida holstii* isolated from OMWW showed the highest germination rate (80%), removed 63% COD and 44% phenolic content. *C. oleophilla* reported 32% higher germination rate, 53 and 80% removal of organic compounds and total phenolic acids [133].

Olive pomace treated with fungi undergoes lignin degradation and solid-state fermentation to improve nutritional availability [134, 135]. Brozzoli *et al.*, [136], reported that when a mixture of olive pomace with wheat bran, middlings, barley grains, crimson clover, wheat flour shorts and field beans was treated with *Pleurotus ostreatus* and *P. pulmonarius*, crude protein content was increased by 7-30% and phenolic content was decreased by 50-90% after 6-weeks of incubation.

Biological treatment of OMWW with plants also showed a significant reduction in genotoxicity and phytotoxicity [6]. According to El Hajjouji *et al.,* [67], *Vicia faba* have shown successful biodegradation of phenolic contents (79%). Mixotrophic microalgae and photosynthetic bacteria can catabolize organic matter during the photosynthesis process and sustainably convert them into lipids, proteins and carbohydrates [137].

Other than fungi, yeast and microalgae, consortia of aerobic bacteria found in agricultural soil, municipal and industrial waste, are also used in OMWW detoxification and their COD reduction [29, 138 - 140]. Aerobic *Lactobacillus* sp. carry out fermentative or co-fermentative dephenolization and decolourization of OMWW [141, 142]. *Azotobacter vinelandii* and *Bacillus pumilis* are reported to remove phytotoxin and decrease phenolic acids by 50% [143, 144].

Composting

Composting is a natural recycling process where microbes biodegrade organic waste (crop residues, leaves, grass trimmings, food waste) into simpler forms under aerobic conditions. Composting produces high quality organic fertilizer to replace harmful chemical fertilizers [3, 22].

Though olive mill solid residue is rich in organic matter (93%) [145], its sticky nature and unsuitable physical (low porosity, moist and acidic condition) and chemical conditions (high concentration of potassium, soluble carbohydrates, lipids, and phenols) make OMSW a poor raw material for direct composting [17]. Complementary residues like animal manure, olive leaves and cereal straw or horticultural residues are added in the composting process to optimize the process, which is known as co-composting [146, 147]. The chemical properties and lignocellulosic content of raw material are critical for reaching maturity. Spectroscopic methods showed that the higher the lignocellulose content, higher the time required to reach maturity [148]. During composting, microbial degradation process of carbohydrates leads to decreased molecular heterogeneity, increased unsaturated structures, aromatic poly-condensation, level of conjugated chromophores, molecular size, and degree of humification [149]. Frequent mechanical turning and forced ventilation are important for reduced maturity time, better humification, higher mineralization, increased germination rate, decreased fat- and water-soluble phenols [150 - 153]. Moderate alkaline pH (7.75-9.7) of the final product is maintained with elemental sulphur and it is an important parameter of composting. Reportedly, a decrease in pH by 1.1 unit can be achieved by adding 0.5% sulphur and 40% moisture [154].

Composting of OMWW is a bio-oxidative process which involves partial humidification and mineralization of organic matter to produce bio-fertilizers

[155]. A study by Tomati *et al*. [156], (1996) was first to report enhanced plant soil activity upon adding OMWW compost. Zenjari *et al.,* [157], showed when OMWW was composted with barley straw, phenolic compounds were reduced by 95% and complete removal of phytotoxicity was achieved after two months.

Agaricus bisporus and *Pleurotus* sp. are reported to grow on olive waste to use it as nutrient sources [158 - 160]. Laccase produced by these fungi drives the ligninolytic activity which results in 65% of decolourization and 75% removal of phenols in OMWW [161]. OMWW (25% v/v) mixed with tap water is used for commercial production of oyster mushrooms using wheat straw as substrate [162].

Production of Bioenergy and Biofuel

Traditional direct combustion of olive solid waste (OSW) yields carbon monoxide and hydrocarbons which are considered as an environmental threat. Very recently, with increasing interest in renewable energy, agricultural waste started gaining importance as a valid resource for clean energy [1]. Different technologies that are implicated to derive energy from olive mill solid waste and olive mill wastewater, are discussed in detail in this section.

Anaerobic treatment of OMWW

OMWW contains volatile acids, fats, polyalcohol, and sugars, making it a suitable substrate for biohydrogen, biomethane and bioethanol production following fermentation and anaerobic digestion [162, 163].

Biohydrogen

Many taxonomically diverse bacteria can produce bio-hydrogen through the metabolic pathway (single or combined) using hydrogenases and nitrogenases as enzymes [164]. Among them, photosynthetic microbes covert solar energy into bio-hydrogen through fermentative catabolic pathways from organic and inorganic substrates [165]. Although the dark colour of OMWW poses a negative impact on the photobiological hydrogen production using photosynthetic bacteria. Eroğlu *et al.,* [166], reported that further dilution of OMWW and nitrogen supplementation can be used to overcome the problem.

Fermentation of water diluted TPOMW yields 2.8-4.5 mmol of bio-hydrogen per gram of carbohydrate. Dark fermentation for bio-hydrogen production can be one or two step process. Eroğlu *et al.,* [166], has demonstrated dark fermentation of diluted OMWW (2-20%) in column photobioreactors with *Rhodobacter sphaeroides* and production of bio-hydrogen was approximately 13.9L from each litre of OMWW. The C/N molar ratio of OMWW is a critical parameter for bio-hydrogen production and studies has shown that OMWW with highest C/N molar ratio and highest organic content (acetate, glutamate, and aspartate) produced highest bio-hydrogen [167]. According to Eroğlu *et al.*, 2011 [168], bio-hydrogen yield can be improved by supplementing OMWW with iron and molybdenum as they are part of nitrogenase enzyme complex. Two step dark fermentation by *Rhodobacter sphaeroides* produced 29L of bio-hydrogen per litre of OMWW (50% v/v). The first step of reaction is directed with bio-hydrogen producing bacteria isolated from thermally treated anaerobic sludge. In the second step, anaerobic mixed culture from hydrogenogenic CSTR-type digester is used to produce methane at 35°C. Though Koutrouli *et al.*, [169], reported that thermophilic condition (55°C) showed higher yield and better production rate. Microalgae such as *Rhodopseudomonas palustris* and *Chlamydomonas reinhardtii* are also reported to produce bio-hydrogen and methane by both photosystem II-driven water splitting and fermentation of carbohydrates [137, 170 - 172,].

Biomethane

Production of methane from olive mill waste is a two-step process; anaerobic-aerobic pre-treatment and two-phase anaerobic digestion. *Aspergillus niger* (anaerobic treatment) and *A. terreus* (aerobic treatment) are used to pre-treat the OMWW before anaerobic digestion to improve methane production. Yeasts like *Candida tropicalis* have also been used in the aerobic treatment of OMWW to reduce phenolic content and COD. During two-step anaerobic digestion process, at the first step, acidogenic bacteria degrade carbohydrates, proteins, lipids, amino acids and fatty acids into mainly volatile fatty acids and alcohols. In the second step, these metabolites are transformed into methane and CO_2 by archaea or methanogens. A study by [173, 174] showed that the hydraulic retention time (HRT) of the first step (acidification) and the second step ranges from 14 to 24 days and 24-36 days respectively. The organic load at the first step was higher (5.54 to 14 g COD/ L) than in the second step (2.2 to 9.1 g COD/L). The methane production is reported to be 32L per liter of OMWW.

Bioethanol

Bioethanol production by bacteria and yeast is favored by the high organic content of OMWW. The bioconversion of different sugars in olive mill waste into ethanol involves the first step of enzymatic hydrolysis and the second step of generation of ethanol. Enzymatic hydrolysis (with Celluclast or Novozyme 88) [175] or physicochemical pre-treatment of OMWW increases the reducing sugar content, which then is converted into ethanol by yeast or bacteria [6]. El Asli and Qatibi [176], have reported that pre-treatment of olive cake with sulphuric acids releases lignocellulosic components and fermentation inhibitors are eliminated by precipitation and filtration. The available soluble sugar (18.1g/L) is then fermented by recombinant *Escherichia coli*, *Saccharomyces cerevisiae* and *Thermoanaerobacter mathranii* into ethanol (0.45/g of sugar). Thermal processing and pre-treatment with *Pleurotus sajor-caju* of OMWW were reported to produce 14.2g ethanol per litre of OMWW after 48 hrs of yeast fermentation [177].

Biodiesel

Biodiesel is another renewable source of energy that is non-toxic and bio-degradable. Studies have shown that olive oil by-products are a suitable substrate for biodiesel production. Studies demonstrated that *Lipomyces starkeyi* can use OMWW as substrate, converting the lipid content into biodiesel [178, 179]. Similarly immobilized lipase from *Thermomyces lanuginosus* can produce 93% biodiesel after 24 hrs at 25°C using olive cake powder as substrate.

Pyrolysis

When less oxygen is present than required or under anaerobic conditions, the combustion of materials is known as pyrolysis [180]. Pyrolysis of olive cake is greatly dependent on various physical characteristics and generates three different phases (solid, liquid, and gaseous). This depends on temperature and is independent of the particle size. The loss of humidity and the volatile compound was not dependent on oxygen concentration, though the initial temperature of char decreased while oxygen concentration increased. During oxidative pyrolysis, the emission of CO and CO_2, temperature range for volatilization, char oxidation and remaining ash were found to be dependent on the particle size of the materials [181]. Chouchene *et al.,* [182], showed oxidative pyrolysis (with 10% oxygen) of olive waste of 2-2.8 mm particle size offering significantly higher char yield and lower remaining ash. Uzun *et al.,* [183], (2007) used a fixed bed tubular reactor to produce bio-oil from a rapid pyrolysis of olive cake where at 500°C, maximum oil (41.5%) production was recorded. Encinar *et al.,.* [184, 185], showed that optimum temperature for olive cake pyrolysis ranges from 600 to 700°C.

Gasification

In gasification, carbonaceous materials are converted into CO and H_2 (syngas) under controlled oxygen and at high temperature. This method is useful as the energy efficiency of syngas is higher than original fuel [1]. Gómez-Barea *et al.,* [186], and Skoulou *et al.,* [187], reported that gasification of olive cake at 700-850°C using bubbling fluidized bed reactor was better in gas yield and carbon conversion . In addition, problems like sintering and agglomeration were solved by using ofite and olivine as inert bed materials. Gasification of olive cake using downdraft fixed bed gasifier and circulation fluidized bed reactor at 950°C favoured the gas production. Higher temperatures directed a higher syngas yield but produced less methane, carbon monoxide and light hydrocarbons [188, 189]. Co-gasification is another useful technique where poor-quality coal is mixed with agricultural waste to reduce NOx and SOx levels [190]. Resources utilization and treatment contribute to the production of value-added products [191 - 196]. Treatment of substrates also leads to the production of useful resources [197 - 201]. Co-gasification of olive cake (40%) in a fluidized bed reactor at 850-900°C showed the best result and also reduced the production of tars and heavier hydrocarbons [202, 203].

CONCLUSION

Olive oil production contributes a large portion of the economy of Mediterranean countries. The high organic matter and phenolic acid content of olive oil waste residues impose a serious environmental threat. This chapter discussed different valorization methods of olive waste and their importance. The valorization techniques mainly cured important phytochemical compounds which are used in pharmaceutical and cosmetic industries. The next target of these approaches is to remove or convert the by-products and to reduce environmental pollution. Efficient physical, Physico-chemical and biotechnological treatments are used to reduce the heterogeneity of olive mill wastes and convert them into value-added products like fuels, enzymes, organic acids, and biopolymers. The dilution of OMWW still requires a large amount of water which is critical for some drought-hit Mediterranean regions. It offers scopes for the development of better treatment with fewer water requirements and also recombinant microbial strains for the detoxification process with promising prospects and improvements. Further research can improve economic feasibility, and sustainability and scale up the valorization process to deal with the higher volume of waste at once. Different valorization methods offer different outcomes and their efficiency depends on many factors. As one valorization technique cannot solve all the problems, researchers suggest that the choice of valorization technique should consider the specific needs of a particular area.

REFERENCES

[1]   Kalderis D, Diamadopoulos E. Valorization of solid waste residues from olive oil mills: A review. Terr Aquat Environ Toxicol 2010; 4(1): 7-20.

[2]   Azaizeh H, Tayeh HN, Gerchman Y. Valorisation of olive oil industry solid waste and production of ethanol and high value-added biomolecules.Biovalorisation of Wastes to Renewable Chemicals and Biofuels. Elsevier 2020; pp. 27-40.
[http://dx.doi.org/10.1016/B978-0-12-817951-2.00002-X]

[3]   Khdair A, Abu-Rumman G. Sustainable Environmental Management and Valorization Options for

Olive Mill Byproducts in the Middle East and North Africa (MENA) Region. Processes (Basel) 2020; 8(6): 671.
[http://dx.doi.org/10.3390/pr8060671]

[4]   Schwingshackl L, Lampousi A-M, Portillo MP, Romaguera D, Hoffmann G, Boeing H. Olive oil in the prevention and management of type 2 diabetes mellitus: a systematic review and meta-analysis of cohort studies and intervention trials. Nutr Diabetes 2017; 7(4)e262.
[http://dx.doi.org/10.1038/nutd.2017.12] [PMID: 28394365]

[5]   Miranda I, Simões R, Medeiros B, *et al.* Valorization of lignocellulosic residues from the olive oil industry by production of lignin, glucose and functional sugars. Bioresour Technol 2019; 292121936.
[http://dx.doi.org/10.1016 /j.biortech .2019.121936] [PMID: 31398542]

[6]   Dermeche S, Nadour M, Larroche C, Moulti-Mati F, Michaud P. Olive mill wastes. Biochemical characterizations and valorization strategies. Process Biochem 2013; 48(10): 1532-52.
[http://dx.doi.org/10.1016/j .procbio. 2013.07.010]

[7]   Hajimahmoo M, Sadeghi N, Jannat B, *et al.* Antioxidant activity reducing power and total phenolic content of iranian olive cultivar. J Biol Sci (Faisalabad, Pak) 2008; 8(4): 779-83.
[http://dx.doi.org/10.3923/jbs.2008.779.783]

[8]   Atmani D, Chaher N, Atmani D, Berboucha M, Debbache N, Boudaoud H. Flavonoids in human health: from structure to biological activity. Curr Nutr Food Sci 2009; 5(4): 225-37.
[http://dx.doi.org/10.2174/157340109790218049]

[9]   Cicerale S, Lucas L, Keast R. Biological activities of phenolic compounds present in virgin olive oil. Int J Mol Sci 2010; 11(2): 458-79.
[http://dx.doi.org/10.3390/ijms11020458] [PMID: 20386648]

[10]   Atmani D, Begontilde M, Ruiz-Sanz JI, Bakkali F, Atmani D. Antioxidant potential, cytotoxic activity and phenolic content of Clematis flammula leaf extracts. J Med Plants Res 2011; 5(4): 589-98.

[11]   Faiza I, Wahiba K, Nassira G, Chahrazed B, Atik BF. Antibacterial and antifungal activities of olive (Olea europaea L.) from Algeria. J Microbiol Biotechnol Res 2011; 1(2): 69-73.

[12]   Loizzo MR, Lecce GD, Boselli E, Menichini F, Frega NG. Inhibitory activity of phenolic compounds from extra virgin olive oils on the enzymes involved in diabetes, obesity and hypertension. J Food Biochem 2011; 35(2): 381-99.
[http://dx.doi.org/10.1111/j.1745-4514.2010.00390.x]

[13]   Ohno T, Inoue M, Ogihara Y, Saracoglu I. Antimetastatic activity of acteoside, a phenylethanoid glycoside. Biol Pharm Bull 2002; 25(5): 666-8.
[http://dx.doi.org/10.1248/bpb.25.666] [PMID: 12033512]

[14]   Ciriminna R, Meneguzzo F, Fidalgo A, Ilharco LM, Pagliaro M. Extraction, benefits and valorization of olive polyphenols. Eur J Lipid Sci Technol 2016; 118(4): 503-11.
[http://dx.doi.org/10.1002/ejlt.201500036]

[15]   Lesage-Meessen L, Navarro D, Maunier S, *et al.* Simple phenolic content in olive oil residues as a function of extraction systems. Food Chem 2001; 75(4): 501-7.
[http://dx.doi.org/10.1016/S0308-8146(01)00227-8]

[16]   Kapellakis IE, Tsagarakis KP, Crowther JC. Olive oil history, production and by-product management. Rev Environ Sci Biotechnol 2008; 7(1): 1-26.
[http://dx.doi.org/10.1007/s11157-007-9120-9]

[17]   Niaounakis M, Halvadakis CP. Olive processing waste management: literature review and patent survey. Elsevier 2006.

[18]   Laufenberg G, Kunz B, Nystroem M. Transformation of vegetable waste into value added products. Bioresour Technol 2003; 87(2): 167-98.
[http://dx.doi.org/10.1016/S0960-8524(02)00167-0] [PMID: 12765356]

[19]   Wyman CE. Potential synergies and challenges in refining cellulosic biomass to fuels, chemicals, and power. Biotechnol Prog 2003; 19(2): 254-62.
[http://dx.doi.org/10.1021/bp025654l] [PMID: 12675557]

[20]   Vandermeersch T, Alvarenga RAF, Ragaert P, Dewulf J. Environmental sustainability assessment of food waste valorization options. Resour Conserv Recycling 2014; 87: 57-64.
[http://dx.doi.org/10.1016/j.resconrec.2014.03.008]

[21]   Roels K, Van Gijseghem D. Loss and waste in the food chain. Department of Agriculture and Fisheries, Monitoring and Study Department 2011.

[22]   Roig A, Cayuela ML, Sánchez-Monedero MA. An overview on olive mill wastes and their valorisation methods. Waste Manag 2006; 26(9): 960-9.
[http://dx.doi.org/10.1016/j.wasman.2005.07.024] [PMID: 16246541]

[23]   Azbar N, Bayram A, Filibeli A, Muezzinoglu A, Sengul F, Ozer A. A review of waste management options in olive oil production. Crit Rev Environ Sci Technol 2004; 34(3): 209-47.
[http://dx.doi.org/10.1080/10643380490279932]

[24]   Di Giovacchino L, Sestili S, Di Vincenzo D. Influence of olive processing on virgin olive oil quality. Eur J Lipid Sci Technol 2002; 104(9-10): 587-601.
[http://dx.doi.org/10.1002/1438-9312(200210)104:9/10<587::AID-EJLT587>3.0.CO;2-M]

[25]   Sánchez Moral P, Ruiz Méndez MV. Production of pomace olive oil. Grasas Aceites 2006; 57(1): 47-55.
[http://dx.doi.org/10.3989/gya.2006.v57.i1.21]

[26]   Alburquerque J, Gonzálvez J, García D, Cegarra J. Agrochemical characterisation of "alperujo", a solid by-product of the two-phase centrifugation method for olive oil extraction. Bioresour Technol 2004; 91(2): 195-200.
[http://dx.doi.org/10.1016/S0960-8524(03)00177-9] [PMID: 14592750]

[27]   Morillo JA, Antizar-Ladislao B, Monteoliva-Sánchez M, Ramos-Cormenzana A, Russell NJ. Bioremediation and biovalorisation of olive-mill wastes. Appl Microbiol Biotechnol 2009; 82(1): 25-39.
[http://dx.doi.org/10.1007/ s00253 -008-1801-y] [PMID: 19082586]

[28]   Brunetti G, Plaza C, Senesi N. Olive pomace amendment in Mediterranean conditions: effect on soil and humic acid properties and wheat (Triticum turgidum L.) yield. J Agric Food Chem 2005; 53(17): 6730-7.
[http://dx.doi.org/10.1021/jf050152j] [PMID: 16104792]

[29]   Benitez J, Beltran-Heredia J, Torregrosa J, Acero JL, Cercas V. Aerobic degradation of olive mill wastewaters. Appl Microbiol Biotechnol 1997; 47(2): 185-8.
[http://dx.doi.org/10.1007/s002530050910] [PMID: 9077005]

[30]   McNamara CJ, Anastasiou CC, O'Flaherty V, Mitchell R. Bioremediation of olive mill wastewater. Int Biodeterior Biodegradation 2008; 61(2): 127-34.
[http://dx.doi.org/10.1016/j.ibiod.2007.11.003]

[31]   Caputo AC, Scacchia F, Pelagagge PM. Disposal of by-products in olive oil industry: waste-to-energy solutions. Appl Therm Eng 2003; 23(2): 197-214.
[http://dx.doi.org/10.1016/S1359-4311(02)00173-4]

[32]   Obied HK, Allen MS, Bedgood DR, Prenzler PD, Robards K, Stockmann R. Bioactivity and analysis of biophenols recovered from olive mill waste. J Agric Food Chem 2005; 53(4): 823-37.
[http://dx.doi.org/10.1021/jf048569x] [PMID: 15712986]

[33]   Paixão SM, Mendonça E, Picado A, Anselmo AM. Acute toxicity evaluation of olive oil mill wastewaters: A comparative study of three aquatic organisms. Environ Toxicol 1999; 14(2): 263-9.
[http://dx.doi.org/10.1002/(SICI)1522-7278(199905)14:2<263::AID-TOX7>3.0.CO;2-D]

[34] Zabaniotou A, Stavropoulos G, Skoulou V. Activated carbon from olive kernels in a two-stage process: Industrial improvement. Bioresour Technol 2008; 99(2): 320-6.
[http://dx.doi.org/10.1016/j.biortech.2006.12.020] [PMID: 17307355]

[35] Blázquez G, Calero M, Hernáinz F, Tenorio G, Martín-Lara MA. Equilibrium biosorption of lead(II) from aqueous solutions by solid waste from olive-oil production. Chem Eng J 2010; 160(2): 615-22.
[http://dx.doi.org/10.1016/j.cej.2010.03.085]

[36] Tortosa G, Alburquerque JA, Ait-Baddi G, Cegarra J. The production of commercial organic amendments and fertilisers by composting of two-phase olive mill waste ("alperujo"). J Clean Prod 2012; 26: 48-55.
[http://dx.doi.org/10.1016/j.jclepro.2011.12.008]

[37] Tortosa G, Alburquerque JA, Bedmar EJ, Ait-Baddi G, Cegarra J. Strategies to produce commercial liquid organic fertilisers from "alperujo" composts. J Clean Prod 2014; 82: 37-44.
[http://dx.doi.org/10.1016/j.jclepro.2014.06.083]

[38] Ballesteros I, Oliva JM, Saez F, Ballesteros M. Ethanol production from lignocellulosic byproducts of olive oil extraction Appl Biochem Biotechnol 2001; 91(1-9): 237-52.

[39] Battista F, Mancini G, Ruggeri B, Fino D. Selection of the best pretreatment for hydrogen and bioethanol production from olive oil waste products. Renew Energy 2016; 88: 401-7.
[http://dx.doi.org/10.1016/j.renene.2015.11.055]

[40] Christoforou EA, Fokaides PA, Kyriakides I. Monte Carlo parametric modeling for predicting biomass calorific value. J Therm Anal Calorim 2014; 118(3): 1789-96.
[http://dx.doi.org/10.1007/s10973-014-4027-5]

[41] Paraskeva P, Diamadopoulos E. Technologies for olive mill wastewater (OMW) treatment: a review. J Chem Technol Biotechnol: International Research in Process. Environmental & Clean Technology 2006; 81(9): 1475-85.

[42] Galanakis CM, Tornberg E, Gekas V. The effect of heat processing on the functional properties of pectin contained in olive mill wastewater. Lebensm Wiss Technol 2010; 43(7): 1001-8.
[http://dx.doi.org/10.1016/j.lwt.2010.01.004]

[43] Rahmanian N, Jafari SM, Galanakis CM. Recovery and removal of phenolic compounds from olive mill wastewater. J Am Oil Chem Soc 2014; 91(1): 1-18.
[http://dx.doi.org/10.1007/s11746-013-2350-9]

[44] Jarboui R, Sellami F, Kharroubi A, Gharsallah N, Ammar E. Olive mill wastewater stabilization in open-air ponds: Impact on clay–sandy soil. Bioresour Technol 2008; 99(16): 7699-708.
[http://dx.doi.org/10.1016/j.biortech.2008.01.074] [PMID: 18337092]

[45] Borja R, Sánchez E, Raposo F, Rincón B, Jiménez AM, Martín A. A study of the natural biodegradation of two-phase olive mill solid waste during its storage in an evaporation pond. Waste Manag 2006; 26(5): 477-86.
[http://dx.doi.org/10.1016/j.wasman.2005.02.024] [PMID: 15963711]

[46] Rinaldi M, Rana G, Introna M. Olive-mill wastewater spreading in southern Italy: effects on a durum wheat crop. Field Crops Res 2003; 84(3): 319-26.
[http://dx.doi.org/10.1016/S0378-4290(03)00097-2]

[47] Mulinacci N, Romani A, Galardi C, Pinelli P, Giaccherini C, Vincieri FF. Polyphenolic content in olive oil waste waters and related olive samples. J Agric Food Chem 2001; 49(8): 3509-14.
[http://dx.doi.org/10.1021/jf000972q] [PMID: 11513620]

[48] Sierra J, Martí E, Montserrat G, Cruañas R, Garau MA. Characterisation and evolution of a soil affected by olive oil mill wastewater disposal. Sci Total Environ 2001; 279(1-3): 207-14.
[http://dx.doi.org/10.1016/S0048-9697(01)00783-5] [PMID: 11712597]

[49] Alba-Mendoza J, Ruiz-Gomez A, Hidalgo-Casado F. Technological evolution of the different

processes for olive oil extraction Edible fats and oils processing: basic principles and modern practices. Champaign, Illinois, USA: Am Oil Chem. Soc. 1990; pp. 341-7.

[50]   Fadil K, Chahlaoui A, Ouahbi A, Zaid A, Borja R. Aerobic biodegradation and detoxification of wastewaters from the olive oil industry. Int Biodeterior Biodegradation 2003; 51(1): 37-41.
[http://dx.doi.org/10.1016/S0964-8305(02)00073-2]

[51]   Bravo L. Polyphenols: Chemistry, dietary sources, metabolism, and nutritional significance. Nutr Rev 1998; 56(11): 317-33.
[http://dx.doi.org/10.1111/j.1753-4887.1998.tb01670.x] [PMID: 9838798]

[52]   Shahidi F, Naczk M. Phenolics in food and nutraceuticals. Boca Raton, FL, USA: CRC Press 2003.
[http://dx.doi.org/10.1201/9780203508732]

[53]   Bianco A, Buiarelli F, Cartoni G, Coccioli F, Jasionowska R, Margherita P. Analysis by liquid chromatography-tandem mass spectrometry of biophenolic compounds in olives and vegetation waters, Part I. J Sep Sci 2003; 26(5): 409-16.
[http://dx.doi.org/10.1002/jssc.200390053]

[54]   Visioli F, Poli A, Gall C. Antioxidant and other biological activities of phenols from olives and olive oil. Med Res Rev 2002; 22(1): 65-75.
[http://dx.doi.org/10.1002/med.1028] [PMID: 11746176]

[55]   DellaGreca M, Monaco P, Pinto G, Pollio A, Previtera L, Temussi F. Phytotoxicity of low-molecula--weight phenols from olive mill waste waters Bull Environ Contamin Toxicol 2001; 67(3): 0352-9.

[56]   Capasso R, Evidente A, Scognamiglio F. A simple thin layer chromatographic method to detect the main polyphenols occurring in olive oil vegetation waters. Phytochem Anal 1992; 3(6): 270-5.
[http://dx.doi.org/10.1002/pca.2800030607]

[57]   Aramendía MA, Boráu V, García I, *et al.* Qualitative and Quantitative Analyses of Phenolic Compounds by High-performance Liquid Chromatography and Detection with Atmospheric Pressure Chemical Ionization Mass Spectrometry. Rapid Commun Mass Spectrom 1996; 10(13): 1585-90.
[http://dx.doi.org/10.1002/(SICI)1097-0231(199610)10:13<1585::AID-RCM673>3.0.CO;2-O]

[58]   Maestro Durán R, León Cabello R, Ruiz-Gutiérrez V, Fiestas P, Vázquez-Roncero A. Bitter phenolic glycosides from olive seeds (Olea europea). Fat Oils 1994; 45(5): 332-5.

[59]   Ryan D, Prenzler PD, Lavee S, Antolovich M, Robards K. Quantitative changes in phenolic content during physiological development of the olive (Olea europaea) cultivar Hardy's Mammoth. J Agric Food Chem 2003; 51(9): 2532-8.
[http://dx.doi.org/10.1021/jf0261351] [PMID: 12696932]

[60]   Fernández-Bolaños J, Felizón B, Brenes M, Guillén R, Heredia A. Hydroxytyrosol and tyrosol as the main compounds found in the phenolic fraction of steam-exploded olive stones. J Am Oil Chem' Soc 1998; 75(11):1643-9.

[61]   García ML, Cáceres E, Selgas MD. Utilisation of fruit fibres in conventional and reduced-fat cooked-meat sausages. J Sci Food Agric 2007; 87(4): 624-31.
[http://dx.doi.org/10.1002/jsfa.2753]

[62]   Kennedy JF, Meullenet F. Olive-Mill Waste Management, Literature Review and Patent Survey, M. Niaounakis, CP Halvadakis, Typophyto-George Dardanos Publications, Athens, Greece (2003),(xv+ 430 pp., Currently Not For Sale—Free under Request (500 copies only), ISBN 960-402-123-0).

[63]   Jiménez A, Guillén R, Fernández-Bolaños J, Heredia A. GUILLÉN R, Fernandez-Bolaños JU, HEREDIA A. Cell wall composition of olives. J Food Sci 1994; 59(6): 1192-6.
[http://dx.doi.org/10.1111/j.1365-2621.1994.tb14674.x]

[64]   Vierhuis E, Korver M, Schols HA, Voragen AGJ. Structural characteristics of pectic polysaccharides from olive fruit (*Olea europaea* cv moraiolo) in relation to processing for oil extraction. Carbohydr Polym 2003; 51(2): 135-48.
[http://dx.doi.org/10.1016/S0144-8617(02)00158-3]

[65]   Paredes MJ, Moreno E, Ramos-Cormenzana A, Martinez J. Characteristics of soil after pollution with waste waters from olive oil extraction plants. Chemosphere 1987; 16(7): 1557-64.
[http://dx.doi.org/10.1016/0045-6535(87)90096-8]

[66]   Kavvadias V, Doula MK, Komnitsas K, Liakopoulou N. Disposal of olive oil mill wastes in evaporation ponds: Effects on soil properties. J Hazard Mater 2010; 182(1-3): 144-55.
[http://dx.doi.org/10.1016/j.jhazmat.2010.06.007] [PMID: 20580156]

[67]   El Hajjouji H, Pinelli E, Guiresse M, Merlina G, Revel JC, Hafidi M. Assessment of the genotoxicity of olive mill waste water (OMWW) with the *Vicia faba* micronucleus test. Mutat Res Genet Toxicol Environ Mutagen 2007; 634(1-2): 25-31.
[http://dx.doi.org/10.1016/j.mrgentox.2007.05.015] [PMID: 17851113]

[68]   Mechri B, Cheheb H, Boussadia O, *et al.* Effects of agronomic application of olive mill wastewater in a field of olive trees on carbohydrate profiles, chlorophyll a fluorescence and mineral nutrient content. Environ Exp Bot 2011; 71(2): 184-91.
[http://dx.doi.org/10.1016/j.envexpbot.2010.12.004]

[69]   Spandre R, Dellomonaco G. Polyphenols pollution by olive mill waste waters, Tuscany, Italy. J Environ Hydrol 1996; 4: 1-3.

[70]   Danellakis D, Ntaikou I, Kornaros M, Dailianis S. Olive oil mill wastewater toxicity in the marine environment: Alterations of stress indices in tissues of mussel *Mytilus galloprovincialis*. Aquat Toxicol 2011; 101(2): 358-66.
[http://dx.doi.org/10.1016/j.aquatox.2010.11.015] [PMID: 21216346]

[71]   Berbel J, Posadillo A. Review and analysis of alternatives for the valorisation of agro-industrial olive oil by-products. Sustainability (Basel) 2018; 10(1): 237.
[http://dx.doi.org/10.3390/su10010237]

[72]   Peri C, Proietti P. Olive mill waste and by-products The extra-virgin olive oil handbook 2014; 22: 283-02.

[73]   Haddadin MS, Abdulrahim SM, Al-Khawaldeh GY, Robinson RK. Solid state fermentation of waste pomace from olive processing. J Chem Technol Biotechnol International Research in Process, Environmental & Clean Technology. 74(7): 613-8.

[74]   Salameh WK. Treatment of olive mill wastewater by ozonation and electrocoagulation processes. Civil Environ Res 2015; 7(2): 80-91.

[75]   Lafi WK, Shannak B, Al-Shannag M, Al-Anber Z, Al-Hasan M. Treatment of olive mill wastewater by combined advanced oxidation and biodegradation. Separ Purif Tech 2009; 70(2): 141-6.
[http://dx.doi.org/10.1016/j.seppur.2009.09.008]

[76]   Cooperband L. 2002. The art and science of composting. Center for Integrated agricultural systems. WI, USA, 2002; 1–14.

[77]   Sánchezarias V, Fernández F, Villaseñor J, Rodríguez L. Enhancing the co-composting of olive mill wastes and sewage sludge by the addition of an industrial waste. Bioresour Technol 2008; 99(14): 6346-53.
[http://dx.doi.org/10.1016/j.biortech.2007.12.013] [PMID: 18194860]

[78]   Vuppala S, Bavasso I, Stoller M, Di Palma L, Vilardi G. Olive mill wastewater integrated purification through pre-treatments using coagulants and biological methods: Experimental, modelling and scale-up. J Clean Prod 2019; 236117622.
[http://dx.doi.org/10.1016/j.jclepro.2019.117622]

[79]   Rodríguez G, Lama A, Rodríguez R, Jiménez A, Guillén R, Fernández-Bolaños J. Olive stone an attractive source of bioactive and valuable compounds. Bioresour Technol 2008; 99(13): 5261-9.
[http://dx.doi.org/10.1016/j.biortech.2007.11.027] [PMID: 18160280]

[80]   Rodríguez G, Lama A, Trujillo M, Espartero JL, Fernández-Bolaños J. Isolation of a powerful

antioxidant from Olea europaea fruit-mill waste: 3,4-Dihydroxyphenylglycol. Lebensm Wiss Technol 2009; 42(2): 483-90.
[http://dx.doi.org/10.1016/j.lwt.2008.08.015]

[81] Obied HK, Allen MS, Bedgood DR Jr, Prenzler PD, Robards K. Investigation of Australian olive mill waste for recovery of biophenols. J Agric Food Chem 2005; 53(26): 9911-20.
[http://dx.doi.org/10.1021/jf0518352] [PMID: 16366674]

[82] Obied HK, Bedgood DR Jr, Prenzler PD, Robards K. Bioscreening of Australian olive mill waste extracts: Biophenol content, antioxidant, antimicrobial and molluscicidal activities. Food Chem Toxicol 2007; 45(7): 1238-48.
[http://dx.doi.org/10.1016/j.fct.2007.01.004] [PMID: 17329005]

[83] Obied HK, Prenzler PD, Robards K. Potent antioxidant biophenols from olive mill waste. Food Chem 2008; 111(1): 171-8.
[http://dx.doi.org/10.1016/j.foodchem.2008.03.058]

[84] Suárez M, Romero MP, Ramo T, Macià A, Motilva MJ. Methods for preparing phenolic extracts from olive cake for potential application as food antioxidants. J Agric Food Chem 2009; 57(4): 1463-72.
[http://dx.doi.org/10.1021/jf8032254] [PMID: 19178195]

[85] Roselló-Soto E, Koubaa M, Moubarik A, et al. Emerging opportunities for the effective valorization of wastes and by-products generated during olive oil production process: Non-conventional methods for the recovery of high-added value compounds. Trends Food Sci Technol 2015; 45(2): 296-310.
[http://dx.doi.org/10.1016/j.tifs.2015.07.003]

[86] Lozano-Sánchez J, Castro-Puyana M, Mendiola J, Segura-Carretero A, Cifuentes A, Ibáez E. Recovering bioactive compounds from olive oil filter cake by advanced extraction techniques. Int J Mol Sci 2014; 15(9): 16270-83.
[http://dx.doi.org/10.3390/ijms150916270] [PMID: 25226536]

[87] Zhang B, Yang R, Liu CZ. Microwave-assisted extraction of chlorogenic acid from flower buds of Lonicera japonica Thunb. Separ Purif Tech 2008; 62(2): 480-3.
[http://dx.doi.org/10.1016/j.seppur.2008.02.013]

[88] Chemat F, Vian MA, Cravotto G. Green extraction of natural products: concept and principles. Int J Mol Sci 2012; 13(7): 8615-27.
[http://dx.doi.org/10.3390/ijms13078615] [PMID: 22942724]

[89] Barba FJ, Grimi N, Vorobiev E. New approaches for the use of non-conventional cell disruption technologies to extract potential food additives and nutraceuticals from microalgae. Food Eng Rev 2015; 7(1): 45-62.
[http://dx.doi.org/10.1007/s12393-014-9095-6]

[90] Camel V. Microwave-assisted solvent extraction of environmental samples. TrAC 2000; 19(4): 229-48.

[91] Zulueta A, Barba FJ, Esteve MJ, Frígola A. Changes in quality and nutritional parameters during refrigerated storage of an orange juice–milk beverage treated by equivalent thermal and non-thermal processes for mild pasteurization. Food Bioprocess Technol 2013; 6(8): 2018-30.
[http://dx.doi.org/10.1007/s11947-012-0858-x]

[92] Donsì F, Ferrari G, Pataro G. Applications of pulsed electric field treatments for the enhancement of mass transfer from vegetable tissue. Food Eng Rev 2010; 2(2): 109-30.
[http://dx.doi.org/10.1007/s12393-010-9015-3]

[93] Abenoza M, Benito M, Saldaña G, Álvarez I, Raso J, Sánchez-Gimeno AC. Effects of pulsed electric field on yield extraction and quality of olive oil. Food Bioprocess Technol 2013; 6(6): 1367-73.
[http://dx.doi.org/10.1007/s11947-012-0817-6]

[94] Barba FJ, Grimi N, Vorobiev E. Evaluating the potential of cell disruption technologies for green selective extraction of antioxidant compounds from *Stevia rebaudiana Bertoni* leaves. J Food Eng

2015; 149: 222-8.
[http://dx.doi.org/10.1016/j.jfoodeng.2014.10.028]

[95]   Deng Q, Zinoviadou KG, Galanakis CM, *et al*. The effects of conventional and non-conventional processing on glucosinolates and its derived forms, isothiocyanates: extraction, degradation, and applications. Food Eng Rev 2015; 7(3): 357-81.
[http://dx.doi.org/10.1007/s12393-014-9104-9]

[96]   Barba FJ, Brianceau S, Turk M, Boussetta N, Vorobiev E. Effect of alternative physical treatments (ultrasounds, pulsed electric fields, and high-voltage electrical discharges) on selective recovery of bio-compounds from fermented grape pomace. Food Bioprocess Technol 2015; 8(5): 1139-48.
[http://dx.doi.org/10.1007/s11947-015-1482-3]

[97]   Lebovka N, Vorobiev E, Chemat F, Eds. Enhancing extraction processes in the food industry. CRC Press 2016.
[http://dx.doi.org/10.1201/b11241]

[98]   Oliveira JV, Oliveira D. Kinetics of the enzymatic alcoholysis of palm kernel oil in supercritical CO_2. Ind Eng Chem Res 2000; 39(12): 4450-4.
[http://dx.doi.org/10.1021/ie990865p]

[99]   Roselló-Soto E, Galanakis CM, Brnčić M, *et al*. Clean recovery of antioxidant compounds from plant foods, by-products and algae assisted by ultrasounds processing. Modeling approaches to optimize processing conditions. Trends Food Sci Technol 2015; 42(2): 134-49.
[http://dx.doi.org/10.1016/j.tifs.2015.01.002]

[100]   Herrero M, Temirzoda TN, Segura-Carretero A, Quirantes R, Plaza M, Ibañez E. New possibilities for the valorization of olive oil by-products. J Chromatogr A 2011; 1218(42): 7511-20.
[http://dx.doi.org/10.1016/j.chroma.2011.04.053] [PMID: 21600577]

[101]   Turner C, Ibañez E. Pressurized hot water extraction and processing. 2011.
[http://dx.doi.org/10.1201/b11241-9]

[102]   Mustafa A, Turner C. Pressurized liquid extraction as a green approach in food and herbal plants extraction: A review. Anal Chim Acta 2011; 703(1): 8-18.
[http://dx.doi.org/10.1016/j.aca.2011.07.018] [PMID: 21843670]

[103]   Aljundi IH, Jarrah N. A study of characteristics of activated carbon produced from Jordanian olive cake. J Anal Appl Pyrolysis 2008; 81(1): 33-6.
[http://dx.doi.org/10.1016/j.jaap.2007.07.006]

[104]   Zabaniotou A, Stavropoulos G, Skoulou V. Activated carbon from olive kernels in a two-stage process: Industrial improvement. Bioresour Technol 2008; 99(2): 320-6.
[http://dx.doi.org/10.1016/j.biortech.2006.12.020] [PMID: 17307355]

[105]   Juárez-Galán JM, Silvestre-Albero A, Silvestre-Albero J, Rodríguez-Reinoso F. Synthesis of activated carbon with highly developed "mesoporosity". Microporous Mesoporous Mater 2009; 117(1-2): 519-21.
[http://dx.doi.org/10.1016/j.micromeso.2008.06.011]

[106]   Bouzid J, Elouear Z, Ksibi M, Feki M, Montiel A. A study on removal characteristics of copper from aqueous solution by sewage sludge and pomace ashes. J Hazard Mater 2008; 152(2): 838-45.
[http://dx.doi.org/10.1016/j.jhazmat.2007.07.092] [PMID: 17822842]

[107]   Banat F, Al-Asheh S, Al-Ahmad R, Bni-Khalid F. Bench-scale and packed bed sorption of methylene blue using treated olive pomace and charcoal. Bioresour Technol 2007; 98(16): 3017-25.
[http://dx.doi.org/10.1016/j.biortech.2006.10.023] [PMID: 17158045]

[108]   Gondar D, Bernal MP. Copper binding by olive mill solid waste and its organic matter fractions. Geoderma 2009; 149(3-4): 272-9.
[http://dx.doi.org/10.1016/j.geoderma.2008.12.005]

[109]   Kolokassidou K, Szymczak W, Wolf M, Obermeier C, Buckau G, Pashalidis I. Hydrophilic olive cake

extracts: Characterization by physicochemical properties and Cu(II) complexation. J Hazard Mater 2009; 164(2-3): 442-7.
[http://dx.doi.org/10.1016/j.jhazmat.2008.08.016] [PMID: 18801615]

[110] Konstantinou M, Kolokassidou K, Pashalidis I. Studies on the interaction of olive cake and its hydrophylic extracts with polyvalent metal ions (Cu(II), Eu(III)) in aqueous solutions. J Hazard Mater 2009; 166(2-3): 1169-73.
[http://dx.doi.org/10.1016/j.jhazmat.2008.12.016] [PMID: 19135296]

[111] Fernández-Bolaños J, Felizón B, Heredia A, Guillén R, Jiménez A. Characterization of the lignin obtained by alkaline delignification and of the cellulose residue from steam-exploded olive stones. Bioresour Technol 1999; 68(2): 121-32.
[http://dx.doi.org/10.1016/S0960-8524(98)00134-5]

[112] Heredia-Moreno A, Guillén-Bejarano R, Fernández-Bolaños J, Rivas-Moreno M. Olive stones as a source of fermentable sugars. Biomass 1987; 14(2): 143-8.
[http://dx.doi.org/10.1016/0144-4565(87)90016-3]

[113] Valiente C, Arrigoni E, Esteban RM, Amadò R. Chemical composition of olive by-product and modifications through enzymatic treatments. J Sci Food Agric 1995; 69(1): 27-32.
[http://dx.doi.org/10.1002/jsfa.2740690106]

[114] Fernández-Bolaños J, Felizón B, Heredia A, Rodríguez R, Guillén R, Jiménez A. Steam-explosion of olive stones: hemicellulose solubilization and enhancement of enzymatic hydrolysis of cellulose. Bioresour Technol 2001; 79(1): 53-61.
[http://dx.doi.org/10.1016/S0960-8524(01)00015-3] [PMID: 11396908]

[115] El Asli A, Qatibi AI. Ethanol production from olive cake biomass substrate. Biotechnol Bioprocess Eng; BBE 2009; 14(1): 118-22.
[http://dx.doi.org/10.1007/s12257-008-0071-y]

[116] Rodríguez R, Jiménez A, Fernández-Bolaños J, Guillén R, Heredia A. Dietary fibre from vegetable products as source of functional ingredients. Trends Food Sci Technol 2006; 17(1): 3-15.
[http://dx.doi.org/10.1016/j.tifs.2005.10.002]

[117] Cardoso SM, Coimbra MA, Lopes da Silva JA. Temperature dependence of the formation and melting of pectin–Ca^{2+} networks: a rheological study. Food Hydrocoll 2003; 17(6): 801-7.
[http://dx.doi.org/10.1016/S0268-005X(03)00101-2]

[118] Cardoso SM, Coimbra MA, Lopes da Silva JA. Calcium-mediated gelation of an olive pomace pectic extract. Carbohydr Polym 2003; 52(2): 125-33.
[http://dx.doi.org/10.1016/S0144-8617(02)00299-0]

[119] Cardoso SM, Silva AMS, Coimbra MA. Structural characterisation of the olive pomace pectic polysaccharide arabinan side chains. Carbohydr Res 2002; 337(10): 917-24.
[http://dx.doi.org/10.1016/S0008-6215(02)00082-4] [PMID: 12007474]

[120] Galanakis CM, Tornberg E, Gekas V. Clarification of high-added value products from olive mill wastewater. J Food Eng 2010; 99(2): 190-7.
[http://dx.doi.org/10.1016/j.jfoodeng.2010.02.018]

[121] Piotrowska A, Rao MA, Scotti R, Gianfreda L. Changes in soil chemical and biochemical properties following amendment with crude and dephenolized olive mill waste water (OMW). Geoderma 2011; 161(1-2): 8-17.
[http://dx.doi.org/10.1016/j.geoderma.2010.11.011]

[122] Mekki A, Dhouib A, Sayadi S. Evolution of several soil properties following amendment with olive mill wastewater. Prog Nat Sci 2009; 19(11): 1515-21.
[http://dx.doi.org/10.1016/j.pnsc.2009.04.014]

[123] Sayadi S, Ellouz R. Decolourization of olive mill waste-waters by the white-rot fungus Phanerochaete chrysosporium: involvement of the lignin-degrading system. Appl Microbiol Biotechnol 1992; 37(6):

813-7.
[http://dx.doi.org/10.1007/BF00174851]

[124] Sayadi S, Ellouz R. Roles of lignin peroxidase and manganese peroxidase from *Phanerochaete chrysosporium* in the decolorization of olive mill wastewaters. Appl Environ Microbiol 1995; 61(3): 1098-103.
[http://dx.doi.org/10.1128/aem.61.3.1098-1103.1995] [PMID: 16534959]

[125] García García I, Jiménez Peña PR, Bonilla Venceslada JL, Martín Martín A, Martín Santos MA, Ramos Gómez E. Removal of phenol compounds from olive mill wastewater using *Phanerochaete chrysosporium, Aspergillus niger, Aspergillus terreus* and *Geotrichum candidum.* Process Biochem 2000; 35(8): 751-8.
[http://dx.doi.org/10.1016/S0032-9592(99)00135-1]

[126] Sampedro I, Marinari S, D'Annibale A, Grego S, Ocampo JA, García-Romera I. Organic matter evolution and partial detoxification in two-phase olive mill waste colonized by white-rot fungi. Int Biodeterior Biodegradation 2007; 60(2): 116-25.
[http://dx.doi.org/10.1016/j.ibiod.2007.02.001]

[127] D'Annibale A, Casa R, Pieruccetti F, Ricci M, Marabottini R. Lentinula edodes removes phenols from olive-mill wastewater: impact on durum wheat (Triticum durum Desf.) germinability. Chemosphere 2004; 54(7): 887-94.
[http://dx.doi.org/10.1016/j.chemosphere.2003.10.010] [PMID: 14637346]

[128] Lakhtar H, Ismaili-Alaoui M, Philippoussis A, Perraud-Gaime I, Roussos S. Screening of strains of *Lentinula edodes* grown on model olive mill wastewater in solid and liquid state culture for polyphenol biodegradation. Int Biodeterior Biodegradation 2010; 64(3): 167-72.
[http://dx.doi.org/10.1016/j.ibiod.2009.10.006]

[129] Saavedra M, Benitez E, Cifuentes C, Nogales R. Enzyme activities and chemical changes in wet olive cake after treatment with *Pleurotus ostreatus* or *Eisenia fetida.* Biodegradation 2006; 17(1): 93-102.
[http://dx.doi.org/10.1007/s10532-005-4216-9] [PMID: 16453175]

[130] Fadil K, Chahlaoui A, Ouahbi A, Zaid A, Borja R. Aerobic biodegradation and detoxification of wastewaters from the olive oil industry. Int Biodeterior Biodegradation 2003; 51(1): 37-41.
[http://dx.doi.org/10.1016/S0964-8305(02)00073-2]

[131] Aissam H, Penninckx MJ, Benlemlih M. Reduction of phenolics content and COD in olive oil mill wastewaters by indigenous yeasts and fungi. World J Microbiol Biotechnol 2007; 23(9): 1203-8.
[http://dx.doi.org/10.1007/s11274-007-9348-0]

[132] Assas N, Ayed L, Marouani L, Hamdi M. Decolorization of fresh and stored-black olive mill wastewaters by *Geotrichum candidum.* Process Biochem 2002; 38(3): 361-5.
[http://dx.doi.org/10.1016/S0032-9592(02)00091-2]

[133] Amaral C, Lucas MS, Sampaio A, *et al.* Biodegradation of olive mill wastewaters by a wild isolate of Candida oleophila. Int Biodeterior Biodegradation 2012; 68: 45-50.
[http://dx.doi.org/10.1016/j.ibiod.2011.09.013]

[134] Galanakis CM, Tornberg E, Gekas V. Dietary fiber suspensions from olive mill wastewater as potential fat replacements in meatballs. Lebensm Wiss Technol 2010; 43(7): 1018-25.
[http://dx.doi.org/10.1016/j.lwt.2009.09.011]

[135] Valiente C, Arrigoni E, Esteban RM, Amadò R. Chemical composition of olive by-product and modifications through enzymatic treatments. J Sci Food Agric 1995; 69(1): 27-32.
[http://dx.doi.org/10.1002/jsfa.2740690106]

[136] Brozzoli V, Bartocci S, Terramoccia S, *et al.* Stoned olive pomace fermentation with *Pleurotus* species and its evaluation as a possible animal feed. Enzyme Microb Technol 2010; 46(3-4): 223-8.
[http://dx.doi.org/10.1016/j.enzmictec.2009.09.008]

[137] Pinto G, Pollio A, Previtera L, Stanzione M, Temussi F. Removal of low molecular weight phenols

from olive oil mill wastewater using microalgae. Biotechnol Lett 2003; 25(19): 1657-9.
[http://dx.doi.org/10.1023/A:1025667429222] [PMID: 14584924]

[138] Zouari N, Ellouz R. Microbial consortia for the aerobic degradation of aromatic compounds in olive oil mill effluent. J Ind Microbiol 1996; 16(3): 155-62.
[http://dx.doi.org/10.1007/BF01569998]

[139] Ranalli A. Microbiological treatment of oil mill waste waters Fats and oils 1992; 43(1): 16-9.

[140] Borja R, Alba J, Garrido SE, *et al.* Comparative study of anaerobic digestion of olive mill wastewater (OMW) and OMW previously fermented with *Aspergillus terreus.* Bioprocess Eng 1995; 13(6): 317-22.
[http://dx.doi.org/10.1007/BF00369564]

[141] Lamia A, Moktar H. Fermentative decolorization of olive mill wastewater by *Lactobacillus plantarum.* Process Biochem 2003; 39(1): 59-65.
[http://dx.doi.org/10.1016/S0032-9592(02)00314-X]

[142] Aouidi F, Gannoun H, Ben Othman N, Ayed L, Hamdi M. Improvement of fermentative decolorization of olive mill wastewater by *Lactobacillus paracasei* by cheese whey's addition. Process Biochem 2009; 44(5): 597-601.
[http://dx.doi.org/10.1016/j.procbio.2009.02.014]

[143] Papadelli M, Roussis A, Papadopoulou K, *et al.* Biochemical and molecular characterization of an *Azotobacter vinelandii* strain with respect to its ability to grow and fix nitrogen in olive mill wastewater. Int Biodeterior Biodegradation 1996; 38(3-4): 179-81.
[http://dx.doi.org/10.1016/S0964-8305(96)00048-0]

[144] Ramos-Cormenzana A, Juárez-Jiménez B, Garcia-Pareja MP. Antimicrobial activity of olive mill wastewaters (alpechin) and biotransformed olive oil mill wastewater. Int Biodeterior Biodegradation 1996; 38(3-4): 283-90.
[http://dx.doi.org/10.1016/S0964-8305(96)00061-3]

[145] García-Gómez A, Roig A, Bernal MP. Composting of the solid fraction of olive mill wastewater with olive leaves: organic matter degradation and biological activity. Bioresour Technol 2003; 86(1): 59-64.
[http://dx.doi.org/10.1016/S0960-8524(02)00106-2] [PMID: 12421010]

[146] Canet R, Pomares F, Cabot B, *et al.* Composting olive mill pomace and other residues from rural southeastern Spain. Waste Manag 2008; 28(12): 2585-92.
[http://dx.doi.org/10.1016/j.wasman.2007.11.015] [PMID: 18262780]

[147] Montemurro F, Diacono M, Vitti C, Debiase G. Biodegradation of olive husk mixed with other agricultural wastes. Bioresour Technol 2009; 100(12): 2969-74.
[http://dx.doi.org/10.1016/j.biortech.2009.01.038] [PMID: 19261466]

[148] Sellami F, Hachicha S, Chtourou M, Medhioub K, Ammar E. Maturity assessment of composted olive mill wastes using UV spectra and humification parameters. Bioresour Technol 2008; 99(15): 6900-7.
[http://dx.doi.org/10.1016/j.biortech.2008.01.055] [PMID: 18328696]

[149] Droussi Z, D'Orazio V, Hafidi M, Ouatmane A. Elemental and spectroscopic characterization of humic-acid-like compounds during composting of olive mill by-products. J Hazard Mater 2009; 163(2-3): 1289-97.
[http://dx.doi.org/10.1016/j.jhazmat.2008.07.136] [PMID: 18804912]

[150] Cegarra J, Alburquerque JA, Gonzálvez J, Tortosa G, Chaw D. Effects of the forced ventilation on composting of a solid olive-mill by-product ("alperujo") managed by mechanical turning. Waste manage 2006; 26(12): 1377-83.

[151] Alburquerque JA, Gonzálvez J, García D, Cegarra J. Measuring detoxification and maturity in compost made from "alperujo", the solid by-product of extracting olive oil by the two-phase centrifugation system. Chemosphere 2006; 64(3): 470-7.
[http://dx.doi.org/10.1016/j.chemosphere.2005.10.055] [PMID: 16337988]

[152] Alburquerque JA, Gonzálvez J, García D, Cegarra J. Composting of a solid olive-mill by-product ("alperujo") and the potential of the resulting compost for cultivating pepper under commercial conditions. Waste Manag 2006; 26(6): 620-6.
[http://dx.doi.org/10.1016/j.wasman.2005.04.008] [PMID: 16005202]

[153] Cayuela ML, Sánchez-Monedero MA, Roig A. Evaluation of two different aeration systems for composting two-phase olive mill wastes. Process Biochem 2006; 41(3): 616-23.
[http://dx.doi.org/10.1016/j.procbio.2005.08.007]

[154] Roig A, Cayuela ML, Sánchez-Monedero MA. The use of elemental sulphur as organic alternative to control pH during composting of olive mill wastes. Chemosphere 2004; 57(9): 1099-105.
[http://dx.doi.org/10.1016/j.chemosphere.2004.08.024] [PMID: 15504468]

[155] Haddadin MSY, Haddadin J, Arabiyat OI, Hattar B. Biological conversion of olive pomace into compost by using *Trichoderma harzianum* and *Phanerochaete chrysosporium*. Bioresour Technol 2009; 100(20): 4773-82.
[http://dx.doi.org/10.1016/j.biortech.2009.04.047] [PMID: 19467866]

[156] Tomati U, Galli E, Fiorelli F, Pasetti L. Fertilizers from composting of olive-mill wastewaters. Int Biodeterior Biodegradation 1996; 38(3-4): 155-62.
[http://dx.doi.org/10.1016/S0964-8305(96)00044-3]

[157] Zenjari B, El Hajjouji H, Ait Baddi G, *et al*. Eliminating toxic compounds by composting olive mill wastewater–straw mixtures. J Hazard Mater 2006; 138(3): 433-7.
[http://dx.doi.org/10.1016/j.jhazmat.2006.05.071] [PMID: 16973264]

[158] Altieri R, Esposito A, Parati F, Lobianco A, Pepi M. Performance of olive mill solid waste as a constituent of the substrate in commercial cultivation of *Agaricus bisporus*. Int Biodeterior Biodegradation 2009; 63(8): 993-7.
[http://dx.doi.org/10.1016/j.ibiod.2009.06.008]

[159] Zervakis G, Yiatras P, Balis C. Edible mushrooms from olive oil mill wastes. Int Biodeterior Biodegradation 1996; 38(3-4): 237-43.
[http://dx.doi.org/10.1016/S0964-8305(96)00056-X]

[160] Kalmıs E, Azbar N, Yıldız H, Kalyoncu F. Feasibility of using olive mill effluent (OME) as a wetting agent during the cultivation of oyster mushroom, *Pleurotus ostreatus*, on wheat straw. Bioresour Technol 2008; 99(1): 164-9.
[http://dx.doi.org/10.1016/j.biortech.2006.11.042] [PMID: 17239585]

[161] Lakhtar H, Ismaili-Alaoui M, Philippoussis A, Perraud-Gaime I, Roussos S. Screening of strains of *Lentinula edodes* grown on model olive mill wastewater in solid and liquid state culture for polyphenol biodegradation. Int Biodeterior Biodegradation 2010; 64(3): 167-72.
[http://dx.doi.org/10.1016/j.ibiod.2009.10.006]

[162] Fedorak PM, Hrudey SE. The effects of phenol and some alkyl phenolics on batch anaerobic methanogenesis. Water Res 1984; 18(3): 361-7.
[http://dx.doi.org/10.1016/0043-1354(84)90113-1]

[163] Rincón B, Borja R, Martín MA, Martín A. Evaluation of the methanogenic step of a two-stage anaerobic digestion process of acidified olive mill solid residue from a previous hydrolytic–acidogenic step. Waste Manag 2009; 29(9): 2566-73.
[http://dx.doi.org/10.1016/j.wasman.2009.04.009] [PMID: 19450962]

[164] Meherkotay S, Das D. Biohydrogen as a renewable energy resource—Prospects and potentials. Int J Hydrogen Energy 2008; 33(1): 258-63.
[http://dx.doi.org/10.1016/j.ijhydene.2007.07.031]

[165] Hawkes F, Hussy I, Kyazze G, Dinsdale R, Hawkes D. Continuous dark fermentative hydrogen production by mesophilic microflora: Principles and progress. Int J Hydrogen Energy 2007; 32(2): 172-84.

[http://dx.doi.org/10.1016/j.ijhydene.2006.08.014]

[166] Eroğlu E, Gündüz U, Yücel M, Türker L, Eroğlu İ. Photobiological hydrogen production by using olive mill wastewater as a sole substrate source. Int J Hydrogen Energy 2004; 29(2): 163-71.
[http://dx.doi.org/10.1016/S0360-3199(03)00110-1]

[167] Eroğlu E, Eroğlu İ, Gündüz U, Yücel M. Comparison of physicochemical characteristics and photofermentative hydrogen production potential of wastewaters produced from different olive oil mills in Western-Anatolia, Turkey. Biomass Bioenergy 2009; 33(4): 706-11.
[http://dx.doi.org/10.1016/j.biombioe.2008.11.001]

[168] Eroglu E, Gunduz U, Yucel M, Eroglu I. Effect of iron and molybdenum addition on photofermentative hydrogen production from olive mill wastewater. Int J Hydrogen Energy 2011; 36(10): 5895-903.
[http://dx.doi.org/10.1016/j.ijhydene.2011.02.062]

[169] Koutrouli EC, Kalfas H, Gavala HN, Skiadas IV, Stamatelatou K, Lyberatos G. Hydrogen and methane production through two-stage mesophilic anaerobic digestion of olive pulp. Bioresour Technol 2009; 100(15): 3718-23.
[http://dx.doi.org/10.1016/j.biortech.2009.01.037] [PMID: 19246194]

[170] Ena A, Pintucci C, Carlozzi P. Production of bioH$_2$ by *Rhodopseudomonas palustris* (strain 6A) grown in pre-treated olive mill waste, under batch or semi-continuous regime. J Biotechnol 2010; 150(150): 180.
[http://dx.doi.org/10.1016/j.jbiotec.2010.08.469]

[171] Sánchez Villasclaras S, Martínez Sancho ME, Espejo Caballero MT, Delgado Pérez A. Production of microalgae from olive mill wastewater. Int Biodeterior Biodegradation 1996; 38(3-4): 245-7.
[http://dx.doi.org/10.1016/S0964-8305(96)00057-1]

[172] Faraloni C, Ena A, Pintucci C, Torzillo G. Enhanced hydrogen production by means of sulfur-deprived *Chlamydomonas reinhardtii* cultures grown in pretreated olive mill wastewater. Int J Hydrogen Energy 2011; 36(10): 5920-31.
[http://dx.doi.org/10.1016/j.ijhydene.2011.02.007]

[173] Fezzani B, Ben Cheikh R. Two-phase anaerobic co-digestion of olive mill wastes in semi-continuous digesters at mesophilic temperature. Bioresour Technol 2010; 101(6): 1628-34.
[http://dx.doi.org/10.1016/j.biortech.2009.09.067] [PMID: 19896368]

[174] Fezzani B, Cheikh RB. Thermophilic anaerobic co-digestion of olive mill wastewater with olive mill solid wastes in a tubular digester. Chem Eng J 2007; 132(1-3): 195-203.
[http://dx.doi.org/10.1016/j.cej.2006.12.017]

[175] Haagensen F, Skiadas IV, Gavala HN, Ahring BK. Pre-treatment and ethanol fermentation potential of olive pulp at different dry matter concentrations. Biomass Bioenergy 2009; 33(11): 1643-51.
[http://dx.doi.org/10.1016/j.biombioe.2009.08.006]

[176] El Asli A, Qatibi AI. Ethanol production from olive cake biomass substrate. Biotechnol Bioprocess Eng; BBE 2009; 14(1): 118-22.
[http://dx.doi.org/10.1007/s12257-008-0071-y]

[177] Massadeh MI, Modallal N. Ethanol production from olive mill wastewater (OMW) pretreated with *Pleurotus sajor-caju*. Energy Fuels 2008; 22(1): 150-4.
[http://dx.doi.org/10.1021/ef7004145]

[178] Kessler B, Weusthuis R, Witholt B, Eggink G. Production of microbial polyesters: fermentation and downstream processes. 2001.
[http://dx.doi.org/10.1007/3-540-40021-4_5]

[179] Lopez MJ, Ramos-Cormenzana A. Xanthan production from olive-mill wastewaters. Int Biodeterior Biodegradation 1996; 38(3-4): 263-70.
[http://dx.doi.org/10.1016/S0964-8305(96)00059-5]

[180]   Mohan D, Pittman CU Jr, Steele PH. Pyrolysis of wood/biomass for bio-oil: a critical review. Energy Fuels 2006; 20(3): 848-89.
[http://dx.doi.org/10.1021/ef0502397]

[181]   Chouchene A, Jeguirim M, Khiari B, Zagrouba F, Trouvé G. Thermal degradation of olive solid waste: Influence of particle size and oxygen concentration. Resour Conserv Recycling 2010; 54(5): 271-7.
[http://dx.doi.org/10.1016/j.resconrec.2009.04.010]

[182]   Chouchene A, Jeguirim M, Khiari B, Trouvé G, Zagrouba F. Study on the emission mechanism during devolatilization/char oxidation and direct oxidation of olive solid waste in a fixed bed reactor. J Anal Appl Pyrolysis 2010; 87(1): 168-74.
[http://dx.doi.org/10.1016/j.jaap.2009.11.008]

[183]   Uzun BB, Pütün AE, Pütün E. Rapid pyrolysis of olive residue. 1. Effect of heat and mass transfer limitations on product yields and bio-oil compositions. Energy Fuels 2007; 21(3): 1768-76.
[http://dx.doi.org/10.1021/ef060171a]

[184]   Encinar JM, González JF, Martínez G, González JM. Two stages catalytic pyrolysis of olive oil waste. Fuel Process Technol 2008; 89(12): 1448-55.
[http://dx.doi.org/10.1016/j.fuproc.2008.07.005]

[185]   Encinar JM, González JF, Martínez G, Román S. Catalytic pyrolysis of exhausted olive oil waste. J Anal Appl Pyrolysis 2009; 85(1-2): 197-203.
[http://dx.doi.org/10.1016/j.jaap.2008.11.018]

[186]   Gómez-Barea A, Arjona R, Ollero P. Pilot-plant gasification of olive stone: a technical assessment. Energy Fuels 2005; 19(2): 598-605.
[http://dx.doi.org/10.1021/ef0498418]

[187]   Skoulou V, Koufodimos G, Samaras Z, Zabaniotou A. Low temperature gasification of olive kernels in a 5-kW fluidized bed reactor for H2-rich producer gas. Int J Hydrogen Energy 2008; 33(22): 6515-24.
[http://dx.doi.org/10.1016/j.ijhydene.2008.07.074]

[188]   Skoulou V, Zabaniotou A, Stavropoulos G, Sakelaropoulos G. Syngas production from olive tree cuttings and olive kernels in a downdraft fixed-bed gasifier. Int J Hydrogen Energy 2008; 33(4): 1185-94.
[http://dx.doi.org/10.1016/j.ijhydene.2007.12.051]

[189]   García-Ibañez P, Cabanillas A, Sánchez JM. Gasification of leached orujillo (olive oil waste) in a pilot plant circulating fluidised bed reactor. Preliminary results. Biomass Bioenergy 2004; 27(2): 183-94.
[http://dx.doi.org/10.1016/j.biombioe.2003.11.007]

[190]   Demirbaş A. Sustainable cofiring of biomass with coal. Energy Convers Manage 2003; 44(9): 1465-79.
[http://dx.doi.org/10.1016/S0196-8904(02)00144-9]

[191]   Awasthi MK, Kumar V, Yadav V, *et al.* Current state of the art biotechnological strategies for conversion of watermelon wastes residues to biopolymers production: A review. Chemosphere 2022; 290133310.
[http://dx.doi.org/10.1016/j.chemosphere.2021.133310] [PMID: 34919909]

[192]   Liu H, Kumar V, Jia L, *et al.* Biopolymer poly-hydroxyalkanoates (PHA) production from apple industrial waste residues: A review. Chemosphere 2021; 284131427.
[http://dx.doi.org/10.1016/j.chemosphere.2021.131427] [PMID: 34323796]

[193]   Awasthi SK, Kumar M, Kumar V, *et al.* A comprehensive review on recent advancements in biodegradation and sustainable management of biopolymers. Environ Pollut 2022; 307119600.
[http://dx.doi.org/10.1016/j.envpol.2022.119600] [PMID: 35691442]

[194]   Duan Y, Tarafdar A, Kumar V, *et al.* Sustainable biorefinery approaches towards circular economy for conversion of biowaste to value added materials and future perspectives. Fuel 2022; 325124846.

[http://dx.doi.org/10.1016/j.fuel.2022.124846]

[195] Kumar V, Sharma N, Umesh M, *et al.* Emerging challenges for the agro-industrial food waste utilization: A review on food waste biorefinery. Bioresour Technol 2022; 362127790.
[http://dx.doi.org/10.1016/j.biortech.2022.127790] [PMID: 35973569]

[196] Awasthi SK, Sarsaiya S, Kumar V, *et al.* Processing of municipal solid waste resources for a circular economy in China: An overview. Fuel 2022; 317123478.
[http://dx.doi.org/10.1016/j.fuel.2022.123478]

[197] Kumar V, Sharma N, Maitra SS. *In vitro* and *in vivo* toxicity assessment of nanoparticles. Int Nano Lett 2017; 7(4): 243-56.
[http://dx.doi.org/10.1007/s40089-017-0221-3]

[198] Vinay K, Neha S, Maitra S. Protein and Peptide Nanoparticles: Preparation and Surface Modification, in Functionalized Nanomaterials I. CRC Press 2020; pp. 191-204.

[199] Vallinayagam S, *et al.* Recent developments in magnetic nanoparticles and nano-composites for wastewater treatment. J Environ Chem Eng 2021; 9(6)106553.
[http://dx.doi.org/10.1016/j.jece.2021.106553]

[200] Kumar V, Sharma N, Lakkaboyana SK, Maitra SS. "Silver nanoparticles in poultry health: Applications and toxicokinetic effects," in Silver Nanomaterials for Agri-Food Applications. Elsevier 2021; pp. 685-704.
[http://dx.doi.org/10.1016/B978-0-12-823528-7.00005-6]

[201] Egbosiuba T C. Biochar and bio-oil fuel properties from nickel nanoparticles assisted pyrolysis of cassava peel 2022.
[http://dx.doi.org/10.1016/j.heliyon.2022.e10114]

[202] André R, Pinto F, Franco C, *et al.* Fluidised bed co-gasification of coal and olive oil industry wastes. Fuel 2005; 84(12-13): 1635-44.
[http://dx.doi.org/10.1016/j.fuel.2005.02.018]

[203] Pinto F, André RN, Franco C, Lopes H, Gulyurtlu I, Cabrita I. Co-gasification of coal and wastes in a pilot-scale installation 1: Effect of catalysts in syngas treatment to achieve tar abatement. Fuel 2009; 88(12): 2392-402.
[http://dx.doi.org/10.1016/j.fuel.2008.12.012]

CHAPTER 5

Organic Residues Valorization For Value-added Chemicals Production

Charumathi Jayachandran[1,*], **Sowmiya Balasubramanian**[2] and **R. Kamatchi**[2]

[1] *Department of Biotechnology, Bhupat and Jyoti Metha School of Biosciences, Indian Institute of Technology Madras, Chennai 600036, India*

[2] *Centre for Biotechnology, Anna University, Chennai, Tamil Nadu, India*

Abstract: In recent years, more studies on waste valorization are emerging due to excessive accumulation in the land, foul-smelling, and lack of conventional disposal practices to sustain a proper ecosystem. The decline in the supply of fossil fuels and their high-cost led to finding alternative technologies that use renewable resources as raw materials to manufacture value-added goods. The waste contains organic residues like carbohydrates, proteins, and fats, which are helpful in producing bio-based chemicals. However, several roadblocks ought to be crossed for adopting organic waste as nutrients for microbes to obtain high yields of desired products. Many studies have shown potential ways to solve these problems and have achieved high yields. Nevertheless, this technology has not been globally explored to manufacture commercial products, as many other issues are associated with biorefinery and product costs. This chapter addresses the organic residues present in the wastes, their use in manufacturing platform chemicals, methods for the pretreatment process, and ways to overcome the challenges.

Keywords: *Aspergillus terreus*, Acid catalyst hydrolysis, Building blocks chemicals, Cellulose, Detoxification, Food and fruit waste, Gluconic acid, Itaconic acid, Levulinic acid, Lignocellulose biomass, Microwave-assisted heating method, Organic wastes, Succinic acid, Sugar alcohols, Sugarcane bagasse, Spent aromatic wastes, Transesterification, Xylose, 5-HMF SSF.

INTRODUCTION

In this growing population, enormous amount of waste is created as a result of human lifestyle, and industrial development. The waste produced poses challenges to waste management technology. The conventional waste disposal practice is incompetent to build a sustainable system to maintain a healthy ecosys-

* **Corresponding author Charumathi Jayachandran:** Department of Biotechnology, Bhupat and Jyoti Metha School of Biosciences, Indian Institute of Technology Madras, Chennai 600036, India; E-mail: mathij28@gmail.com

Vinay Kumar, Sivarama Krishna Lakkaboyana & Neha Sharma (Eds.)

tem and human well-being [1]. Indeed, a large part of the waste constitutes organic residues collected from the household (food, fruits, and vegetables), municipal waste, animal excretes, agricultural waste, public places (shops, hotels, and office activity), and industrial by-products that are usually processed by incineration [2].

Introducing a new process to utilize organic waste not only replaces the traditional waste disposal method but also reduces the accumulation of toxic substances in the environment by generating various bio-based products. Considering environmental, economic, and social perspectives, the application of recycled waste for agricultural purposes is more promising as it lessens environmental contamination [3]. However, the direct application of organic waste as manure in agricultural land is harmful due to the presence of lignocellulose material. Composting organic waste into stabilized manure is a good fertilizer for plant cultivation and agriculture [4].

The evolution of new microorganisms and the development of synthetic biotechnology can utilize recalcitrant waste as carbon and nitrogen substrates to produce various bio-based chemicals. For instance, a biodiesel refinery generates about 10% crude glycerol as the main by-product. The glycerol is utilized as a feedstock by microbes to produce potential chemicals (1,3- propanediol, citric acid, poly hydroxyalkonates, phytase, *etc.*) and as animal feed [5]. The characterization of the waste material is a prerequisite to segregating the potential organic compounds from undesirable hazardous materials. Agricultural waste contains cellulose (40%), hemicellulose (30%), lignin (20%), proteins (5%), and minerals (5%) [6]. Food waste hydrolysate contains polymers such as starch (30-60%), cellulose, lignin, proteins (5-10%), lipids (10-40%), organic acids, and inorganic compounds. They serve as a rich nutrient medium for microorganisms' growth [7]. The biochemical conversion of wastes into simple sugars as a hydrolysate requires chemical, enzymatic, or hydrothermal treatment. Depending on the type of biomass, the pretreatment methods vary. A simple enzyme hydrolysis step is adequate for recovering nutrients from food wastes, starch, sucrose, *etc.* Lignocellulose biomass requires harsh treatment due to its complex heterogeneous structure [8].

Therefore, building a biorefinery to produce multiple products from one raw material, like a petroleum refinery, has remarkable strength. Current technologies can make industrial products derived from fossil fuel resources from organic waste biomass [9]. However, certain limitations must be overcome to achieve this transition of using biodegradable raw materials from waste to value-added chemicals. Herein, the chapter gives detailed information about organic waste valorization, process strategies, and difficulties overlooked for value-added chemicals. The chapter covers industrial organic acids reported in the top 12 building block chemicals by the U.S. Department of Energy [10].

3-Hydroxypropionic Acid

3-Hydroxypropionic acid (3-HP) is the third most crucial chemical among the top twelve value-added platform chemicals produced from biomass [10]. It serves as a versatile precursor for diverse high-value compounds such as 1,3-propanediol, acrylic acid, acrylamide, malonic acid, 3-hydroxypropionaldehyde, and acryl-based polymers by a slight modification of chemical reactions [11, 12]. These compounds are extensively used in food preservations, as a crosslinking agent for polymer coatings, medical sutures, *etc*. The global market potential of 3-HP was projected to be >1 million tons per year. Commercially 3-HP is produced by chemical processes. However, fermentative routes have also been extensively studied [13, 14]. Until now, glucose and glycerol have been the major sources of renewable raw materials to produce 3HP. Besides, several other organic residues were successfully investigated, such as sucrose from sugar beet, corn starch, and pre-treated lignocellulose materials from hydrolyzed food waste, forest biomass, agricultural industry, and municipal waste [15, 16].

A recent study has shown to achieve 3-HP production at a low concentration from CO_2 and xylose, although considerable research is in progress. Kildegaard and co-workers [17] reported the feasibility of producing 3-HP in *S. cerevisiae* strains from xylose through the β-alanine pathway that achieved a high titer of 6.09 ± 0.33 g/L. A recent study focused on *L. reuteri* growth in wheat and sugar beet by-products [18]. The suspended solid particles in low-purity sugar beetroot syrup and wheat extract were filtered and used directly as sugar sources. The bacteria in this medium displayed a high product yield of 0.40 g/g compared to the conventional MRS medium. However, it is to be noted that almost very few attempts have been made at producing 3-HP from organic waste (except glycerol). Substrate and recovery costs play a critical role in the commercialization of bio-based compounds, especially in the case of 3-HP. Though significant progress has been made in improving fermentative production of 3-HP using crude glycerol, establishing this on a commercial scale remains insignificant mainly due to the toxicity of 3-HP and the regeneration of NAD^+. Thus, intensive research efforts are required to enhance 3-HP synthesis from available renewable resources.

Succinic Acid

Succinic acid is recognized as one of the top twelve potential chemical building blocks used in synthesizing various high-value commodity derivatives such as 1,4-butanediol tetrahydrofuran, polybutylene succinate, and polyurethanes that find extensive application in the field of pharmaceutical, food, antibiotics, bioplastic sectors, surfactants, and detergents. Commercial succinic acid is produced by catalytic hydrogenation of petrochemically derived maleic acid or maleic anhydride [19, 20]. But the current demand has shifted towards bio-based succinic acid. Besides metabolic engineering strategies, renewable feedstock from in-situ enzymatic hydrolysis is extensively studied to achieve economically feasible fermentative succinic acid production.

Actinobacillus succinogenes is one of the most promising succinic acid-producing strains. Fermentation using *Actinobacillus succinogenes* 130Z from cheese whey produced succinic acid and by-products such as acetic acid and formic acid [21]. Succinic acid production was investigated from orange peel rich in citrus acid. Orange fruits are cultivated in large quantities, and the fruit peel will be discarded as waste except for essential oil preparation. The microbe *F. succinogenes* S85 efficiently converts cellulose in the orange peel to succinate. The presence of the essential oil D-limonene in orange peel could inhibit microbial growth, which is pretreated by steam distillation to minimize the level. Yet deep understanding at the process level is necessary to improve the yield as fermentation from orange peel enhanced succinic acid at the volumetric level [22]. Leung and co-workers [23] employed bakery waste as a raw material for succinic acid fermentation by *A. succinogenes*, which led to the production of 24.8 and 31.7 g/L with a yield of 0.80 and 0.67 g/g, respectively. Solid-state fermentation of the bread waste by *Aspergillus awamori* and *Aspergillus oryzae* produced enzyme complexes resulting in hydrolysate used as the sole feedstock for fermentative succinic acid production. Similarly, Zhang and co-workers [24] examined the feasibility of using enzymatic hydrolysis and fungal autolysis of bakery waste in succinic acid production by *Actinobacillus succinogenes*. The resulting cake and pastry waste hydrolysate was rich in glucose and free amino nitrogen. The titer of about 47.3 g/L succinic acid with a yield of 1.16 g/g sugar was achieved. This work led to the development of a biorefinery set-up to utilize bakery waste to produce succinic acid and other value-added chemicals. Also, the possibility of fermentative succinic acid production using corncob hydrolysate with yeast extract was investigated. The predominant monosaccharide xylose released from hemicellulose from diluted acid-treated corncob was used directly as the carbon source in the fermentation that produced 23.6 g/L succinic acids with a yield of 0.58 g/g sugar [25].

Meanwhile, corn stover and pinewood-derived pretreated lignocellulosic biomass were employed, which led to the succinic acid production of 20.7 g/ L, with an average yield of 0.37 g per g of biomass *A. succinogenes* [26]. Biotin and pantothenic acid in cane molasses play an important role in succinic acid biosynthesis. Molasses pretreated by 150 kDa ultra filtrate membrane produced a succinic acid concentration of 45.6 g/L with a productivity of 1.27 g/L·h and yielded 0.76 g /g sugars [27]. Dessie and co-workers [28] have optimized hydrolyzed fruit and vegetable waste as fermentable feedstock that achieved the succinic acid concentration of 7.03 g/L of SA with 1.18 g−1g−1 yield and 1.28 gL−1 h−1 productivity. In a recent work, hydrolysate derived from fruit and vegetable waste (apples, pears, oranges, potatoes, cabbage, lettuce, and taros) has been shown to produce the highest succinic acid concentration of 140.6 g/L from an engineered *Yarrowia lipolytic* strain. The study signifies the importance of nitrogen sources during fermentation for succinic acid production. Further addition of corn steep liquor enhanced cell growth and production. The highest yield was achieved from *in situ* fibrous bed bioreactor and fed-batch fermentation [29]. Furthermore, current research focuses on the development of innovative food waste utilization strategies and to increase volumetric productivity and yield to make fermentative succinic acid production more advantageous.

Maleic Acid

Maleic acid (MA) is an important chemical intermediate that finds applications in food and beverages as an acidulant, in metal cleaning, textile finishing, pharmaceuticals, and in agriculture [30]. The global market size varies from 40,000–60,000 metric tons with the growth rate increasing by 4% annually [31]. The market value of malic acid has been estimated to be $130 million. Currently, the industrial production of malic acid is *via* the petroleum-based method. However, studies have shown bio-based malic acid production from organic residue valorization as an effective competitor to petroleum-based products. The utilization of alternative cheap raw materials from industry, food, and agricultural waste for low-cost microbial malic acid production is being extensively studied. Miscanthus, beechwood, corn straw, sweet potato, or Jerusalem artichoke was ideal substrate for malic acid production. Several species of the fungus *Aspergillus* have been shown to support microbial malic acid production from thin silage, bio-oil from agricultural biomass, and treated lignocellulosic biomass [32]. *A. flavus* was identified as the first producer of malic acid (≤58.4g/L), which was synthesized within 9 days in a shake flask from glucose [33]. Production of poly (β-L-malic acid) by strains of *A. pullulans* from agricultural biomass has been reported where polymalic acid was subsequently hydrolyzed to malic acid. The strain A. *pullulans* YJ6-11 that could utilize thin stillage, the residue recovered after dry milling of corn supported malic acid production of 32.4 g/L from 90 g/L of mixed sugars, namely glucose, xylose, and arabinose obtained from pretreated corncob hydrolysate [34]. Thin stillage was primarily generated during the dry-milling process for ethanol production, mainly consisting of glycerol and polysaccharides. This is good as well as a cheap substrate for malic acid production as the dry-milling process can generate more than ethanol produced [35].

Li and co-workers [36] have shown the highest titers of about 120 g/L malic acid production from the bioconversion of corn straw hydrolysate using an isolated *Rhizopus delemar* strain. In another study, malic acid production by *A. pullulans*, ZX-10 capable of utilizing soybean hull hydrolysate (26.8 g/L carbohydrates) supplemented with corn steep liquor (10 g/L, *w/v*) produced 31.3 g/L malic acids [37]. In the same study, *A. pullulans* ZX-10 also fermented soybean molasses (26.8 g/L carbohydrates) to produce 71.9 g/L malic acids. Studies have also shown the advantage of utilizing bio-oil from wheat straw hydrolysate that supports malic acid production by *A. oryzae* DSM 1863 [38]. Using molasses, malic acid concentrations in the range of 71.9 to 109.7 g/L and yields between 0.28 and 0.69 g/g have been reported [39]. An engineered strain of the thermophilic actinobacterium *Thermobifida fusca* has been shown to produce malic acid from cellulose and treated lignocellulosic biomass. The metabolically engineered soil bacterial *T. fusca* muC-16 strain could ferment cellulose and corn stover into malic acid. The strain was engineered with a heterologous gene pyruvate carboxylase from *Corynebacterium glutamicum* to increase the flux of pyruvate to oxaloacetate. The strain produced about 63 g/L of malic acid from 100 g/L of cellulose and 21.5 g/L from corn stover [31].

Furthermore, waste xylose mother liquor originating from the production of xylitol based on lignocellulosic feedstock has been successfully used to produce malic acid of about 66 g/L and a yield of 0.77 g/g making the process competitive to that of glucose [40]. Crude glycerol from biodiesel production was also used as a substrate for malic acid. Though the yield was less than pure glycerol, further optimization can increase productivity [35]. Though malic acid can be synthesized by microorganisms from lignocellulosic biomass, extensive research is required to focus on the utilization of other hydrolyzed biomass, such as agricultural and bakery waste, and wood, for microbial malic production. Also, the high solubility of malic acid would be a challenging task during downstream processing, strategies must be re-evaluated when using crude industrial side streams.

Fumaric Acid

Fumaric acid is another important raw material that finds applications in nearly every field of industrial chemistry. It is used as a food acidulent and as a raw material in the manufacture of unsaturated polyester resins, quick-setting inks, furniture lacquers, and paper-sizing chemicals [41]. It is generally produced by fermentation. Fumaric acid was initially produced *via* the petrochemical route from maleic acid by isomerization. Chemical reactions always have limitations like side reactions due to the high temperature or the reaction attaining equilibrium that often reduces the yield. A biological system for fumaric acid production was identified that has the maleic isomerase enzyme. The organisms like *Pseudomonas* sp. and *Bacillus* sp. were initially used for fumaric acid production and later were shifted to *Rhizopus* sp. as this was found to be the best producer [42].

Xylose, one of the most abundant monosaccharide fractions of lignocellulose biomass and agro-waste, has proved to be an excellent fumaric acid producer. Enormous lignocellulose wastes from paper and pulp were used after proper treatment, producing 41.65 g of fumaric acid per kg of dry solid waste [43]. In a study, *R. oryzae* efficiently produced 41.32 g/L fumaric acids from hydrolyzed corncob by simultaneous saccharification and fermentation processes. Enzymatically hydrolyzed yucca bagasse and potato residue have been shown to ferment the resulting starchy material to fumaric acid of about 21.28 g/L and yield 0.23 g/g [44]. Apple pomace, which represents about 20% to 35% of the fruit weight, represents about 20% to 35% of the fruit weight and is an essential waste in the juice industry. The fermentation process post ultrafiltration utilizing high sugar content of apple pomace has been optimized using *R. oryzae* 1526 that produced fumaric acid titer of about 25.2 g/L and about 52 g/kg of solid when apple pomace was used directly without prior treatment [45]. With the same strain as a biocatalyst and utilizing brewery wastewaters as raw material under appropriate fermentation conditions, about 31.3 g/L fumaric acid was obtained, which indicates that brewery wastewaters are also a promising and cheap substrate for fumaric acid production [46]. Food wastes obtained from hotels and restaurants were ground, heated at 100°C, and centrifuged into three parts. The fractions are divided into waste oil, liquid part, and solid part. Preheated and separated solid food waste was shown to produce fumaric acid of about 31.65 g/L using *R. arrhizus* RH-07–13; but the liquid fraction of food waste has been shown to produce better yield and titer of 32.68 g/L than that obtained from glucose [47]. Treated biomass wastes such as corn steep liquor and/or soybean meal hydrolysate should also be analyzed as an alternate source of nitrogen to develop eco-friendly processes.

Gluconic Acid

Gluconic acid (GA) belongs to the aldonic acid family. It is a stereoisomer of 2,3,4,5,6- pentahydroxyhexanoic acid. It is obtained from specific oxidation of the aldehyde group at C1 in D-glucose to a carboxyl group by chemical, electrochemical, and catalytic biotransformation. Gluconic acid has found applications in the food, pharmaceutical, textile, and building industries. GA production is estimated to amount to approximately 100000 tons/year [48], with production costs ranging from 1.20 US$/kg for GA to 8.50 US$/kg for calcium gluconate and gluconolactone [49]. GA is commercially available as 50% aqueous solution (pH of 1.82 and 1.23 g/cm3 density) [50]. To produce GA, surface fermentation with *Penicillium* fungi has been superseded by the use of submerged cultures of filamentous fungi such as *Aspergillus niger* or *Gluconobacter oxydans* [51]. *Acetobacter diazotrophicus* and *Zymomonas mobilis* have also been identified to metabolize glucose to GA [52].

Table **1** lists the main agro-industrial byproducts and fermentation techniques used for GA production, sample pretreatments, and the yields obtained in each case. Glucose and sucrose substrates were replaced with cheaper agro-industrial by-products, such as starch, sugarcane molasses, fig, banana, grape, pear, whey, and paper waste. Hydrolysates from lignocellulosic biomass were used as natural sources of carbohydrates. The use of lignocellulosic-based biomass needs serious conditions (high temperature and pressure, low pH) for hydrolysis. It produces a wide range of compounds like carboxylic acids, furan aldehydes, and aromatic compounds, which are toxic and inhibitory to microorganisms [53]. Enzymatic production of GA seems to be uneconomical at the industrial scale because of the instability of the enzymes and the high costs. Multi-enzyme system to produce GA using sucrose from sugarcane as raw material and soluble enzymes (invertase, GOD, and catalase) has been studied in an airlift reactor [54]; in comparison with other alternatives, a pretty high GA productivity per gram of glucose was obtained (0.266 g GA/g·h).

Table 1. Main agro-industrial wastes used as substrates for the microbiological production of gluconic acid [55].

Carbon source	Microorganism or enzyme used	Culturing method	Substrate pretreatment	GA yield (%)
Corn starch	*A. niger* ORS-4	Batch culture/surface fermentation	None	24,39
Starch	Immobilized glucoamylase and glucose oxidase (from *A. niger*)	Batch culture/submerged fermentation Chemically reduced graphene oxide (CRGO)	Hydrolysis	82,00
Lignocellulosic biomass	*A. niger* SIIM M276	Batch culture/submerged fermentation	Hydrolysis	94,83
Sugarcane molasses	*A. niger* ORS-4	Batch culture/surface fermentation	None Clarification	8,27 38,47
Whey	*A. niger* NCIM 548	Batch culture/submerged fermentation	Deproteination	69,00
Cheese whey/glucose	*P. taetrolens* LMG 2336	Fed-batch culture/submerged fermentation	Sweetening	-
Wastepaper	*A. niger* IAM 2094	Batch culture/submerged fermentation	Hydrolysis	84,84

(Table 1) cont.....

Carbon source	Microorganism or enzyme used	Culturing method	Substrate pretreatment	GA yield (%)
Fig	*A. niger* ATCC 1057	Batch culture/solid-state fermentation	Extraction and sterilization	63,00
Banana must	*A. niger* ORS-4	Batch culture/surface fermentation	Clarification	79,70
Grape must	*A. niger* ORS-4410	Batch culture/Submerged fermentation	Rectification	60,40
Strawberry purée	*G. japonicus* CECT 8443	Batch culture/submerged fermentation	Pasteurization	95,00

The revalorization of agro-industrial by-products as fermentation substrates can be absorbing with regard to perishables, for which fermentation significantly extends their shelf life. *A. niger* showed a promising result in producing GA from waste like grape must, fig fruits, deproteinized whey, and waste paper hydrolysate. The critical part of the work is clarifying the substrates with heavy metals in varying amounts, which is expected to affect the process. The grape and banana extracts were clarified using enzymes such as cytolase, klerzyme, and rapidase. A 20-25% increase in GA production was observed from clarified hydrolysates compared to crude extracts [49]. Solid substrates like tea wastes require an SSF process to support the microbes for cultivation. Tea waste contains phenols, sugars, and amino acids suitable for microbial growth and the production of acids. The SSF type is more advantageous than submerged fermentation as it consumes low energy, and water, with no foam formation, and high yield that reduces enzymes' cost. By using this technique, tea waste along with other supplements namely yeast extract, sugar bagasse, and soya oil further enhanced the GA production (76 – 82 g/L) [56].

In case of fruits and garden vegetables, products are hygienically and nutritionally acceptable. Still, they are discarded based on aesthetic grounds (*viz.*, size, weight, and appearance) because they fail to meet quality regulations for marketing because of their low market value. The use of fruit surpluses poses the problem of working with complex matrices that may imply many restrictions on fermentation alternatives. For example, strawberry surpluses have been used to obtain new non-alcoholic fermented beverages containing GA as a significant ingredient, which also retains the nutritional and sensory properties of the fruit. Strawberry puree was used to make a glucose-free, naturally sweet drink by turning glucose into GA while retaining the puree's original fructose [57]. Pasteurization helped to preserve the original properties of the fruit instead of sterilization. Thus, to ensure the prevalence of the *Gluconobacter* strain over undesirable microorganisms, the inoculation procedure had to be carefully observed. A two-stage procedure, a pre-cultivation step in sterilized strawberry purée which yields a pretty active inoculum of *G. japonicus* that can transform the glucose content before the yeast activity was started [58]. In addition to this, due to the complexity of the *Gluconobacter* metabolism, there was a need to preserve the formed GA in the fermented end-product, avoiding its conversion into keto-gluconates. This was achieved by maintaining proper pH throughout fermentation [59].

Itaconic Acid

Itaconic acid (IA) is a single unsaturated (C5) dicarboxylic acid known as methylene succinic acid or 2- methylidenebutanedioic acid. IA was included in the list of top 12 building blocks chemicals due to its potential transition into different derivatives such as itaconic diamide, 2-methyl-1,4-butane diamine, *etc* [10]. It has two carboxyl groups which make it an alternative in synthesizing biodegradable polymers like plastics and unsaturated polyester resins [60]. The chemical price varies between US$ 1.80 and US$ 2.00/ kg depending on the substrate cost, production system, and quality of the product [61 - 63]. Fig. (**1**) summarizes the IA's production route, applications, and derivatives (A road map of Itaconic acid). Itaconic acid was first discovered as a thermal decomposition product of citric acid [64]. The biological pathway of IA from glucose involves the conversion of citric acid into cis-aconitate, which will be decarboxylated to IA by a cytoplasmic enzyme *cis*- aconitic acid decarboxylase (CadA). *Aspergillus itaconicus*, a filamentous fungus, was identified to produce IA *via* the tricarboxylic acid (TCA) cycle or *via* the citramalate pathway [65]. To date, *A. terreus* is considered the best producer of IA (130- 160g/L) from glucose [66 - 68]. Wild-type strains like U. maydis, Candida sp., Rodotorula sp., Pseudozyma sp., Helicobasidium sp., *etc*. are also IA producers. However, the fungal production of IA was less compared to citric acid production, which was around 200g/L [69].

Furthermore, genetically engineered organisms were developed to increase the production of IA. The wild-type *A. terreus* strain was engineered to express the pfkA gene of *A. niger* (encodes 6-phosphofructo-1-kinase). The modified *A. terreus* could produce 45.5 g/L of IA, higher than the wild type [70]. *A. niger,* a native citric acid producer, was engineered with cadA, mttA, mfsA genes to produce IA. The resulting strain enhanced the IA production level by 25 times compared to a cadA-expressing strain [71]. Other engineered organisms such as *E. coli* [72, 73], *Corynebacterium glutamicum* [74], *Saccharomyces cerevisiae* [75], *Yarrowia lipolytica* [76] and *Pichia kudriavzevii* [77] were also studied to produce IA. So far, glucose is used as the primary substrate for the industrial production of IA. Other substrates like sucrose, xylose, arabinose, mannose, galactose, and rhamnose were also reported to produce IA. However, the yield was lower than glucose's [61, 78]. Anyhow, exploring various sugars to produce IA by microbes is still in progress to achieve a better yield and profit presents the methodological flow of itaconic acid preparation.

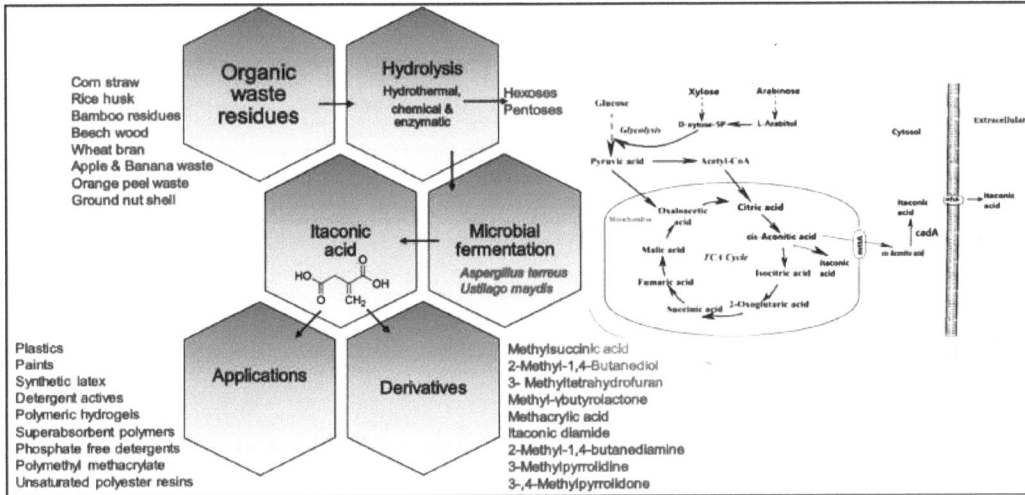

Fig. (1). A road map for Itaconic acid

Biological Production of IA from Waste Residues

In recent years, several types of research have been carried out to utilize the organic residues generated from food, agriculture, and industrial waste to produce IA. This approach could minimize the waste accumulation in the ecosystem, reduce the substrate cost, and offer a profitable market. Lignocellulose-based materials are widely used to produce many value-added products, including IA due to their plenteous availability. Lignocellulose biomass contains cellulose, hemicellulose, lignin, and ash [79]. In this direction, numerous researchwere research studies reported IA production from waste residues such as sugarcane bagasse, molasses, bamboo, beechwood hydrolysate, corn stover hydrolysate, fruit waste extract, wheat bran, and wheat chaff hydrolysate (Table **2**).

Table 2. Production of IA from organic waste substrates [83, 84, 61]

Microorganism	Substrate	Itaconic acid titer (g L^{-1})
Aspergillus terreus M-8	Starch hydrolysate	55
A. terreus (mutant)	Market refuse apple and banana	30 - 32
A. terreus	Palm oil mill effluent	5.76
A. terreus SKR10	Sago starch	48.2
A. terreus	Jatropha Cake	48.7

(Table 2) cont.....

Microorganism	Substrate	Itaconic acid titer (g L⁻¹)
A. terreus	Olive & beet waste	44
A. terreus (mutant)	Corn stover hydrolysate	19.3
A. terreus DSM 23081	Wheat chaff	27.7
A. terreus ATCC10020	Rice husk	1.9
A. terreus CICC40205	Wheat bran hydrolysate	49.6
A. terreus NRRL1960	Bleached cellulose pulp	37.5

Jatropha seed cake is a by-product in biodiesel production and is disposed of in the land as waste. It is not suitable for of applying livestock feed but utilized as a nutrient source by *A. terreus* for IA production (24.46 g/l – 48.7g/L) [80, 81]. Rafi and co-workers [82] produced IA using agro-waste residues like ground nutshells, groundnut oil cake, rice bran, sugarcane bagasse, sunflower husk, and orange fruit waste. The substrates were steam dried (80°C) and chemically treated to make them accessible for microbial fermentation. The fungi *U. maydis* was efficient in converting all the substrates into IA. The titer of IA was high in orange fruit waste, which could be attributed to the citric acid in the fruit.

Pre-treatment and Detoxification of Waste Residues

Different pre-treatment-treatment methods were investigated to generate monomeric sugars from the recalcitrant mixtures with fewer impurities. Depolymerization or destruction of biomass follows the thermochemical process involving acid hydrolysis or steam explosion and biological treatment using enzyme mixtures like cellulase, β-1-4-endoglucanases, cellobiohydrolase, α - amylase, and amyloglucosidase [85 - 87]. Generally, lignocellulosic biomass requires harsh pre-treatment methods to break the heteropolymer chain into many simple convertible sugars. Rice husk generated after cereal processing is procured for IA production by introducing a different chemical pre-treatment process. In this study, they applied a pressurized acid condition with nitric acid and phosphoric acid at three different parameters: acid concentration, time, and temperature. Phosphoric acid treatment resulted in a 45% yield of sugars (glucose, xylose, and arabinose) in less time, with fewer impurities, and the resulting hydrolysate was compatible with the detoxification process. The fermentation with *A. terreus* produced 1.9 g/L IA without any other nutrient supplementation [88]. Beechwood (plant biomass) on treatment with chemo-catalytic biomass fractionation resulted in three phases: cellulose pulp, hemicellulose, and organic lignin solutions. The cellulose pulp is enzymatically hydrolyzed to glucose, and fermented by *A. terreus* or *U. maydis* for IA synthesis. The beechwood was pretreated by the Organo Cat process, which uses acid treatment to hydrolyze hemicellulose and organic solvent for lignin. This method is economically best due to the recycling of fractions in each step, reducing the loss of organic sugars. The final hemicellulose hydrolysate in the aqueous phase consists of xylose and glucose (65g/L and 11g/L). The conversion efficiency of cellulose pulp to glucose was comparable (60%) to that of pure cellulose convertibility (70%). However, IA yield by *A. terrus* was low compared to genetically modified *U. maydis*. The drawback of the experiment is the impurities present in the hydrolysate, which inhibited cell growth and the residual glucose accumulation at the end of fermentation. Finally, the study signifies the importance of glucose convertibility to IA, which is low, and this can be overcome by process optimization or can be neglected if a high product yield is achieved [89].

The robustness of microorganisms against substrate utilization, growth against impurities, and inhibitors in the pretreated biomass are of paramount importance than the product yield. The impurities may come from the lignin depolymerization during hydrolysis (chemical & thermal treatment), mixed enzyme biochemical reaction, microelements (Fe, Mn, Cu, & Zn), and salts released from the hydrolysis of various substrates [90]. In 2016, a research group provided evidence on the discharge of impurities along with glucose and xylose after the steam explosion and enzyme treatment of corn stover hydrolysate. The impurities in the hydrolysate were identified as formic acid, acetic acid, levulinic acid, furfural, and 5- hydroxymethyl furaldehyde which is known to inhibit *A. terreus* growth. Consequently, the fungal strain exhibits no growth in the undetoxified hydrolysate. However, the mutant *A.terreus* strain showed growth in the undetoxified enzymatic hydrolysate and produced IA (19.3g/L) [91]. *A. terreus* is more sensitive to medium impurities like phosphorous, acids, and metal ions, and *U. maydis* requires a nitrogen-limited fermentation condition for better growth and IA production [90]. Still, cultivation of *A. terreus* displayed retarded growth and no production of IA in the diluted acid hydrolysate of wheat straw [92].

Lignin-derived inhibitors are weak acid, ferulic acid, HMF, furfural, phenol, syringaldehyde, vanillin, 4- hydroxybenzaldehyde, and levulinic acid [93 - 95]. A recent study estimated the inhibition level of acetic acid, furfural, HMF in biomass hydrolysate and, it was found to be above 188 mg/L, 175 mg/L, and 700 mg/L. The presence of traces of acetic acid is sufficient to reduce IA production [87]. Another group experimentally proved that acetic acid and manganese ion concentrations above $\geq$ 0.5 g/L and 0.1 mg/L in bamboo enzymatic hydrolysate inhibited cell growth and product formation. After a 100-fold dilution of the hydrolysate with a suitable nitrogen source (steep corn liquor) and deionized water, replacing the buffer improved cell growth and enhanced IA production [62].

Detoxification is one possible way to remove impurities from the hydrolysate. A 90% removal of acetic acid by liquid-liquid extraction was reported by Roque and co-workers [96]. Biodetoxification is a method that involves the usage of microorganisms capable of utilizing the inhibitors in the hydrolysate. A fungus, namely *Amorphotheca resinase* ZN1 was able to degrade almost all the toxins as a carbon source for its growth in a broad spectrum of lignocellulose feedstock. This is more beneficial as no wastewater is released, and the energy input is low. Zhang and co-workers [97] reported that the strain readily utilized toxins in the corn stover hydrolysate without degrading the cellulose. The fungal metabolism was high in solid-state cultivation rather than liquid fermentation. The disadvantage of this method was that the fungus could utilize xylose in the hydrolysate, and the process was prolonged. However, a remnant of 0.85g/L acetic acid in the bio-detoxified corn stover hydrolysate inhibited the growth of the *A. terreus* M69 and produced no IA. Further detoxification by activated charcoal reduced the acetic acid level to 0.45g/L and showed IA production of 33.6g/L [95]. Activated charcoal-based detoxification also reduced the concentration of furfural and HMF to 99% by adsorption [98]. Kumar and co-workers [99] used other methods like ultrafiltration, neutralization, and adsorption using ion exchange resins to minimize the inhibitory compounds. Deionization of potato starch hydrolysate enhanced IA to 16-fold compared to crude hydrolysate. The presence of ions in the potato starch hydrolysate hinders IA production, which elevated IA levels upon removal using ion exchange resin [86]. Another study produced IA without detoxification by using a simple alkaline pretreatment method. The method reduced the impurities level in the wheat chaff hydrolysate, and the cation exchanger recovered metal traces without any loss in sugar concentration [100].

Thus, the different substrates, varying compositions, and processivity greatly influenced the IA production. For instance, cellulose, starch, sugar bagasse, and crude glycerol are promising substrate for IA, when considering the impurities generation level. Cellulose pulp by enzyme hydrolysis and sterilization by membrane filtration had small traces of impurities (HMF and furfural) and with supplementation of nitrogen source, the fungi *A. terreus* produced 37.5 g/L of IA [84]. Other factors like medium and fermentation process optimization also influence IA production. Process strategies like fed-batch with simultaneous saccharification and fermentation (SSF) displayed many advantages in terms of cost and labor. To improve the microorganism's growth, additional organic sources like nitrogen, amino acids, and vitamins are supplemented in the medium in the form of yeast extract, steep corn liquor, and beef extract [62]. This method improved the culture growth and cell morphology. The general studies gave a conclusive understanding that the IA production was influenced by the type of substrate, biomass pretreatment, the impurities in the hydrolysate, detoxification, fermentation medium and process optimization, and the expediency of the whole process. This ultimately reflects in the final price of the product, which is expected to be less than the current market value.

Levulinic Acid

Levulinic acid (LA) is a C5 platform chemical referred to as a keto acid or 4-oxopentanoic acid. The structure has a carboxylic group and a ketone that enables the formation of different derivatives. It gained more attention from chemical industries as an intermediate and is included in the top 12 building blocks chemicals derived from biomass. A few of its derivatives are acrylic acid, γ-valerolactone, 1,4- pentanediol, levulinate esters, diphenolic acid, δ-aminolevulinate, and β-acetylacrylic acid [10]. δ- aminolevulinate is used as a natural ingredient in herbicides [101], and levulinate ester is used as an additive in gasoline and biodiesel [102]. Diphenolic acid gains an extreme interest in the polycarbonate industry as it can replace the standard bisphenol A which is producing over 3 million tonnes [103]. LA has an application in pharmaceuticals, and plasticizer synthesis and can utilize as a solvent, antifreeze agent, and food flavoring agent [104].

LA is chemically synthesized from a variety of raw materials containing carbohydrates, acid catalysts, or solid catalysts and precursors like 5-hydroxymethylfurfural (5-HMF) and furfural. The chemical reaction involves the dehydration of fructose to 5-HMF in the presence of an acid catalyst, and the latter rehydrates into LA (Fig **2**). There are limitations in the commercialization of LA, as the chemical synthesis process requires high-value raw materials, equipment, and the yields were generally low due to the presence of impurities [105, 106]. Such restrictions urged replacing pure sugars with low-cost renewable resources as raw materials.

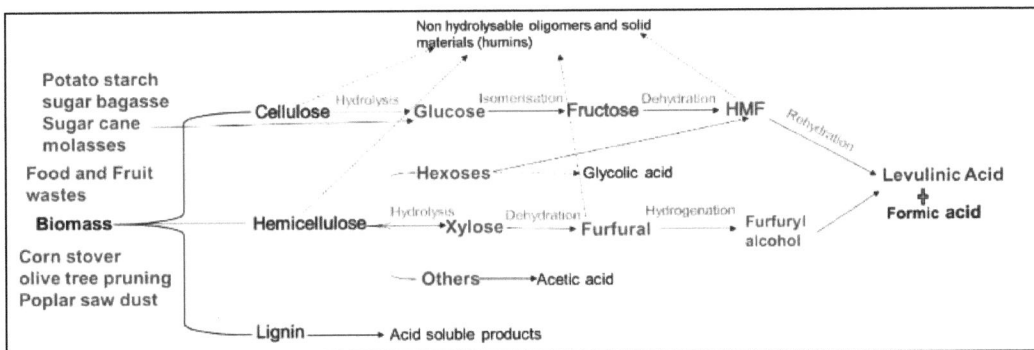

Fig. (2). Chemical synthesis of LA from waste biomass

The choice of the raw material is primarily based on the ability to convert the hexose sugar to the precursor 5-HMF. Monosaccharides such as glucose, fructose, sorbose, galactose, *etc*, and polysaccharides like starch, cellulose, chitin, chitosan, sucrose, inulin, lactose, *etc*, were studied for the synthesis of the precursor 5-HMF [107 - 110]. Fructose is the dominant sugar against others in 5-HMF synthesis, as it could make a maximum of 100% convertibility. The acid hydrolysis allows side reactions generating impurities like formic acid, humins, and a small amount of LA. Thus, selecting appropriate sugar-based biomass and a catalyst is a prime requirement for LA production from 5-HMF [108]. Like 5-HMF, furfural is also considered a good precursor for LA production. Xylose and arabinose present in the hemicellulose biomass are the chief sources of furfural.

Pretreatment Methods for Synthesis of Levulinic Acid from Organic Wastes

Organic residues rich in cellulose will degrade by applying a hydrothermal treatment in the presence of a catalyst for LA production. In general, cellulose on hydrolysis releases glucose, which is either isomerized to fructose or directly converted to precursors 5-HMF or furfuryl alcohol, depending on the type of acid used and may induce side reactions. The fructose is dehydrated to 5- HMF and then rehydrated to LA. The pretreatment conditions may vary according to the biomass, for instance, a milder pretreatment is required for less complex structures like corn starch, potato starch waste, sago waste, *etc*., unlike lignocellulosic biomass (Biofine process) [111]. It is crucial to select an acid or an alkaline treatment and a suitable heating temperature for processing the biomass. Chemical synthesis of LA from organic wastes follows a pre-drying step and heating above 105°C in the presence of a catalyst. Conventional heating type uses an oil bath or a jacketed vessel type apparatus. Recently heating is conducted using a microwave device [112].

The production of LA with microwave heating was reported using raw materials like tobacco chops, olive tree pruning, paper mill sludge, and poplar sawdust. The microwave reactor offers fast heating and cooling with high gas pressure. The advantage is that the microwave radiation uniformly absorbs into the slurry and thus provides homogeneous heating [104]. In 2013, a research group reported LA production from plant-derived cellulose, animal-derived chitin, and chitosan by applying a mono-mode Multisynth microwave reactor. By this method, the maximum yield obtained from the glycans in the presence of 2M H_2SO_4 at 170°C in 30 minutes was between 20 – 37% [113]. Likewise, LA from tomato plant waste showed better yield using HCl as a catalyst than conventional type. The conversion was completed in 2 min irradiation and obtained clean LA in the range 63-95% [114]. Microwave heating along with ionic liquid (alkali metal halides &

Phosphoric acid) assisted a complete conversion of cellulose into LA without pretreatment [115].

Tukacs and co-workers [116] reported LA production from wheat straws, pea straws, shells (nut, pistachio, coconut) and peels (orange, banana, carrot) gathered from household waste, potato peels from the food and canning industry, waste of instant coffee manufacturing and tea waste (dried leaves or tea bags). The study recommended the direct processing of biomass for hydrolysis instead of pre-drying, as it requires more energy and reduces the water content in the biomass. Anyway, there was no change in the LA production, but the orange and carrot peel waste resulted in a two-fold increase. The water content enabled the H^+ ions penetration and diffusion within the complex lignocellulosic structure for hydrolysis, as this could happen in a slow process after drying the biomass. From these studies, it is apparent that microwave-assisted heating allowed a maximum disruption of polymeric structure into monomers in a short time and aided differential yield from different biomass. This difference depends on the hexoses present in the hydrolysate and their convertibility into LA . Table **3** presents the biomass which can be utilized for LA production.

Table 3. List of biomasses utilized for the production of LA

Biomass	Solvent system	Catalyst	Temperature and reaction time	Levulinic acid titer (g L^{-1})
Sugarcane bagasse	Water	HCl	220°C, 45 min	22.8
Paddy straw	Water	HCl	220°C, 45 min	23.7
Rice straw	Water	$S_2O_8^{2-}/ZrO_2$-SiO_2-Sm_2O_3	200°C, 10 min	49
Olive tree pruning	Water	HCl	200°C, 1 h	47.2
Poplar sawdust	Water	HCl	200°C, 1 h	37
Empty fruit bunch	Water	$CrCl_3$ + HY hybrid catalyst	145°C	15.5
Cow dung	Water	HCl	180°C, 150 min	338.9g/kg
Corn stover	Water - GVL	Amberlyst 70	160°C, 16h	54
Carbohydrate- rich potato peel waste	Water	H_2So_4, $CrCl_3/AlCl_3$	180°C, 15 min	49%
Oil palm empty fruit bunch	Water	$InCl_3^-$ [HMIM][HSo4]	177°C, 288 min	17.7%
Corn cob residues	Water	H_2So_4	180°C, 50 min	107.9
Corn stover	Water - GVL	H_2So_4	160°C, 19h	66

Types of Catalysts, and Solvents Used in LA Chemical Synthesis

Catalyst supplement enhances the reaction rate of conversion and product formation. Initially, solid catalysts (strong acid cation exchange resins, zeolite, solid metal phosphates, TiO_2, and ZrO_2) were highly used in the reaction medium. Later liquid catalysts like sulphuric acid, hydrochloric acid, phosphoric acid, and a combination of weak acid-base were employed for the betterment [117, 118]. Solid catalyst often results in low yield, more reaction time, and limitation in mass transfer, and it loses activity after usage for a prolonged period leading to the formation of humin or metal groups [8]. An average of 32% yield w.r.t dry biomass was obtained from most feedstocks using mineral acid as a catalyst [116]. However, mineral acids also exhibit some challenges in LA production, for instance, incompetence in product separation, recycling, corrosion effect, and inconsistent yield [119]. However, mineral acids like HCL and H_2SO_4 had a good effect in all types of substrates, especially lignocellulosic biomass. A well-suited biorefinery set-up is a prerequisite for acid treatments to minimize rusting problems and a different plant for acid recovery. The acid-based chemical process can be replaced with a renewable, biodegradable solvent to reduce the risk of acid contamination in the environment. Dwivedi and co-workers [120] extracted a biodegradable *p*-cymene sulphonic acid dihydrate (*p*-CSA) from essential oil *d*-limonene present in citrus fruits. This chemical and HCl, produced LA from spent aromatic wastes (lemongrass, citronella grass, and palmarosa fibres) generated after essential oil extraction. Almost 6.0 million tons of aromatic crops are generated in India and most of them are discarded as waste or burnt. The pre-treated aromatic waste containing cellulose and *p*-CSA and HCL was treated at 140°C - 180°C in a glass reactor for two hrs, which produced maximum LA. They claimed that the combined action of *P*-CSA and HCL was effective in LA generation. The addition of *p*-CSA in the reaction mixture increased the cellulose conversion efficiency to LA [121].

Water as a solvent is universally used for most biomass for solubilization with a low impact on the environment and less energy requirement for solvent recovery in the downstream process. However, the combination of heterogeneous catalysts and some water-sensitive reagents may pull down the solubility of the polymers. A recent study showed the effectiveness of heterogeneous catalysts and a saltwater system in the production of LA. The presence of salt anions enhanced the conversion of glucose to LA [122]. The choice of solvent is important as the solubility of the substrate in the solvent influences the reaction. Apart from water, other solvents like DMSO, DMF, DMA, THF, MIBK, *etc.*, were also found to be good in solubilization [123]. Wood has potential applications in producing bio-based products and is vulnerable to depolymerization into monomers. Eucalyptus wood chip was used as biomass for LA production by using methanol as a solvent and H_2So_4 as a catalyst. Methanol showed the best result in suppressing humins formation from glucose and enhanced the convertibility of glucose to LA (66 mol%) and methyl levulinate [124]. In the case of solid catalyst, water and GVL combination showed better solubilization of cellulose. For example, the Rackemann group reported that the water & GVL combination improved the Amberlyst 70 catalyst activity and porosity, where water resulted in enhanced LA yield (38.7 wt%) from corn stover [125]. This combination worked well with the mineral catalyst as well. Therefore, only an organic solvent with a high LA partition coefficient about water can be used because it reduces the volume required, decreasing the energy input for LA recovery [123].

Unlike lignocellulose biomass, sugarcane molasses produces LA with fewer side reactions *i.e.*, by-products. A new LA production strategy was investigated by recycling the acid-treated sugarcane molasses for the second acid process after the LA production. Operating with an optimum H_2SO_4 concentration, fixed to 0.2M at 180°C for 180 min, was the most challenging part of the experiment. This method fetched a maximum of 24 – 30% increase in LA from the subsequent cycles [126]. A similar approach was introduced with a heterogeneous catalyst acid cation exchange resin for sugar beet molasses. A pre-treatment step was followed to remove the non-sugar components like cations, proteins, and alkaline compounds in the sugar beet molasses, as it deactivates the catalyst. The sugar beet molasses and resin pellets were packed in a column to precisely remove the impurities. In this way, the LA yield in the ion exchange solution was high and the catalyst was active after several applications [127]. Kiwi fruit and vegetable waste were proven to produce LA using 20% Nb/Al oxide and Amberlyst 36 respectively [128, 129]. Besides crop wastes, cow dung from cattle is also investigated for LA production. Cow dung is accumulated in large quantities and is mostly used as land fertilizer, which may lead to groundwater pollution. The cow dung was initially treated with KOH aqueous solution to separate lignin from other polymers in this study. The solid material contained high concentrations of cellulose and hemicellulose, along with small fractions of lignin and ash. The pre-treated cow dung was treated with HCL and allowed to heat at 180°C in a reactor. The LA yield was high in pre-treated cow dung which is 338.9g/kg than the raw cow dung. A significant amount of formic acid was also formed by the chemical reaction as a by-product along with LA amount [130].

The drawbacks above related to the chemical process and ingredients used in LA production can be fixed by varying the substrate composition or opting for an alternative catalyst. In addition, the routine availability of biomass and transportation could affect the total cost. Moreover, the biomass composition with varying sugar levels greatly influenced the economy. Many studies have shown that the yield from sucrose/starch was comparatively higher than lignocellulose biomass due to the disparity in the hexose composition of raw materials and the process involved in the synthesis. In addition, the recovery of LA from the reactor faces a significant challenge due to the formation of humins and the difficulty in separating LA from mineral acids [8]. Therefore, techno-economic analysis and environmental safety assessment will provide a promising lead for the commercialization of the product.

Sugar Alcohols

Sorbitol

Sorbitol is a natural sugar alcohol, odorless, noncariogenic, white crystalline powder, with a molecular weight of 182.17g/mol, 60% relative sweetness compared to sucrose, 2350g/L of solubility, and a pH around 7.0. It is mainly used in the food industry as a sweetener, moisturizer, texturizer, and softener. As its metabolic pathway is not insulin-dependent, it can also be used in dietetic foods for people with diabetes (low energy and non-metabolizable). The global production of sorbitol has been estimated to be more than 500,000 tons per year [131]. Sorbitol is obtained from glucose by catalytic hydrogenation, which is reported as "Generally Recognized as Safe" [132]. Sorbitol is one of the top 12 high-value-added building block intermediate chemicals that can be produced from renewable biomass resources [10]. The biosynthesis of sorbitol may be affected by the concentration of nutrients in the culture medium, environmental conditions, and new biotechnological processes, like cell permeabilization and immobilization [133].

Sorbitol can be produced from cornstarch-based glucose as a replacement for petroleum feedstocks. Corn steep liquor (CSL) is used in the batch process [134]. Maltose feedstock is also known to produce sorbitol, and the technology of converting cellulose, lignin, and oils (from wood and residues) into sorbitol is well-established [135]. Starch is the primary commercial source of sorbitol. The hydrogenation of fructose and mannose yields mannitol and sorbitol. Therefore, a variety of raw materials resulting from agricultural activities are used to produce sorbitol. Moreover, the use of a renewable feedstock can reduce the carbon footprint of the process [136]. For instance, cellulose is the most abundant source that can be converted into sorbitol. The conversion involves both hydrolysis and hydrogenation, and many catalytic processes were addressed for this purpose. Several studies have reported the catalytic conversion of cellulose into sorbitol using noble metal-based catalysts. The highest yield of sugar alcohols (81%) could be achieved by the combination of heteropoly acids with supported ruthenium catalysts [137]. Direct conversion of cellulose from wood residues to sorbitol or any other sugar alcohols without using acids as a catalyst is challenging today. The rigid crystalline structure of the cellulose prevents the complete conversion to sorbitol. A ball-milling method of waste cellulosic material from cotton wool, cotton textile, and tissue paper resulted in the disruption of the crystalline structure and enabled 100% conversion with sorbitol yields of around 50% in 2 h. Though the chemical reaction works well with the ruthenium catalyst, some waste residues like printing paper may not be competent to produce sorbitol due to the presence of impurities [138]. Cheaper catalysts are available (TiO_2, Al_2O_3, SiO_2, MgO, ZnO, and ZrO_2) for the hydrogenation reaction to convert cellulose into sorbitol and mannitol. Isosorbide is a dehydrated product of sorbitol produced *via* an intermediate 1,4-anhydrosorbitol in the presence of a catalyst. This sequential reaction for the synthesis of isosorbide from Japanese cedar using Pt/C and Ru/C catalyst with Amberlyst 70 was reported [139].

The bacteria *Zymomonas mobilis* is one of the potential candidates for sorbitol bioproduction due to its enzyme glucose-fructose oxidoreductase, which can convert fructose and glucose into sorbitol and gluconolactone (shortly converted into gluconic acid). Many attempts were made to reduce ethanol production and at the same time improve sorbitol yield using the overproduction of the oxidoreductase enzyme and the improvement of bioconversion from fructose and glucose to sorbitol and gluconic acid by a new cell treatment inhibiting Entner–Doudoroff pathway enzymes in order [133]. Moreover, sorbitol can be produced by the metabolically engineered lactic bacteria, when lactose is available as a carbon source; this conversion of high-calorie sugar lactose into low-calorie sugar sorbitol may be of particular interest for the development of dairy products with added functional values. De Boeck and co-workers [140] used an added-value metabolite, whey permeate, a waste product from the dairy industry with a high concentration of lactose. A conversion rate of 9.4% of lactose into sorbitol was obtained using an optimized fed-batch cultivation system. A further economically viable way of producing sorbitol is the use of cheap media, such as CSL, a cheap source of vitamins and nitrogen, to obtain high bioconversion [141]. In this case, the process was carried out in a batch mode. Furthermore, employing 25g/L of CSL is likely to reduce the cost of the medium between 25% and 50%.

Glycerol

Glycerol is traditionally produced as a byproduct of the soap and fatty industries. Glycerol is used as an intermediate chemical for the production of cosmetic and food products. It has applications in pharmaceutical, tobacco, polyurethane, and alkyd resins due to its nontoxicity and biodegradability characteristics. It serves as a feedstock for the production of 1,3-propanediol, acrolein, and hydrogen. Many attempts have been made to produce glycerol by biochemical methods as an alternative to chemical synthesis. However, this route does not fulfill its demand. Microbial production of glycerol has been known for 150 years, and glycerol was produced commercially during World War I. Glycerol production is carried out by the chemical process *via* propylene utilization, hydrolysis of oils (triglycerides and caustic soda), and transesterification of fatty acids with ethanol/methanol. The reactions are catalyzed by mineral acids either homogeneous or heterogeneous [142]. The main production stream is from the biodiesel plant, as crude glycerol is the main by-product of the process (Fig. **3**). In a single run producing 100kg biodiesel, about 10% glycerol is always expected with 50-55% purity [143]. The feedstock of biodiesel accounts for 75% of the product cost.

Alternative feedstocks like used cooking oil, beef fat, and chicken fat were analyzed for biodiesel and glycerol production. The transesterification reaction enabled the production of biodiesel in the range of 87 – 89% with crude glycerol as a by-product [144]. Similarly, microalgae lipids, sunflower oil, and seed oils were investigated for glycerol production. The crude glycerol produced by this method comprised \ 30% (w/w) glycerol, 50% methanol, 13% soap, 2% moisture, 2-3% salts, and 2-3% impurities [145]. The optimization of the glycerol production was carried out with different catalysts to enhance the transesterification reaction with less formation of impurities. Based on the free fatty acid content in the triglycerides, the type of catalysts such as acid catalysts with water or alkali catalysts will be used. However, there have been issues with corrosion, high-energy requirements, and acid disposal in the processing of glycerol with mineral acids. Enzymes such as lipases, Novozym 435, *etc.*, are reported to exhibit a promising transesterification reaction for glycerol production [146]. Glycerol is also produced from saponification and hydrolysis of fats with a yield of 35% and 15% respectively [143].

Glycerol production by fermentation has become more attractive, as the cost of propylene increased and its availability has decreased especially in developing countries. The microorganisms such as yeast (*Saccharomyces cerevisiae* & *Candida sp.*), bacteria (*Bacillus subtilis*), and algae (*Dunaliella tertiolecta*) were studied for glycerol biosynthesis. The possibilities of glycerol overproduction by yeast from monosaccharides can be attained by: (1) By developing a complex between acetaldehyde and bisulfite ions thereby impeding ethanol production and balancing redox cofactors through glycerol synthesis, (2) Growing yeast cultures at pH values near 7 or above; or, (3) Using osmotolerant yeasts. In recent years, osmotolerant yeasts have shown promising growth in glycerol production [147]. Peptones and fish hydrolysates are used for the production of glycerol. Growth substrate costs often make up the major part of the production cost of microbial cells and by-products from the fermentation industry. On the other hand, peptones and fish hydrolysates are made either by acid hydrolysis or enzymatic digestion of proteins. Experiments with ram horn protein hydrolysate and hydrolysate on the crop yield of the mushroom *Agaricus bisporus* were tested for glycerol production [148]. Molasses proved to be a more suitable substrate for glycerol production [149]. *Y. lipolytica* can utilize hydrophobic substrates such as fatty acids, lipids, and alkanes and simple carbon sources, such as glucose and glycerol, which can all be found in food waste. Various related substrates can be used to produce useful products including chemicals [150 - 155]. Transformation of the products also produces value-added products [156 - 160]. Fruits and vegetables, animal waste, meat and derivatives, waste oil, and dairy products can produce glycerol and erucic acid. It should be noted that the metabolic versatility of microbes enables the fermentative production of a wide range of products [161].

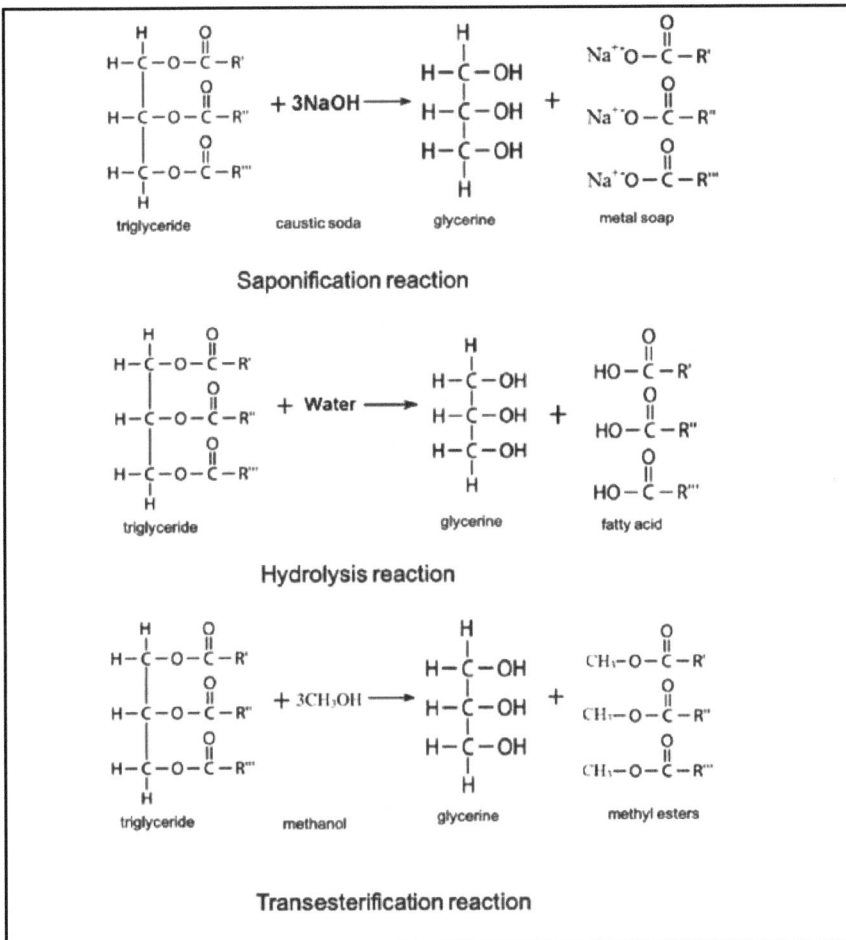

Fig. (3). Basic reaction of glycerol synthesis

CONCLUSION

The valorization of organic waste has become a key to the advancement of waste management systems for the production of bio-based chemicals. The current knowledge of pretreatment techniques makes it possible to understand the disruption of complex waste structures, the feasibility of sugar monomers for conversion into the desired products, and ways to overcome the impurities that block the reaction. This chapter reviewed the potential platform chemicals from waste organic residues. The production of bio-based chemicals from starch, sucrose, and cellulose has made maximum convertibility. More comprehensive research is needed to make better use of whole lignin to produce value-added products with fewer impurities. Building a biorefinery would be an economically advantageous set-up for industrial innovations where all process technology is carried out. To date, successful results have been obtained at the laboratory level, with the goal of large-scale research at the industrial level needing further research to satisfy the economy.

ACKNOWLEDGEMENT

The authors are thankful to their family and friends for their support.

REFERENCES

[1]   Chaher NEH, Hemidat S, Thabit Q, *et al.* Potential of Sustainable Concept for Handling Organic Waste in Tunisia. Sustainability (Basel) 2020; 12(19): 8167.
[http://dx.doi.org/10.3390/su12198167]

[2]   Madejón E, Burgos P, López R, Cabrera F. Agricultural use of three organic residues: effect on orange production and on properties of a soil of the 'Comarca Costa de Huelva' (SW Spain). Nutr Cycl Agroecosyst 2003; 65(3): 281-8.
[http://dx.doi.org/10.1023/A:1022608828694]

[3]   Higashikawa FS, Silva CA, Bettiol W. Chemical and physical properties of organic residues. Rev Bras Ciênc Solo 2010; 34(5): 1742-52.
[http://dx.doi.org/10.1590/S0100-06832010000500026]

[4]   Chatterjee R, Gajjela S, Thirumdasu RK. Recycling of Organic Wastes for Sustainable Soil Health and Crop Growth. Int J Waste Resour 2017; 7(3.)
[http://dx.doi.org/10.4172/2252-5211.1000296]

[5]   Yang F, Hanna MA, Sun R. Value-added uses for crude glycerol--a byproduct of biodiesel production. Biotechnol Biofuels 2012; 5(1): 13.
[http://dx.doi.org/10.1186/1754-6834-5-13] [PMID: 22413907]

[6]   Pleissner D, Venus J. Agricultural Residues as Feedstocks for Lactic Acid Fermentation.Green Technologies for the. Washington, DC: American Chemical Society 2014; pp. 247-63.
[http://dx.doi.org/10.1021/bk-2014-1186.ch013]

[7]   Xiong X, Yu IKM, Tsang DCW, *et al.* Value-added chemicals from food supply chain wastes: State-of-the-art review and future prospects. Chem Eng J 2019; 375121983.
[http://dx.doi.org/10.1016/j.cej.2019.121983]

[8]   Morone A, Apte M, Pandey RA. Levulinic acid production from renewable waste resources:

Bottlenecks, potential remedies, advancements and applications. Renew Sustain Energy Rev 2015; 51: 548-65.
[http://dx.doi.org/10.1016/j.rser.2015.06.032]

[9]   Dessie W, Luo X, Wang M, *et al.* Current advances on waste biomass transformation into value-added products. Appl Microbiol Biotechnol 2020; 104(11): 4757-70.
[http://dx.doi.org/10.1007/s00253-020-10567-2] [PMID: 32291487]

[10]   Werpy T, Petersen G. Top Value Added Chemicals from Biomass: Volume I -- Results of Screening for Potential Candidates from Sugars and Synthesis Gas. Report No.: DOE/GO-102004-1992, 15008859. United States: National Renewable Energy Lab., Golden,CO (US), US Department of Energy August 2004.

[11]   Kumar V, Ashok S, Park S. Recent advances in biological production of 3-hydroxypropionic acid. Biotechnol Adv 2013; 31(6): 945-61.
[http://dx.doi.org/10.1016/j.biotechadv.2013.02.008] [PMID: 23473969]

[12]   Valdehuesa KNG, Liu H, Nisola GM, Chung WJ, Lee SH, Park SJ. Recent advances in the metabolic engineering of microorganisms for the production of 3-hydroxypropionic acid as C3 platform chemical. Appl Microbiol Biotechnol 2013; 97(8): 3309-21.
[http://dx.doi.org/10.1007/s00253-013-4802-4] [PMID: 23494623]

[13]   Rathnasingh C, Raj SM, Lee Y, Catherine C, Ashok S, Park S. Production of 3-hydroxypropionic acid *via* malonyl-CoA pathway using recombinant Escherichia coli strains. J Biotechnol 2012; 157(4): 633-40.
[http://dx.doi.org/10.1016/j.jbiotec.2011.06.008] [PMID: 21723339]

[14]   Ashok S, Raj SM, Rathnasingh C, Park S. Development of recombinant Klebsiella pneumoniae ΔdhaT strain for the co-production of 3-hydroxypropionic acid and 1,3-propanediol from glycerol. Appl Microbiol Biotechnol 2011; 90(4): 1253-65.
[http://dx.doi.org/10.1007/s00253-011-3148-z] [PMID: 21336929]

[15]   Sjöblom M, Matsakas L, Krige A, Rova U, Christakopoulos P. Direct electricity generation from sweet sorghum stalks and anaerobic sludge. Ind Crops Prod 2017; 108: 505-11.
[http://dx.doi.org/10.1016/j.indcrop.2017.06.062]

[16]   Matsakas L, Nitsos C, Vörös D, Rova U, Christakopoulos P. High-Titer Methane from Organosolv-Pretreated Spruce and Birch. Energies 2017; 10(3): 263.
[http://dx.doi.org/10.3390/en10030263]

[17]   Kildegaard KR, Wang Z, Chen Y, Nielsen J, Borodina I. Production of 3-hydroxypropionic acid from glucose and xylose by metabolically engineered *Saccharomyces cerevisiae.* Metab Eng Commun 2015; 2: 132-6.
[http://dx.doi.org/10.1016/j.meteno.2015.10.001] [PMID: 34150516]

[18]   Couvreur J, Teixeira A, Allais F, Spinnler HE, Saulou-Bérion C, Clément T. Wheat and Sugar Beet Coproducts for the Bioproduction of 3-Hydroxypropionic Acid by Lactobacillus reuteri DSM17938. Fermentation (Basel) 2017; 3(3): 32.
[http://dx.doi.org/10.3390/fermentation3030032]

[19]   Nghiem N, Kleff S, Schwegmann S. Succinic Acid: Technology Development and Commercialization. Fermentation (Basel) 2017; 3(2): 26.
[http://dx.doi.org/10.3390/fermentation3020026]

[20]   Beauprez JJ, De Mey M, Soetaert WK. Microbial succinic acid production: Natural versus metabolic engineered producers. Process Biochem 2010; 45(7): 1103-14.
[http://dx.doi.org/10.1016/j.procbio.2010.03.035]

[21]   Wan C, Li Y, Shahbazi A, Xiu S. Succinic acid production from cheese whey using Actinobacillus succinogenes 130 Z. Appl Biochem Biotechnol 2008; 145(1-3): 111-9.
[http://dx.doi.org/10.1007/s12010-007-8031-0] [PMID: 18425617]

[22] Li Q, Siles JA, Thompson IP. Succinic acid production from orange peel and wheat straw by batch fermentations of Fibrobacter succinogenes S85. Appl Microbiol Biotechnol 2010; 88(3): 671-8.
[http://dx.doi.org/10.1007/s00253-010-2726-9] [PMID: 20645087]

[23] Leung CCJ, Cheung ASY, Zhang AYZ, Lam KF, Lin CSK. Utilisation of waste bread for fermentative succinic acid production. Biochem Eng J 2012; 65: 10-5.
[http://dx.doi.org/10.1016/j.bej.2012.03.010]

[24] Zhang AY, Sun Z, Leung CCJ, *et al.* Valorisation of bakery waste for succinic acid production. Green Chem 2013; 15(3): 690.
[http://dx.doi.org/10.1039/c2gc36518a]

[25] Yu J, Li Z, Ye Q, Yang Y, Chen S. Development of succinic acid production from corncob hydrolysate by Actinobacillus succinogenes. J Ind Microbiol Biotechnol 2010; 37(10): 1033-40.
[http://dx.doi.org/10.1007/s10295-010-0750-5] [PMID: 20532948]

[26] Wang C, Yan D, Li Q, Sun W, Xing J. Ionic liquid pretreatment to increase succinic acid production from lignocellulosic biomass. Bioresour Technol 2014; 172: 283-9.
[http://dx.doi.org/10.1016/j.biortech.2014.09.045] [PMID: 25270043]

[27] Cao W, Wang Y, Luo J, Yin J, Xing J, Wan Y. Succinic acid biosynthesis from cane molasses under low pH by Actinobacillus succinogenes immobilized in luffa sponge matrices. Bioresour Technol 2018; 268: 45-51.
[http://dx.doi.org/10.1016/j.biortech.2018.06.075] [PMID: 30071412]

[28] Dessie W, Zhang W, Xin F, *et al.* Succinic acid production from fruit and vegetable wastes hydrolyzed by on-site enzyme mixtures through solid state fermentation. Bioresour Technol 2018; 247: 1177-80.
[http://dx.doi.org/10.1016/j.biortech.2017.08.171] [PMID: 28941663]

[29] Li C, Yang X, Gao S, Chuh AH, Lin CSK. Hydrolysis of fruit and vegetable waste for efficient succinic acid production with engineered Yarrowia lipolytica. J Clean Prod 2018; 179: 151-9.
[http://dx.doi.org/10.1016/j.jclepro.2018.01.081]

[30] Wojcieszak R, Santarelli F, Paul S, Dumeignil F, Cavani F, Gonçalves RV. Recent developments in maleic acid synthesis from bio-based chemicals. Sustainable Chemical Processes 2015; 3(1): 9.
[http://dx.doi.org/10.1186/s40508-015-0034-5]

[31] Deng Y, Mao Y, Zhang X. Metabolic engineering of a laboratory-evolved *Thermobifida fusca* muC strain for malic acid production on cellulose and minimal treated lignocellulosic biomass. Biotechnol Prog 2016; 32(1): 14-20.
[http://dx.doi.org/10.1002/btpr.2180] [PMID: 26439318]

[32] West T. Microbial Production of Malic Acid from Biofuel-Related Coproducts and Biomass. Fermentation (Basel) 2017; 3(2): 14.
[http://dx.doi.org/10.3390/fermentation3020014]

[33] Shigeo A, Akira F, Ichiro TK. Method of producing l-malic acid by fermentation . US Patent 3063910, 1962.

[34] Zou X, Yang J, Tian X, Guo M, Li Z, Li Y. Production of polymalic acid and malic acid from xylose and corncob hydrolysate by a novel Aureobasidium pullulans YJ 6–11 strain. Process Biochem 2016; 51(1): 16-23.
[http://dx.doi.org/10.1016/j.procbio.2015.11.018]

[35] Kövilein A, Kubisch C, Cai L, Ochsenreither K. Malic acid production from renewables: a review. J Chem Technol Biotechnol 2020; 95(3): 513-26.
[http://dx.doi.org/10.1002/jctb.6269]

[36] Li X, Liu Y, Yang Y, *et al.* High levels of malic acid production by the bioconversion of corn straw hydrolyte using an isolated Rhizopus delemar strain. Biotechnol Bioprocess Eng; BBE 2014; 19(3): 478-92.
[http://dx.doi.org/10.1007/s12257-014-0047-z]

[37] Cheng C, Zhou Y, Lin M, Wei P, Yang ST. Polymalic acid fermentation by Aureobasidium pullulans for malic acid production from soybean hull and soy molasses: Fermentation kinetics and economic analysis. Bioresour Technol 2017; 223: 166-74.
[http://dx.doi.org/10.1016/j.biortech.2016.10.042] [PMID: 27792926]

[38] Dörsam S, Kirchhoff J, Bigalke M, Dahmen N, Syldatk C, Ochsenreither K. Evaluation of Pyrolysis Oil as Carbon Source for Fungal Fermentation Front Microbiol 2016 Dec 22 ;07.
[http://dx.doi.org/10.3389/fmicb.2016.02059]

[39] Feng J, Yang J, Yang W, Chen J, Jiang M, Zou X. Metabolome- and genome-scale model analyses for engineering of *Aureobasidium pullulans* to enhance polymalic acid and malic acid production from sugarcane molasses. Biotechnol Biofuels 2018; 11(1): 94.
[http://dx.doi.org/10.1186/s13068-018-1099-7] [PMID: 29632554]

[40] Feng J, Li T, Zhang X, Chen J, Zhao T, Zou X. Efficient production of polymalic acid from xylose mother liquor, an environmental waste from the xylitol industry, by a T-DNA-based mutant of Aureobasidium pullulans. Appl Microbiol Biotechnol 2019; 103(16): 6519-27.
[http://dx.doi.org/10.1007/s00253-019-09974-x] [PMID: 31243500]

[41] Martin-Dominguez V, Estevez J, Ojembarrena F, Santos V, Ladero M. Fumaric Acid Production: A Biorefinery Perspective. Fermentation (Basel) 2018; 4(2): 33.
[http://dx.doi.org/10.3390/fermentation4020033]

[42] Roa Engel CA, Straathof AJJ, Zijlmans TW, van Gulik WM, van der Wielen LAM. Fumaric acid production by fermentation. Appl Microbiol Biotechnol 2008; 78(3): 379-89.
[http://dx.doi.org/10.1007/s00253-007-1341-x] [PMID: 18214471]

[43] Das RK, Brar SK, Verma M. Fumaric Acid: Production and Application Aspects.Platform Chemical Biorefinery. Amsterdam: Elsevier 2016; pp. 133-57.
[http://dx.doi.org/10.1016/B978-0-12-802980-0.00008-0]

[44] Deng Y, Li S, Xu Q, Gao M, Huang H. Production of fumaric acid by simultaneous saccharification and fermentation of starchy materials with 2-deoxyglucose-resistant mutant strains of Rhizopus oryzae. Bioresour Technol 2012; 107: 363-7.
[http://dx.doi.org/10.1016/j.biortech.2011.11.117] [PMID: 22217732]

[45] Das RK, Brar SK, Verma M. A fermentative approach towards optimizing directed biosynthesis of fumaric acid by Rhizopus oryzae 1526 utilizing apple industry waste biomass. Fungal Biol 2015; 119(12): 1279-90.
[http://dx.doi.org/10.1016/j.funbio.2015.10.001] [PMID: 26615750]

[46] Das RK, Brar SK. Enhanced fumaric acid production from brewery wastewater and insight into the morphology of Rhizopus oryzae 1526. Appl Biochem Biotechnol 2014; 172(6): 2974-88.
[http://dx.doi.org/10.1007/s12010-014-0739-z] [PMID: 24469587]

[47] Liu H, Ma J, Wang M, *et al.* Food Waste Fermentation to Fumaric Acid by Rhizopus arrhizus RH7-13. Appl Biochem Biotechnol 2016; 180(8): 1524-33.
[http://dx.doi.org/10.1007/s12010-016-2184-7] [PMID: 27387957]

[48] Climent MJ, Corma A, Iborra S. Converting carbohydrates to bulk chemicals and fine chemicals over heterogeneous catalysts. Green Chem 2011; 13(3): 520.
[http://dx.doi.org/10.1039/c0gc00639d]

[49] Singh OV, Kumar R. Biotechnological production of gluconic acid: future implications. Appl Microbiol Biotechnol 2007; 75(4): 713-22.
[http://dx.doi.org/10.1007/s00253-007-0851-x] [PMID: 17525864]

[50] Anastassiadis S, Morgunov I. Gluconic acid production. Recent Pat Biotechnol 2007; 1(2): 167-80.
[http://dx.doi.org/10.2174/187220807780809472] [PMID: 19075839]

[51] Herrick HT, May OE. The production of gluconic acid by the penicillium luteum-purpurogenum group. II. Some optimal conditions for acid formation. J Franklin Inst 1928; 206(1): 103-4.

[http://dx.doi.org/10.1016/S0016-0032(28)90752-8]

[52]　Rehr B, Wilhelm C, Sahm H. Production of sorbitol and gluconic acid by permeabilized cells of Zymomonas mobilis. Appl Microbiol Biotechnol 1991; 35(2.)
[http://dx.doi.org/10.1007/BF00184677]

[53]　Zhang H, Zhang J, Bao J. High titer gluconic acid fermentation by Aspergillus niger from dry dilute acid pretreated corn stover without detoxification. Bioresour Technol 2016; 203: 211-9.
[http://dx.doi.org/10.1016/j.biortech.2015.12.042] [PMID: 26724553]

[54]　Mafra ACO, Furlan FF, Badino AC, Tardioli PW. Gluconic acid production from sucrose in an airlift reactor using a multi-enzyme system. Bioprocess Biosyst Eng 2015; 38(4): 671-80.
[http://dx.doi.org/10.1007/s00449-014-1306-2] [PMID: 25326720]

[55]　Cañete-Rodríguez AM, Santos-Dueñas IM, Jiménez-Hornero JE, Ehrenreich A, Liebl W, García-García I. Gluconic acid: Properties, production methods and applications—An excellent opportunity for agro-industrial by-products and waste bio-valorization. Process Biochem 2016; 51(12): 1891-903.
[http://dx.doi.org/10.1016/j.procbio.2016.08.028]

[56]　Sharma A, Vivekanand V, Singh RP. Solid-state fermentation for gluconic acid production from sugarcane molasses by Aspergillus niger ARNU-4 employing tea waste as the novel solid support. Bioresour Technol 2008; 99(9): 3444-50.
[http://dx.doi.org/10.1016/j.biortech.2007.08.006] [PMID: 17881224]

[57]　Sainz F, Navarro D, Mateo E, Torija MJ, Mas A. Comparison of d-gluconic acid production in selected strains of acetic acid bacteria. Int J Food Microbiol 2016; 222: 40-7.
[http://dx.doi.org/10.1016/j.ijfoodmicro.2016.01.015] [PMID: 26848948]

[58]　Cañete-Rodríguez AM, Santos-Dueñas IM, Torija-Martínez MJ, Mas A, Jiménez-Hornero JE, García-García I. Preparation of a pure inoculum of acetic acid bacteria for the selective conversion of glucose in strawberry purée into gluconic acid. Food Bioprod Process 2015; 96: 35-42.
[http://dx.doi.org/10.1016/j.fbp.2015.06.005]

[59]　Cañete-Rodríguez AM, Santos-Dueñas IM, Jiménez-Hornero JE, Torija-Martínez MJ, Mas A, García-García I. Revalorization of strawberry surpluses by bio-transforming its glucose content into gluconic acid. Food Bioprod Process 2016; 99: 188-96.
[http://dx.doi.org/10.1016/j.fbp.2016.05.005]

[60]　Ferreira JA, Mahboubi A, Lennartsson PR, Taherzadeh MJ. Waste biorefineries using filamentous ascomycetes fungi: Present status and future prospects. Bioresour Technol 2016; 215: 334-45.
[http://dx.doi.org/10.1016/j.biortech.2016.03.018] [PMID: 26996263]

[61]　Kuenz A, Krull S. Biotechnological production of itaconic acid—things you have to know. Appl Microbiol Biotechnol 2018; 102(9): 3901-14.
[http://dx.doi.org/10.1007/s00253-018-8895-7] [PMID: 29536145]

[62]　Yang J, Xu H, Jiang J, et al. Itaconic acid production from undetoxified enzymatic hydrolysate of bamboo residues using Aspergillus terreus. Bioresour Technol 2020; 307123208.
[http://dx.doi.org/10.1016/j.biortech.2020.123208.] [PMID: 32208342]

[63]　Okabe M, Lies D, Kanamasa S, Park EY. Biotechnological production of itaconic acid and its biosynthesis in Aspergillus terreus. Appl Microbiol Biotechnol 2009; 84(4): 597-606.
[http://dx.doi.org/10.1007/s00253-009-2132-3] [PMID: 19629471]

[64]　Baup S. Ueber eine neue Pyrogen-Citronensäure, und über Benennung der Pyrogen-Säuren überhaupt. Ann Pharm 1836; 19(1): 29-38.
[http://dx.doi.org/10.1002/jlac.18360190107]

[65]　Kinoshita K. Über die Produktion von Itaconsäure und Mannit durch einen neuen Schimmelpilz Aspergillus itaconicus. Acta Phytochim 1932; 5: 271-87.

[66]　Krull S, Hevekerl A, Kuenz A, Prüße U. Process development of itaconic acid production by a natural wild type strain of Aspergillus terreus to reach industrially relevant final titers. Appl Microbiol

Biotechnol 2017; 101(10): 4063-72.
[http://dx.doi.org/10.1007/s00253-017-8192-x] [PMID: 28235991]

[67]   Zhao M, Lu X, Zong H, Li J, Zhuge B. Itaconic acid production in microorganisms. Biotechnol Lett 2018; 40(3): 455-64.
[http://dx.doi.org/10.1007/s10529-017-2500-5] [PMID: 29299715]

[68]   Kumar S, Krishnan S, Samal SK, Mohanty S, Nayak SK. Itaconic acid used as a versatile building block for the synthesis of renewable resource-based resins and polyesters for future prospective: a review. Polym Int 2017; 66(10): 1349-63.
[http://dx.doi.org/10.1002/pi.5399]

[69]   Steiger MG, Blumhoff ML, Mattanovich D, Sauer M. Biochemistry of microbial itaconic acid production. Front Microbiol 2013; 4: 23.
[http://dx.doi.org/10.3389/fmicb.2013.00023] [PMID: 23420787]

[70]   Tevž G, Benčina M, Legiša M. Enhancing itaconic acid production by Aspergillus terreus. Appl Microbiol Biotechnol 2010; 87(5): 1657-64.
[http://dx.doi.org/10.1007/s00253-010-2642-z] [PMID: 20461508]

[71]   van der Straat L, Vernooij M, Lammers M, *et al.* Expression of the Aspergillus terreus itaconic acid biosynthesis cluster in Aspergillus niger. Microb Cell Fact 2014; 13(1): 11.
[http://dx.doi.org/10.1186/1475-2859-13-11] [PMID: 24438100]

[72]   Vuoristo KS, Mars AE, Sangra JV, *et al.* Metabolic engineering of itaconate production in Escherichia coli. Appl Microbiol Biotechnol 2015; 99(1): 221-8.
[http://dx.doi.org/10.1007/s00253-014-6092-x] [PMID: 25277412]

[73]   Yang Z, Gao X, Xie H, Wang F, Ren Y, Wei D. Enhanced itaconic acid production by self-assembly of two biosynthetic enzymes in *Escherichia coli*. Biotechnol Bioeng 2017; 114(2): 457-62.
[http://dx.doi.org/10.1002/bit.26081] [PMID: 27543843]

[74]   Otten A, Brocker M, Bott M. Metabolic engineering of Corynebacterium glutamicum for the production of itaconate. Metab Eng 2015; 30: 156-65.
[http://dx.doi.org/10.1016/j.ymben.2015.06.003] [PMID: 26100077]

[75]   Blazeck J, Miller J, Pan A, *et al.* Metabolic engineering of Saccharomyces cerevisiae for itaconic acid production. Appl Microbiol Biotechnol 2014; 98(19): 8155-64.
[http://dx.doi.org/10.1007/s00253-014-5895-0] [PMID: 24997118]

[76]   Blazeck J, Hill A, Jamoussi M, Pan A, Miller J, Alper HS. Metabolic engineering of Yarrowia lipolytica for itaconic acid production. Metab Eng 2015; 32: 66-73.
[http://dx.doi.org/10.1016/j.ymben.2015.09.005] [PMID: 26384571]

[77]   Sun W, Vila-Santa A, Liu N, *et al.* Metabolic engineering of an acid-tolerant yeast strain *Pichia kudriavzevii* for itaconic acid production. Metab Eng Commun 2020; 10e00124.
[http://dx.doi.org/10.1016/j.mec.2020.e00124] [PMID: 32346511]

[78]   Saha BC, Kennedy GJ, Qureshi N, Bowman MJ. Production of itaconic acid from pentose sugars by *Aspergillus terreus*. Biotechnol Prog 2017; 33(4): 1059-67.
[http://dx.doi.org/10.1002/btpr.2485] [PMID: 28440059]

[79]   Wu X, Liu Q, Deng Y, *et al.* Production of itaconic acid by biotransformation of wheat bran hydrolysate with Aspergillus terreus CICC40205 mutant. Bioresour Technol 2017; 241: 25-34.
[http://dx.doi.org/10.1016/j.biortech.2017.05.080] [PMID: 28550772]

[80]   Dowlathabad M, S MDJH, v P, *et al.* Fermentatative production of itaconic acid by Aspergillus terreus using Jatropha seed cake. Afr J Biotechnol 2007; 6(18): 2140-2.
[http://dx.doi.org/10.5897/AJB2007.000-2333]

[81]   Ahmed El-Imam AM, Kazeem MO, Odebisi MB, Oke MA, Abidoye AO. Production of Itaconic Acid from Jatropha curcas Seed Cake by Aspergillus terreus. Not Sci Biol 2013; 5(1): 57-61.
[http://dx.doi.org/10.15835/nsb518355]

[82]   Rafi MM, Hanumanthu M, Rao DM, Venkateswarlu K. Production of itaconic acid by Ustilago maydis from agro wastes in solid state fermentation. J Biosci Biotechnol 2014; 3(2): 163-8.

[83]   Bafana R, Pandey RA. New approaches for itaconic acid production: bottlenecks and possible remedies. Crit Rev Biotechnol 2018; 38(1): 68-82.
        [http://dx.doi.org/10.1080/07388551.2017.1312268] [PMID: 28425297]

[84]   Kerssemakers AAJ, Doménech P, Cassano M, Yamakawa CK, Dragone G, Mussatto SI. Production of Itaconic Acid from Cellulose Pulp: Feedstock Feasibility and Process Strategies for an Efficient Microbial Performance. Energies 2020; 13(7): 1654.
        [http://dx.doi.org/10.3390/en13071654]

[85]   Moshi AP, Crespo CF, Badshah M, *et al.* Characterisation and evaluation of a novel feedstock, Manihot glaziovii, Muell. Arg, for production of bioenergy carriers: Bioethanol and biogas. Bioresour Technol 2014; 172: 58-67.
        [http://dx.doi.org/10.1016/j.biortech.2014.08.084] [PMID: 25237774]

[86]   Bafana R, Sivanesan S, Pandey RA. Itaconic Acid Production by Filamentous Fungi in Starch-Rich Industrial Residues. Indian J Microbiol 2017; 57(3): 322-8.
        [http://dx.doi.org/10.1007/s12088-017-0661-5] [PMID: 28904417]

[87]   Magalhães AI Jr, de Carvalho JC, Thoms JF, Souza Silva R, Soccol CR. Second-generation itaconic acid: An alternative product for biorefineries? Bioresour Technol 2020; 308123319.
        [http://dx.doi.org/10.1016/j.biortech.2020.123319] [PMID: 32278999]

[88]   Pedroso GB, Montipó S, Mario DAN, Alves SH, Martins AF. Building block itaconic acid from center-over biomass. Biomass Convers Biorefin 2017; 7(1): 23-35.
        [http://dx.doi.org/10.1007/s13399-016-0210-1]

[89]   Regestein L, Klement T, Grande P, *et al.* From beech wood to itaconic acid: case study on biorefinery process integration. Biotechnol Biofuels 2018; 11(1): 279.
        [http://dx.doi.org/10.1186/s13068-018-1273-y] [PMID: 30337958]

[90]   Jing Y, Hao XU, Jianchun J, *et al.* Production of Itaconic Acid Through Microbiological Fermentation of Inexpensive Materials. J Bioresour Bioprod 2019; 4(3): 135-42.

[91]   Li X, Zheng K, Lai C, Ouyang J, Yong Q. Improved Itaconic Acid Production from Undetoxified Enzymatic Hydrolysate of Steam-Exploded Corn Stover using an Aspergillus terreus Mutant Generated by Atmospheric and Room Temperature Plasma. BioResources 2016; 11(4): 9047-58.
        [http://dx.doi.org/10.15376/biores.11.4.9047-9058]

[92]   Saha BC, Kennedy GJ. Ninety six well microtiter plate as microbioreactors for production of itaconic acid by six Aspergillus terreus strains. J Microbiol Methods 2018; 144: 53-9.
        [http://dx.doi.org/10.1016/j.mimet.2017.11.002] [PMID: 29109012]

[93]   He Y, Zhang J, Bao J. Acceleration of biodetoxification on dilute acid pretreated lignocellulose feedstock by aeration and the consequent ethanol fermentation evaluation. Biotechnol Biofuels 2016; 9(1): 19.
        [http://dx.doi.org/10.1186/s13068-016-0438-9] [PMID: 26816529]

[94]   Zhang J, Wang X, Chu D, He Y, Bao J. Dry pretreatment of lignocellulose with extremely low steam and water usage for bioethanol production. Bioresour Technol 2011; 102(6): 4480-8.
        [http://dx.doi.org/10.1016/j.biortech.2011.01.005] [PMID: 21277774]

[95]   Liu Y, Liu G, Zhang J, Balan V, Bao J. Itaconic acid fermentation using activated charcoal-treated corn stover hydrolysate and process evaluation based on Aspen plus model. Biomass Convers Biorefin 2020; 10(2): 463-70.
        [http://dx.doi.org/10.1007/s13399-019-00423-3]

[96]   Roque LR, Morgado GP, Nascimento VM, Ienczak JL, Rabelo SC. Liquid-liquid extraction: A promising alternative for inhibitors removing of pentoses fermentation. Fuel 2019; 242: 775-87.
        [http://dx.doi.org/10.1016/j.fuel.2018.12.130]

[97]   Zhang J, Zhu Z, Wang X, Wang N, Wang W, Bao J. Biodetoxification of toxins generated from lignocellulose pretreatment using a newly isolated fungus, Amorphotheca resinae ZN1, and the consequent ethanol fermentation. Biotechnol Biofuels 2010; 3(1): 26.
[http://dx.doi.org/10.1186/1754-6834-3-26] [PMID: 21092158]

[98]   Santana NB, Dias JCT, Rezende RP, Franco M, Oliveira LKS, Souza LO. Production of xylitol and bio-detoxification of cocoa pod husk hemicellulose hydrolysate by Candida boidinii XM02G. Arora PK, editor. PLOS ONE. 2018 Apr 11;13(4):e0195206.

[99]   Kumar V, Yadav SK, Kumar J, Ahluwalia V. A critical review on current strategies and trends employed for removal of inhibitors and toxic materials generated during biomass pretreatment. Bioresour Technol 2020; 299122633.
[http://dx.doi.org/10.1016/j.biortech.2019.122633] [PMID: 31918972]

[100] Krull S, Eidt L, Hevekerl A, Kuenz A, Prüße U. Itaconic acid production from wheat chaff by Aspergillus terreus. Process Biochem 2017; 63: 169-76.
[http://dx.doi.org/10.1016/j.procbio.2017.08.010]

[101] Rebeiz CA, Montazer-Zouhoor A, Hopen HJ, Wu SM. Photodynamic herbicides: 1. Concept and phenomenology. Enzyme Microb Technol 1984; 6(9): 390-6.
[http://dx.doi.org/10.1016/0141-0229(84)90012-7]

[102] Erner WE. Synthetic liquid fuel and fuel mixtures for oil-burning devices. US Patent 4364743A 1982.

[103] Isoda Y, Azuma M. Preparation of bis(hydroxyaryl)pentanoic acids. JP08053390, 1996.

[104] Raspolli Galletti AM, Antonetti C, De Luise V, Licursi D, Nassi N. Levulinic acid production from waste biomass. BioResources 2012; 7(2): 1824-35.
[http://dx.doi.org/10.15376/biores.7.2.1824-1835]

[105] Moens L. Sugar cane as a renewable feedstock for the chemical industry: challenges and opportunities.Advances in the chemistry and processing of beet and cane sugar. New Orleans, Louisiana, USA: Sugar Processing Research, Institute, Inc. 2002; pp. 26-41.

[106] Yan L, Yang N, Pang H, Liao B. Production of Levulinic Acid from Bagasse and Paddy Straw by Liquefaction in the Presence of Hydrochloride Acid. Clean (Weinh) 2008; 36(2): 158-63.
[http://dx.doi.org/10.1002/clen.200700100]

[107] Binder JB, Cefali AV, Blank JJ, Raines RT. Mechanistic insights on the conversion of sugars into 5-hydroxymethylfurfural. Energy Environ Sci 2010; 3(6): 765.
[http://dx.doi.org/10.1039/b923961h]

[108] de Souza RL, Yu H, Rataboul F, Essayem N. 5-Hydroxymethylfurfural (5-HMF) Production from Hexoses: Limits of Heterogeneous Catalysis in Hydrothermal Conditions and Potential of Concentrated Aqueous Organic Acids as Reactive Solvent System. Challenges 2012; 3(2): 212-32.
[http://dx.doi.org/10.3390/challe3020212]

[109] Omari KW, Besaw JE, Kerton FM. Hydrolysis of chitosan to yield levulinic acid and 5-hydroxymethylfurfural in water under microwave irradiation. Green Chem 2012; 14(5): 1480.
[http://dx.doi.org/10.1039/c2gc35048c]

[110] Wang Y, Pedersen CM, Deng T, Qiao Y, Hou X. Direct conversion of chitin biomass to 5-hydroxymethylfurfural in concentrated ZnCl2 aqueous solution. Bioresour Technol 2013; 143: 384-90.
[http://dx.doi.org/10.1016/j.biortech.2013.06.024] [PMID: 23819974]

[111] Karim AA, Tie APL, Manan DMA, Zaidul ISM. Starch from the Sago (*Metroxylon sagu*) Palm TreeProperties, Prospects, and Challenges as a New Industrial Source for Food and Other Uses. Compr Rev Food Sci Food Saf 2008; 7(3): 215-28.
[http://dx.doi.org/10.1111/j.1541-4337.2008.00042.x] [PMID: 33467803]

[112] Mukherjee A, Dumont MJ. Levulinic Acid Production from Starch Using Microwave and Oil Bath Heating: A Kinetic Modeling Approach. Ind Eng Chem Res 2016; 55(33): 8941-9.

[http://dx.doi.org/10.1021/acs.iecr.6b02468]

[113] Szabolcs Á, Molnár M, Dibó G, Mika LT. Microwave-assisted conversion of carbohydrates to levulinic acid: an essential step in biomass conversion. Green Chem 2013; 15(2): 439-45.
[http://dx.doi.org/10.1039/C2GC36682G]

[114] Tabasso S, Montoneri E, Carnaroglio D, Caporaso M, Cravotto G. Microwave-assisted flash conversion of non-edible polysaccharides and post-harvest tomato plant waste to levulinic acid. Green Chem 2014; 16(1): 73-6.
[http://dx.doi.org/10.1039/C3GC41103F]

[115] Qin K, Yan Y, Zhang Y, Tang Y. Direct production of levulinic acid in high yield from cellulose: joint effect of high ion strength and microwave field. RSC Advances 2016; 6(45): 39131-6.
[http://dx.doi.org/10.1039/C6RA00448B]

[116] Tukacs JM, Holló AT, Rétfalvi N, *et al.* Microwave-Assisted Valorization of Biowastes to Levulinic Acid. ChemistrySelect 2017; 2(4): 1375-80.
[http://dx.doi.org/10.1002/slct.201700037]

[117] De S, Dutta S, Saha B. Critical design of heterogeneous catalysts for biomass valorization: current thrust and emerging prospects. Catal Sci Technol 2016; 6(20): 7364-85.
[http://dx.doi.org/10.1039/C6CY01370H]

[118] Parshetti GK, Suryadharma MS, Pham TPT, Mahmood R, Balasubramanian R. Heterogeneous catalyst-assisted thermochemical conversion of food waste biomass into 5-hydroxymethylfurfural. Bioresour Technol 2015; 178: 19-27.
[http://dx.doi.org/10.1016/j.biortech.2014.10.066] [PMID: 25453435]

[119] Jeong H, Jang SK, Hong CY, *et al.* Levulinic acid production by two-step acid-catalyzed treatment of Quercus mongolica using dilute sulfuric acid. Bioresour Technol 2017; 225: 183-90.
[http://dx.doi.org/10.1016/j.biortech.2016.11.063] [PMID: 27889477]

[120] Dwivedi P, Singh M, Sehra N, Pandey N, Sangwan RS, Mishra BB. Processing of wet Kinnow mandarin (Citrus reticulata) fruit waste into novel Brønsted acidic ionic liquids and their application in hydrolysis of sucrose. Bioresour Technol 2018; 250: 621-4.
[http://dx.doi.org/10.1016/j.biortech.2017.11.100] [PMID: 29220805]

[121] Singh M, Pandey N, Dwivedi P, Kumar V, Mishra BB. Production of xylose, levulinic acid, and lignin from spent aromatic biomass with a recyclable Brønsted acid synthesized from d-limonene as renewable feedstock from citrus waste. Bioresour Technol 2019; 293122105.
[http://dx.doi.org/10.1016/j.biortech.2019.122105] [PMID: 31514116]

[122] Pyo SH, Glaser SJ, Rehnberg N, Hatti-Kaul R. Clean Production of Levulinic Acid from Fructose and Glucose in Salt Water by Heterogeneous Catalytic Dehydration. ACS Omega 2020; 5(24): 14275-82.
[http://dx.doi.org/10.1021/acsomega.9b04406] [PMID: 32596564]

[123] Antonetti C, Licursi D, Fulignati S, Valentini G, Raspolli Galletti A. New Frontiers in the Catalytic Synthesis of Levulinic Acid: From Sugars to Raw and Waste Biomass as Starting Feedstock. Catalysts 2016; 6(12): 196.
[http://dx.doi.org/10.3390/catal6120196]

[124] Kang S, Yu J. Effect of Methanol on Formation of Levulinates from Cellulosic Biomass. Ind Eng Chem Res 2015; 54(46): 11552-9.
[http://dx.doi.org/10.1021/acs.iecr.5b03512]

[125] Rackemann DW, Doherty WOS. The conversion of lignocellulosics to levulinic acid. Biofuels Bioprod Biorefin 2011; 5(2): 198-214.
[http://dx.doi.org/10.1002/bbb.267]

[126] Kang S, Fu J, Zhou N, Liu R, Peng Z, Xu Y. Concentrated Levulinic Acid Production from Sugar Cane Molasses. Energy Fuels 2018; 32(3): 3526-31.
[http://dx.doi.org/10.1021/acs.energyfuels.7b03987]

[127] Kang S, Yu J. Maintenance of a Highly Active Solid Acid Catalyst in Sugar Beet Molasses for Levulinic Acid Production. Sugar Tech 2018; 20(2): 182-93.
[http://dx.doi.org/10.1007/s12355-017-0543-5]

[128] Wang R, Xie X, Liu Y, *et al.* Facile and Low-Cost Preparation of Nb/Al Oxide Catalyst with High Performance for the Conversion of Kiwifruit Waste Residue to Levulinic Acid. Catalysts 2015; 5(4): 1636-48.
[http://dx.doi.org/10.3390/catal5041636]

[129] Chen SS, Yu IKM, Tsang DCW, *et al.* Valorization of cellulosic food waste into levulinic acid catalyzed by heterogeneous Brønsted acids: Temperature and solvent effects. Chem Eng J 2017; 327: 328-35.
[http://dx.doi.org/10.1016/j.cej.2017.06.108]

[130] Su J, Shen F, Qiu M, Qi X. High-Yield Production of Levulinic Acid from Pretreated Cow Dung in Dilute Acid Aqueous Solution. Molecules 2017; 22(2): 285.
[http://dx.doi.org/10.3390/molecules22020285] [PMID: 28216587]

[131] Ortiz ME, Bleckwedel J, Raya RR, Mozzi F. Biotechnological and in situ food production of polyols by lactic acid bacteria. Appl Microbiol Biotechnol 2013; 97(11): 4713-26.
[http://dx.doi.org/10.1007/s00253-013-4884-z] [PMID: 23604535]

[132] Williams M. The Merck Index: An Encyclopedia of Chemicals, Drugs, and Biologicals, 15th Edition Edited by M.J.O'Neil, Royal Society of Chemistry, Cambridge, UK ISBN 9781849736701; 2708 pages. Drug Dev Res. 2013 Aug;74(5):339–339.

[133] Liu C, Dong H, Zhong J, Ryu DDY, Bao J. Sorbitol production using recombinant Zymomonas mobilis strain. J Biotechnol 2010; 148(2-3): 105-12.
[http://dx.doi.org/10.1016/j.jbiotec.2010.04.008] [PMID: 20438775]

[134] M S, R J. The biotechnological production of sorbitol. Appl Microbiol Biotechnol 2002; 59(4-5): 400-8.
[http://dx.doi.org/10.1007/s00253-002-1046-0] [PMID: 12172602]

[135] Anand A, Kulkarni RD, Gite VV. Preparation and properties of eco-friendly two pack PU coatings based on renewable source (sorbitol) and its property improvement by nano ZnO. Prog Org Coat 2012; 74(4): 764-7.
[http://dx.doi.org/10.1016/j.porgcoat.2011.09.031]

[136] Acton A. Advances in Sorbitol Research and Application: 2013 Edition: ScholarlyBrief. ScholarlyEditions 2013.

[137] Palkovits R, Tajvidi K, Ruppert AM, Procelewska J. Heteropoly acids as efficient acid catalysts in the one-step conversion of cellulose to sugar alcohols. Chem Commun (Camb) 2011; 47(1): 576-8.
[http://dx.doi.org/10.1039/C0CC02263B] [PMID: 21103493]

[138] Ribeiro LS, Órfão JJM, Pereira MFR. Direct catalytic production of sorbitol from waste cellulosic materials. Bioresour Technol 2017; 232: 152-8.
[http://dx.doi.org/10.1016/j.biortech.2017.02.008] [PMID: 28222384]

[139] Kohli K, Prajapati R, Sharma B. Bio-Based Chemicals from Renewable Biomass for Integrated Biorefineries. Energies 2019; 12(2): 233.
[http://dx.doi.org/10.3390/en12020233]

[140] De Boeck R, Sarmiento-Rubiano LA, Nadal I, Monedero V, Pérez-Martínez G, Yebra MJ. Sorbitol production from lactose by engineered Lactobacillus casei deficient in sorbitol transport system and mannitol-1-phosphate dehydrogenase. Appl Microbiol Biotechnol 2010; 85(6): 1915-22.
[http://dx.doi.org/10.1007/s00253-009-2260-9] [PMID: 19784641]

[141] Silveira MM, Wisbeck E, Hoch I, Jonas R. Production of glucose-fructose oxidoreductase and ethanol by Zymomonas mobilis ATCC 29191 in medium containing corn steep liquor as a source of vitamins. Appl Microbiol Biotechnol 2001; 55(4): 442-5.

[http://dx.doi.org/10.1007/s002530000569] [PMID: 11398924]

[142] Bagnato G, Iulianelli A, Sanna A, Basile A. Glycerol Production and Transformation: A Critical Review with Particular Emphasis on Glycerol Reforming Reaction for Producing Hydrogen in Conventional and Membrane Reactors. Membranes (Basel) 2017; 7(2): 17.
[http://dx.doi.org/10.3390/membranes7020017] [PMID: 28333121]

[143] Tan HW, Abdul Aziz AR, Aroua MK. Glycerol production and its applications as a raw material: A review. Renew Sustain Energy Rev 2013; 27: 118-27.
[http://dx.doi.org/10.1016/j.rser.2013.06.035]

[144] Hoque ME, Singh A, Chuan YL. Biodiesel from low cost feedstocks: The effects of process parameters on the biodiesel yield. Biomass Bioenergy 2011; 35(4): 1582-7.
[http://dx.doi.org/10.1016/j.biombioe.2010.12.024]

[145] R.g SW, Nomura N, Sato S, Matsumura M. Pre-treatment and utilization of raw glycerol from sunflower oil biodiesel for growth and 1,3-propanediol production by Clostridium butyricum. J Chem Technol Biotechnol 2008; 83(7): 1072-80.
[http://dx.doi.org/10.1002/jctb.1917]

[146] Shimada Y, Watanabe Y, Samukawa T, *et al.* Conversion of vegetable oil to biodiesel using immobilized *Candida antarctica* lipase. J Am Oil Chem Soc 1999; 76(7): 789-93.
[http://dx.doi.org/10.1007/s11746-999-0067-6]

[147] Wang Z, Zhuge J, Fang H, Prior BA. Glycerol production by microbial fermentation. Biotechnol Adv 2001; 19(3): 201-23.
[http://dx.doi.org/10.1016/S0734-9750(01)00060-X] [PMID: 14538083]

[148] Kurbanoglu EB, Kurbanoglu NI. Utilization as peptone for glycerol production of ram horn waste with a new process. Energy Convers Manage 2004; 45(2): 225-34.
[http://dx.doi.org/10.1016/S0196-8904(03)00148-1]

[149] Mostafa NA, Magdy YH. Utilization of molasses and akalona hydrolyzate for continuous glycerol production in a packed bed bioreactor. Energy Convers Manage 1998; 39(7): 671-7.
[http://dx.doi.org/10.1016/S0196-8904(97)00052-6]

[150] Awasthi MK, Kumar V, Yadav V, *et al.* Current state of the art biotechnological strategies for conversion of watermelon wastes residues to biopolymers production: A review. Chemosphere 2022; 290133310.
[http://dx.doi.org/10.1016/j.chemosphere.2021.133310] [PMID: 34919909]

[151] Liu H, Kumar V, Jia L, *et al.* Biopolymer poly-hydroxyalkanoates (PHA) production from apple industrial waste residues: A review. Chemosphere 2021; 284131427.
[http://dx.doi.org/10.1016/j.chemosphere.2021.131427] [PMID: 34323796]

[152] Awasthi SK, Kumar M, Kumar V, *et al.* A comprehensive review on recent advancements in biodegradation and sustainable management of biopolymers. Environ Pollut 2022; 307119600.
[http://dx.doi.org/10.1016/j.envpol.2022.119600] [PMID: 35691442]

[153] Duan Y, Tarafdar A, Kumar V, *et al.* Sustainable biorefinery approaches towards circular economy for conversion of biowaste to value added materials and future perspectives. Fuel 2022; 325124846.
[http://dx.doi.org/10.1016/j.fuel.2022.124846]

[154] Kumar V, Sharma N, Umesh M, *et al.* Emerging challenges for the agro-industrial food waste utilization: A review on food waste biorefinery. Bioresour Technol 2022; 362127790.
[http://dx.doi.org/10.1016/j.biortech.2022.127790] [PMID: 35973569]

[155] Awasthi SK, Sarsaiya S, Kumar V, *et al.* Processing of municipal solid waste resources for a circular economy in China: An overview. Fuel 2022; 317123478.
[http://dx.doi.org/10.1016/j.fuel.2022.123478]

[156] Kumar V, Sharma N, Maitra SS. *In vitro* and *in vivo* toxicity assessment of nanoparticles. Int Nano Lett 2017; 7(4): 243-56.

[http://dx.doi.org/10.1007/s40089-017-0221-3]

[157]   Vinay K, Neha S, Maitra S. Protein and Peptide Nanoparticles: Preparation and Surface Modification, in Functionalized Nanomaterials I. CRC Press 2020; pp. 191-204.

[158]   Vallinayagam S, *et al.* Recent developments in magnetic nanoparticles and nano-composites for wastewater treatment. J Environ Chem Eng 2021; 9(6)106553.
[http://dx.doi.org/10.1016/j.jece.2021.106553]

[159]   Kumar V, Sharma N, Lakkaboyana SK, Maitra SS. Silver nanoparticles in poultry health: Applications and toxicokinetic effects, in Silver Nanomaterials for Agri-Food Applications. Elsevier 2021; pp. 685-704.
[http://dx.doi.org/10.1016/B978-0-12-823528-7.00005-6]

[160]   Egbosiuba T C. Biochar and bio-oil fuel properties from nickel nanoparticles assisted pyrolysis of cassava peel Heliyon, vol 8, no 8, p e10114,2022/08/01/ 2022.
[http://dx.doi.org/10.1016/j.heliyon.2022.e10114]

[161]   Saygün A, Şahin-Yeşilçubuk N, Aran N. Effects of Different Oil Sources and Residues on Biomass and Metabolite Production by Yarrowia lipolytica YB 423-12. J Am Oil Chem Soc 2014; 91(9): 1521-30.
[http://dx.doi.org/10.1007/s11746-014-2506-2]

Waste Valorization, 2023, Vol. 1, 147-160

CHAPTER 6

Use of Date Palm Fruit Processing Wastes to Produce High-Value Products

Shefali Patel[1,*], **Susmita Sahoo**[1], **Vinay Kumar**[2], **Sivarama Krishna Lakkaboyana**[3] and **Ritu Pasrija**[4]

[1] *Department of Biological and Environment Sciences, N. V. Patel College of Pure and Applied Sciences, V. V. Nagar, Gujarat – 388120, India*

[2] *Department of Community Medicine, Saveetha Medical College and Hospital, Saveetha Institute of Medical and Technical Sciences (SIMATS), Chennai, Thandalam-602105, India*

[3] *Vel Tech Rangarajan Dr. Sagunthala R&D Institute of Science and Technology,Chennai, Tamil Nadu, 600062, India*

[4] *Department of Biochemistry, Maharshi Dayanand University, Rohtak, Haryana-121002, India*

Abstract: Fruits of the date have found great value in human nutrition because of their rich content of essential nutrients. Tons of palm fruit waste are being discarded daily. Waste such as date holes represents 10% of date fruit. Within the framework of the bio-economy, there is a high potential for date waste use in ligne-cellulosic products in a broad spectrum of bio-industries. Extensive and varied biomolecules may capture energy for use in the pharmaceutical industry as an active pharmaceutical ingredient (API), or in the development of nutraceuticals without using them as substrates for mass production of bacteria, phenolic, sterols, carotenoids, anthocyanins., procyanidin, flavonoids, minerals, various vitamins, economically beneficial amino acids, organic acids, biosurfactants, biopolymers, biofuels, exopolysaccharides, probiotics with date flavors, *etc.* Date fruits are commonly used to prepare many kinds of products such as date juice concentrate (distribution, syrup, and liquid sugar), date products (wine, alcohol, vinegar, organic acids) and date pastes for different uses (*e.g.*, bakery and -confectionery) without the direct use. Date seeds can be converted into high-value liquids (bio-oil), gas, and solid products (bio-char) by pyrolysis, and coal and activated carbon can be produced from date seeds. Significant progress has been made in developing specific date fruit products and using products from packaging and processing. Additional economic benefits will also increase so far as farmers increase the number of commodities they produce, as well as diversify their sources of income.

Keywords: Date fruit waste, Industrial and medical applications, Traditional use.

* **Corresponding author Shefali Patel:** Department of Biological and Environment Sciences, N. V. Patel College of Pure and Applied Sciences, V. V. Nagar, Gujarat, India; E-mail: shefalipatel1312@gmail.com

INTRODUCTION

Date palm is undoubtedly the world's oldest cultivated tree [1], with a history dating back 10,000 years, and is one of the most widely cultivated trees. It is an ancient grown crop in tropical and subtropical regions, as its production, use and industrial development are increasing mainly in Arabia. Conquerors brought palms to the conquered lands for example, Alexander to western India (now Pakistan) and the Moors to Spain [2]. Next, traders and explorers spread the word to other lands, including Mexico and North America. Palm tree is currently planted across a large belt that covers most of the ancient regions, 8,000 miles from east to west and 2000 Km from north to south [1]. North Africa and South Asia are the major producers of date and global date production is increasing day by day in recent decades [3]. Date fruit is considered as an essential source of livelihood in the desert. This is due to its various useful properties. The palm tree has a special place in economic and social life. Some of the best products based on palm tree residues are now available and affordable. These developments have led to the gradual use of palm fossils in manufacturing traditional, handmade products. In addition, palm plantations have increased recently. The processing of palm is a cost-intensive process that generates a huge amount of waste with no economical applications [4]. Therefore, it would be helpful to develop an effective and economical way to use these dates in producing value-added fermented products. The presented chapter discusses the use of date palm for value-added products such as antibiotics, organic acids, biofuels, *etc*.

Arecaceae is the palm family composed of several genera and thousands of species [5, 6]. Five significant varieties of palm are cultivated as commercial varieties. These include palm, coconut palm, palm oil, nut palm, and areca palm. The oasis area of the date palm is different from the other four species, which are wet palm trees in tropical areas. All five of them were cultivated mainly for their fruit. In the subsistence economy of the farm, all parts of the five palm trees would be carefully processed to obtain any utility or product. Considering the need for survival in the arid region, the palm tree can regenerate the burned parts lost in a fire accident. This ability may have led to the naming of the palm tree by the mythical bird Phoenix, which is said to have lived for 500 years and to have risen with renewed vigor after being burned to ashes [7]. Fruits of different dates weigh 2 to 60 g, 8 to 110 mm long, 8 to 32 mm wide, and yellow to black Table **1** presents the top date producers in the world.

The development of date fruit can be classified into three steps Rutab, Tamr and Khalal. Dates are usually harvested in the fully mature Tamr phase, following the formation of the Total Soluble Solids (TSS) 60–70. Brix is consumed at this phase. Rutab and Tamr are the ripe and fully ripe stages Therefore, the fruit can be eaten at these stages without processing. The large amount of waste produced from the Kabkab date can be used to produce syrup [8]. There are various reports available for palm oil cultivation. But most of the studies are focused on the pharmacology and chemistry of date fruit [9]. Date palm has a range of health benefits, making it a great fruit [10]. The disposal rate of the date palm industries is too high in various countries [11]. These large figures, which are obtained annually in a sustainable manner, provide ample opportunities for the emergence of new bio-entrepreneurs and commercial entrepreneurs in developing countries to fully use palm in addition to better management of palm fruit waste. The date processing wastes can be used to produce various valuable products [10].

Table 1. The significant producers countries of date fruit [3].

Country name	Total annual production per 1000 metric tons
Egypt	1502
Algeria	848
Iran	1084
Iraq	676
Saudi Arabia	1065
Pakistan 527	-
South Sudan	432
Sudan	438
United Arab Emirates	245
Oman	269

Price of Nutrition and Organic Chemicals

Date contains many other vitamins and minerals and is considered a complete diet [1]. Currently, a small portion of waste is used as animal feed [12]. They are widely used and can be used as food for future generations because of their fantastic nutrition, health, and economy. Date contains a high amount of sugar, magnesium, potassium, calcium and vitamins [13]. In addition, it contains various fatty acids [14]. Due to its high sugar content, it can be stored in dried form for longer periods. It has beneficial properties such as anti-inflammatory, antioxidant and antimicrobial [15]. In recent years, the dates have attracted much attention because of their several health benefits [9]. Table **2** presents the date flesh and date seed chemical composition.

Table 2. Date seed and flesh chemical composition.

Nutrient components	Composition flesh of date	Composition of fresh date pits	Amount of date pits	Amount of date pits (roasted)
Fat	0.5-3.3	5.7-8.8	10.2-12.7	8.1
Ash	1.4-2.6	0.8-1.1	1.1-1.2	1.0
Protein	1.1-3.0	4.8-6.9	5.2-5.6	7.1
Carbohydrate	72.85-85.0	2.4-4.7	81.0-83.1	62.3
Moisture	9.7-17.7	8.6-12.5	-	-
Dietary fiber	5.9-18.4	67.6-74.2	-	-

Dietary / Diet Supplements: The dates are soft and dry and can be eaten by hand. They can be packed or stuffed like other dry fruits. In the United States during the holidays, a cake called date nut bread is much famous. The dates can also be cut into small cubes and pasted as honey, *etc.* Daily vinegar is a traditional Middle Eastern product [16]. Nutritious stock food can be made such as dehydrated and digested dates with cereals.

Dry or soft dates are eaten with hand, or they can be canned and stuffed such as almonds, walnuts, pecans, candied oranges, and peel a lemon, tahini, marzipan, or cream ushizi. Dates are also processed into cubes, pastes labeled "'frozen", spread, date syrup, or "honey" called "dibs" or "rub" in Libya, powder (date sugar), vinegar, or alcohol. Dates can also be dehydrated, digested, and mixed with cereals to create nutritious stock food. Dry dates are fed to camels, horses, and dogs in the Sahara. In Israel, date syrup is used to cook chicken, sweets, and desserts [16].

Waste Management-Current Practice and Problems

Date processing and Palm oil industries produce date press cakes and pits.

Due to their insufficient (very soft) composition, lost dates, commonly referred to as products of the dates, are inedible [12]. Within the agricultural systems, the dates fall from the palms of the hands before ripening [8, 17].

Date Palm Fruits Processing Products and By-Products

Dates are used to prepare products such as syrups, sugars, spreads, alcohol, wine, organic acids, vinegar, *etc.* The date products can be used in digestible food or as a thickener. Juice extraction process of syrup was widely studied by researchers [8, 18] (Fig. **1**).

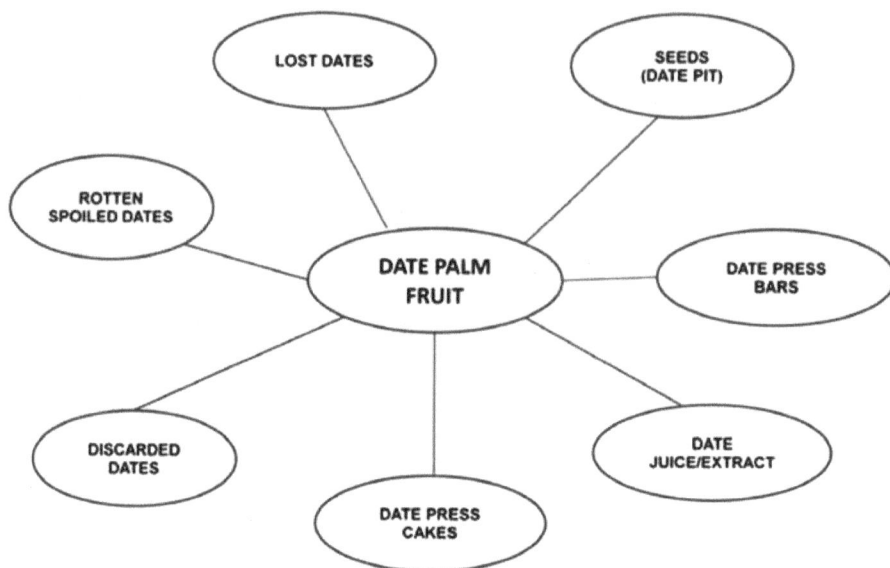

Fig. (1). Products obtained from the date palm processing [50].

After processing, the material contains about 23% humidity and too low water activity [10]. Date products are used to produce sugar substitutes [19]. The syrup of the dates (dibs), which is the primary and standard product, is used in the preparation of foods such as jam, marmalade, strong drinks, and chocolate [20]. Dates are stored in a bag that accumulates for several months to produce syrup. In industry, a large portion of sugar is removed. It is produced at home by extracting boiling the juice [3].

Few products are manufactured and produced by chemical conversion. Polyols have many active hydroxyl groups available to combine with organisms and are often added to food due to their low calories [21]. A study promoted the use of daily pit oil as a non-traditional oil in mayonnaise products [22].

Dates and their products contain dense nutrients fermentation products. Table **3** presents some critical fermentation products originating from dates.

Table 3. Date products and by-Products.

Product	Microorganisms	Substrate	References
Biopolymers	-	-	-
Xanthan gum	*X. campestris* NRRLB-1459	Date palm juice	Salah et. al. (2010) [23]

(Table 3) cont.....

Product	Microorganisms	Substrate	References
	X. campestris	Date syrup	Besbes et. al. (2006) [24]
Poly 3-hydroxybutyrate	*Bacillus megaterium*	Date syrup	Nansib et. al. (2001) [25]
Curdlan	*Rhizobium radiobacter* ATCC 6466	Date palm juice	Salah et. al. (2010) [23]
Carotenoid	*Lactobacillus planetarium* Q53	Date syrup	Elsanhoty et. al. (2012) [26]
Biofuel	-	-	-
Ethanol	*Saccharomyces cerevisae* ATCC 36858	Date extract	Gupta and Kushwaha (2011) [28]
	S. cerevisiae STAR brand	-	Gupta and Kushwaha (2011) [28]
	S. cerevisae SDB	Date waste	Acourene and Ammouche (2012) [29]
Hydrogen	*E. Coli* EGY	Rotten date	Abd-Alla et. al. (2011) [27]
	Clostridium acetobutylicum ATCC 824	-	Abd-Alla et. al. (2011) [27]
	Rhodobacter capsulatus DSM 1710	-	Abd-Alla et. al. (2011) [27]
Butanol	*Clostridium acetobutylicum* ATCC 824	Spoiled date fruit	Abd-Alla and El-Enany (2012) [30]
	Bacillus subtilis DSM 4451	-	Abd-Alla and El-Enany (2012) [30]
Biosurfactant	*Bacillus subtilis* DSM 20B	Date molasses	Al-bahry et. al. (2013) [31]
Amino acids	-	-	-
Glutamic acid	*Corynebacterium glutamicum* CECT 690 & CECT77	Date sugars	Demirbas (2017) [34]
	-	-	-
Organic acids	-	-	-
Citric acid	*Aspergillus niger* ATCC 6275 & 9642	Date extract/molasses	Mehaia and Cheryan (1991) [32]
	niger ANSS –B5	Date waste	Acourene and Amouche (2012) [29]

(Table 3) cont.....

Product	Microorganisms	Substrate	References
Lactic acid	*Lactobacillus casei*	Date fruit waste	Nancib et. al. (2001) [25]
	Lactobacilus delbrucki	Date fruit waste	Yadav et. al. (2011) [33]
Antibiotics	-	-	-
Bleomycin	*Streptomyces mobaraensis*	Date syrup	Manickavasagan et. al. (2012) [36]
Oxytetracycline	*Streptomyces rimosus*	Date-Coatsugar extract; Date-seed hydrolysate	Abou-Zeid et. al. (1993) [17]
Biomass	-	-	-
Baker's yeast	*Saccharomyces cerevisae* (I) *S. dastorianus* NRRL YI2693 *S. cerevisae* (III) *S. dayanus* NRRL Y-12624,	Date Sugars	Al-Jasass et. al. (2010) [35]
Enzymes	-	-	-
Pectinase	Bacillus subtilis EFRL 01	Date syrup	Qureshi et. al. (2008) [37]
Alpha amylase	*Candicla guiliemondii* CGL-A10	Dae waste	Acourene and Amouche (2012) [29]
Probiotics	-	-	-
Lactobacilli	*Lactobacillus casei* ATCC 334	Date powder	Sani et. al. (2016) [16]; Hartati et. al. (2012) [49]

Biopolymers

Xanthan gum: Xanthum gum is one of the products produced from date fruit [24]. The major problem in xanthan gum production is the cost of substrates. Therefore, a search for a low-cost substrate for xanthan gum production is required. The date palm's by-product can be considered an alternative carbon feedstock substrate [24].

Poly (3-hydroxybutyrate)(PHA): PHAs are alternatives to plastics that can change petroleum-based plastics. *Bacillus megaterium* can be used for PHA production using beet molasses or date syrup. It was noted that when dietary syrup and beet molasses are used alone without any additional nitrogen source, the bacterium produced approximately 3 gL-1 cell density and a PHB content of

50% (w / w). In addition, NH_4NO_3 supplementation followed by ammonium acetate and then NH_4Cl was found to support cell growth to 4.8 gL-1, while PHB concentrations increased with NH_4Cl followed by ammonium acetate, NH_4NO_3, and $(NH_4)_2SO_4$ to PHB content approximately 42% (w / w) [12].

Curdlan: It is a polysaccharide with high molecular weight and contains C-(1,3) glucose linkages. It can be produced by a bacterium called Agrobacterium biovar.The bacterium contains β- (1,3) -descriptive glucose residues and has an unusual ability to form a gel that expands when liquid suspension heats up. It is a dietary supplement used as a stabilizing agent. Curdlan can also be produced from palm juice using bacterium such as *Rhizobium radiobacter* ATCC 6466 [10].

Carotenoids: Proper modification of intermediate parts can be done using date palms [26]. It was observed that the production of carotenoids can be increased by adding syrup with a sugar concentration of 5% [10].

Biosurfactants: Molasses can be used as a substrate to produce the biosurfactant [31]. It was observed that the surfactant produced could reduce the friction and surface tension between the surfaces. The surfactant does not have significant stability in different pH ranges and temperatures. Biosurfactants have the potential to produce oil which was evidenced in flood studies [10].

Biofuels

Hydrogen Production:Hydrogen produced from rotten dates through successive fermentation of three phases was studied using three different strains [30]. In a study, *Escherichia coli* EGY was used in three stages for hydrogen production. In the first stage, oxygen was absorbed to maintain anaerobic conditions. In the second stage, hydrogen is produced with anaerobic *Clostridium acetobutylicum* ATCC 824. In the last stage, fermentation with *Rhodobacter capsulatus* DSM 1710 was carried out. The hydrogen production was maximum with sucrose at 5 g/L. The results suggest that rotten dates are more effective in hydrogen production.

Bio-ethanol, acetone, and Butanol: The most widely used compounds are the bio-energy liquids such as ethanol and butanol. Three main feedstock categories are used to produce ethanol sugar, starch, and lignocellulose. Date palm sap is very nutritious and high in sugar. Therefore, it was used for ethanol production [38]. It has been observed that S. cerevisiae ATCC 36858 can selectively produce ethanol from glucose with the nominal conversion of fructose [28]. It can be produced using different resources [44 - 49]. In strict anaerobic conditions, *Clostridium* spp. produces butanol with traces of ethanol and acetone [37]. Treatment of feedstock produce the raw material for production [50 - 54]. Dried date fruit has been tested as a potential component of fermentation to produce butanol, acetone, and ethanol using mixed cultures of *Bacillus subtilis* DSM 4451 and *C. Acetotylicum* ATCC 824 [30]. In the growth period, the mixed culture utilized all the components, such as amino acids, sugar, and NPK. The aerobic bacterium, *B. subilis* DSM 5541, can be used to maintain the anaerobic condition. Ammonium acetate or yeast extract was added to the fruit of the damaged homogenate date and it significantly increased the production rate of ABE. The study demonstrated that rotten dates can be used as feedstock without any pretreatment or adding any reducing agent [30]. Fruits can also be used to produce bioethanol by *C. acetobutylicum* NCIMB 13357 [39].

Theoretical production of bioethanol was obtained at 261.5 kg/ton of date palms. Sugar production cost from date palms is cheaper than from sugar feedstocks. Vegetable oil in dates ranges from 5 to 13. Date seeds contain biodiesel. The seed of the product contains varying ranges of moisture, carbon, and ash content. Pyrolysis of the seed can produce gas, liquids, and biochar. Chemicals and phenolics can be produced by producing pyrolysis oils [40].

Antibiotics

Bleomycin (BLM)

Streptomyces mobaraensis was used to produce bleomycin using date syrup as a carbon source. The polynomial regression improved the concentration and fluid formation. Compared with glucose, date syrup can produce a high amount of antibiotics. Soybean acts as an excellent source of nitrogen. Therefore, date syrup can act as an additional carbon source for increased bleomycin production [10].

Oxytetracycline (OTC)

In the fermentation medium, cheaper carbon sources such as seed hydrolysate, sugar extract, and seed ash can produce antibiotics such as oxytetracycline by using *Streptomyces rimosus*. In the process, the sugar extract worked as the carbon source, while the seed hydrolysate acted as the nitrogen source. The seed ash replaced the nutrients such as iron, zinc, manganese, and magnesium in the culture medium. In the natural environment, the date materials can be an excellent source to produce oxytetracycline using *S. rimosus* [10].

Organic Acids

Citric acid: Can be produced using strains of *Aspergillus* such as *Aspergillus* ATCC 9642 and 6375 in a nutrient medium that contains molasses supplemented with methanol, whey and tricalcium phosphate. A study utilized date syrup to produce citric acid in immobile and free cell systems. [38]. Citric acid production by *Aspergillus niger* ANSS-B5 was influenced by various parameters such as sugar content, temperature, and concentration of methanol. Therefore, the citric acid production under waste syrup fermentation was enhanced [29].

Lactic acid: Although a few substrates are tested for lactic acid production. It was observed in lactic acid production that yeast extract acts as an essential source of nitrogen. But to replace the yeast extract, various nitrogen sources were tested. In this context, date syrup has been used as a source of nitrogen to produce lactic acid, and the yield was comparable to that of yeast extract as a supplement [25]. The combinatorial nitrogen source and date juice can also produce citric acid [41]. Lactic acid was produced by using the Burman design using the bacterium *Lactobacillus* sp.KCP01 [42]. Lactic acid is produced by the bacterium *Lactobacillus delbruckii,* and it was analyzed in a productive manner consisting of 10% of the carbon source found in sweet cereals [33].

CONCLUSION

Date palm is considered one of the most nutritious foods. It has been known for its various health benefits. In addition to that, the date palm has been popular to produce high-value products. The date palm can produce various value-added products such as biopolymers, carotenoids, biofuel, ethanol, hydrogen, organic acids, *etc*. There are various processes where different parts of date palm can be used. Therefore, date palm presents an essential feedstock for producing value-added products.

REFERENCES

[1]   Zaid A, de Wet PF. Date palm cultivation. Abdelouahhab Zaid (ed.) Chapter II. FAO Plant production and protection 2002

[2]   Al-Hooti S, Sidhu JS, Qabazard H. Physicochemical characteristics of five date fruit cultivars grown in the United Arab Emirates. Plant Foods Hum Nutr 1997; 50(2): 101-13.
[http://dx.doi.org/10.1007/BF02436030] [PMID: 9201745]

[3]   FAO, Food and Agriculture Organization statistical database (FAOSTAT), Retrieved from http://faostat3.fao.org/, 2013

[4]   Besbes S, Drira L, Blecker C, Deroanne C, Attia H. Adding value to hard date (Phoenix dactylifera L.): Compositional, functional and sensory characteristics of date jam. Food Chem 2009; 112(2): 406-11.
[http://dx.doi.org/10.1016/j.foodchem.2008.05.093]

[5]   Govaerts R, Dransfield J. World checklist of palms. Royal Botanic Gardens 2005.

[6]   Dransfield J, Uhl NW, Lange CBA, Baker WJ, Harley MM, Lewis CE. Genera Palmarum: the evolution and classification of palms. Kew Publishing 2008.

[7]   Johnson DV. Enhancement of Date Palm As A Source of Multiple Products: Examples from Fromother Industrialized Palms. Emir J Food Agric 2012; 408-17.

[8]   Al-Hooti SN, Sidhu JS, Al-Saqer JM, Al-Othman A. Chemical composition and quality of date syrup as affected by pectinase/cellulase enzyme treatment. Food Chem 2002; 79(2): 215-20.
[http://dx.doi.org/10.1016/S0308-8146(02)00134-6]

[9]   Vayalil PK. Date fruits (Phoenix dactylifera Linn): an emerging medicinal food. Crit Rev Food Sci Nutr 2012; 52(3): 249-71.
[http://dx.doi.org/10.1080/10408398.2010.499824] [PMID: 22214443]

[10]   Chandrasekaran M, Bahkali AH. Valorization of date palm (Phoenix dactylifera) fruit processing by-products and wastes using bioprocess technology – Review. Saudi J Biol Sci 2013; 20(2): 105-20.
[http://dx.doi.org/10.1016/j.sjbs.2012.12.004] [PMID: 23961227]

[11]   Badawi MA. Production of Biochar from Date Palm Fronds and its Effects on Soil Properties By-Products of Palm Trees and Their Applications 2019; 11-159.

[12]   Besbes S, Blecker C, Deroanne C, Lognay G, Drira NE, Attia H. Quality characteristics and oxidative stability of date seed oil during storage. Food Sci Technol Int 2004; 10(5): 333-8.
[http://dx.doi.org/10.1177/1082013204047777]

[13]   Lambiote B. Some aspects of the role of dates in human nutrition. In Proceedings of the first international symposium on date palm. Saudi Arabia: King Faisal University. 1982; pp.16.

[14]   Al-shahib W, Marshall RJ. The fruit of the date palm: its possible use as the best food for the future? Int J Food Sci Nutr 2003; 54(4): 247-59.
[http://dx.doi.org/10.1080/09637480120091982] [PMID: 12850886]

[15]   Baliga MS, Baliga BRV, Kandathil SM, Bhat HP, Vayalil PK. A review of the chemistry and pharmacology of the date fruits (Phoenix dactylifera L.). Food Res Int 2011; 44(7): 1812-22.
[http://dx.doi.org/10.1016/j.foodres.2010.07.004]

[16]   Sani MN, Abdulkadir F, Salim FB, Abubakar MM, Kutama AS. Date Palm (Phoenix dactylifera) as a food supplement and antimicrobial Agent in the 21st Century – A review. IOSR Journal of Pharmacy and Biological Sciences (IOSR-JPBS) 2016; 11(4): 46-51.
[http://dx.doi.org/10.9790/3008-1104034651]

[17]   Abou-zied MM. Biological studies on some bivalves from the Suez Canal 1991.

[18]   Ramadan BR. Biochemical, nutritional, and technological studies on dates. Assiut University 1995.

[19]   Alhamdan AM, Hassan BH. Water sorption isotherms of date pastes as influenced by date cultivar and storage temperature. J Food Eng 1999; 39(3): 301-6.
[http://dx.doi.org/10.1016/S0260-8774(98)00170-8]

[20]   Al-Showiman SS. Date, food and health. Qassim, Saudi Arabia: Dar Al-Khareji Press 1998.

[21]   Briones R, Serrano L, Younes RB, Mondragon I, Labidi J. Polyol production by chemical modification of date seeds. Ind Crops Prod 2011; 34(1): 1035-40.
[http://dx.doi.org/10.1016/j.indcrop.2011.03.012]

[22]   Basuny AMM, AL-Marzooq MA. Production of mayonnaise from date pit oil. Food Nutr Sci 2011; 2(9): 938-43.
[http://dx.doi.org/10.4236/fns.2011.29128]

[23]   Salah RB, Chaari K, Besbes S, Blecker C, Attia H. Production of xanthan gum from *Xanthomonas campestris* NRRL B-1459 by fermentation of date juice palm by-products (Phoenix dactylifera L.). J Food Process Eng 2011; 34(2): 457-74.
[http://dx.doi.org/10.1111/j.1745-4530.2009.00369.x]

[24] Besbes S, Cheikh Rouhou S, Blecker C, Derouanne C, Lognay G, Drira NE. Voies de valorisation des sous produits de dattes: Valorisation de la pulpe. Microbiologie Hygiène Alimentaire 2006; 18: 3-7.

[25] Nancib N, Nancib A, Boudjelal A, Benslimane C, Blanchard F, Boudrant J. The effect of supplementation by different nitrogen sources on the production of lactic acid from date juice by Lactobacillus casei subsp. rhamnosus. Bioresour Technol 2001; 78(2): 149-53.
[http://dx.doi.org/10.1016/S0960-8524(01)00009-8] [PMID: 11333033]

[26] Elsanhoty RM, Al-Turki IA, Ramadan MF. Screening of medium components by Plackett–Burman design for carotenoid production using date (*Phoenix dactylifera*) wastes. Ind Crops Prod 2012; 36(1): 313-20.
[http://dx.doi.org/10.1016/j.indcrop.2011.10.013]

[27] Abd-Alla MH, Morsy FM, El-Enany AWE. Hydrogen production from rotten dates by sequential three stages fermentation. Int J Hydrogen Energy 2011; 36(21): 13518-27.
[http://dx.doi.org/10.1016/j.ijhydene.2011.07.098]

[28] Gupta N, Kushwaha H. Date Palm as a Source of Bioethanol Producing Microorganisms.Springer Netherlands Date Palm Biotechnology. 2011; pp. 711-27.
[http://dx.doi.org/10.1007/978-94-007-1318-5_33]

[29] Acourene S, Ammouche A. Optimization of ethanol, citric acid, and α-amylase production from date wastes by strains of *Saccharomyces cerevisiae*, *Aspergillus niger*, and *Candida guilliermondii*. J Ind Microbiol Biotechnol 2012; 39(5): 759-66.
[http://dx.doi.org/10.1007/s10295-011-1070-0] [PMID: 22193823]

[30] Abd-Alla MH, Elsadck El-Enany A-W. Production of acetone-butanol-ethanol from spoilage date palm (Phoenix dactylifera L.) fruits by mixed culture of *Clostridium acetobutylicum* and *Bacillus subtilis*. Biomass Bioenergy 2012; 42: 172-8.
[http://dx.doi.org/10.1016/j.biombioe.2012.03.006]

[31] Al-Bahry SN, Al-Wahaibi YM, Elshafie AE, *et al.* Biosurfactant production by *Bacillus subtilis* B20 using date molasses and its possible application in enhanced oil recovery. Int Biodeterior Biodegradation 2013; 81: 141-6.
[http://dx.doi.org/10.1016/j.ibiod.2012.01.006]

[32] Mehaia MA, Cheryan M. Fermentation of date extracts to ethanol and vinegar in batch and continuous membrane reactors. Enzyme Microb Technol 1991; 13(3): 257-61.
[http://dx.doi.org/10.1016/0141-0229(91)90138-Z]

[33] Yadav AK, Bipinraj NK, Chaudhari AB, Kothari RM. Production of L(+) lactic acid from sweet sorghum, date palm, and golden syrup as alternative carbon sources. Stärke 2011; 63(10): 632-6.
[http://dx.doi.org/10.1002/star.201100006]

[34] Demirbas A. Utilization of date biomass waste and date seed as bio-fuels source. Energy Sources A Recovery Util Environ Effects 2017; 39(8): 754-60.
[http://dx.doi.org/10.1080/15567036.2016.1261208]

[35] Al-Jasass FM, Al-Eid SM, Ali SHH. A comparative study on date syrup (dips) as the substrate for the production of baker's yeast (Saccharomyces cerevisiae). Acta Hortic 2010; 8(882): 699-704.
[http://dx.doi.org/10.17660/ActaHortic.2010.882.76]

[36] Manickavasagan A, Essa MM, Sukumar E. Dates: production, processing, food, and medicinal values. 2012.
[http://dx.doi.org/10.1201/b11874]

[37] Qureshi N, Ezeji TC, Ebener J, Dien BS, Cotta MA, Blaschek HP. Butanol production by *Clostridium beijerinckii*. Part I: Use of acid and enzyme hydrolyzed corn fiber. Bioresour Technol 2008; 99(13): 5915-22.
[http://dx.doi.org/10.1016/j.biortech.2007.09.087] [PMID: 18061440]

[38] Besbes S, Hentati B, Blecker C, Derouanne C, Lognay G, Drira NE. Voies de valorisation des sous

produits de dattes: Valorisation du noyau. Microbiologie Hygiène Alimentaire 2005; 18: 3-11.

[39]   Khamaiseh EI, Olujimi Dada , Ibrahim El-Shawabkeh , Ibrahim ES, Wan M. Date fruit as a carbon source in RCM-Modified medium to produce biobutanol by *Clostridium acetobutylicum* NCIMB 13357. J Appl Sci (Faisalabad) 2012; 12(11): 1160-5.
[http://dx.doi.org/10.3923/jas.2012.1160.1165]

[40]   Gaily MH, Sulieman AK, Zeinelabdeen MA, AlZahrani SM, Atiyeh HK, Abasaeed AE. The effects of activation time on the production of fructose and bioethanol from date extract. Afr J Biotechnol 2012; 11(33): 8212-7.
[http://dx.doi.org/10.5897/AJB12.082]

[41]   Nancib A, Nancib N, Meziane-Cherif D, Boubendir A, Fick M, Boudrant J. Joint effect of nitrogen sources and B vitamin supplementation of date juice on lactic acid production by *Lactobacillus casei* subsp. rhamnosus. Bioresour Technol 2005; 96(1): 63-7.
[http://dx.doi.org/10.1016/j.biortech.2003.09.018] [PMID: 15364082]

[42]   Chauhan K, Trivedi U, Patel K. Statistical screening of medium components by Plackett–Burman design for lactic acid production by *Lactobacillus sp.* KCP01 using date juice. Bioresour Technol 2007; 98(1): 98-103.
[http://dx.doi.org/10.1016/j.biortech.2005.11.017] [PMID: 16386897]

[43]   Hartati AI, Pramono YB, Legowo AM. Lactose and reduction sugar concentrations, pH, and the sourness of date flavored yogurt drink as a probiotic beverage. J Appl Food Technol 2012; 1: 1-3.

[44]   Awasthi MK, Kumar V, Yadav V, *et al.* Current state of the art biotechnological strategies for conversion of watermelon wastes residues to biopolymers production: A review. Chemosphere 2022; 290133310.
[http://dx.doi.org/10.1016/j.chemosphere.2021.133310] [PMID: 34919909]

[45]   Liu H, Kumar V, Jia L, *et al.* Biopolymer poly-hydroxyalkanoates (PHA) production from apple industrial waste residues: A review. Chemosphere 2021; 284131427.
[http://dx.doi.org/10.1016/j.chemosphere.2021.131427] [PMID: 34323796]

[46]   Awasthi SK, Kumar M, Kumar V, *et al.* A comprehensive review on recent advancements in biodegradation and sustainable management of biopolymers. Environ Pollut 2022; 307119600.
[http://dx.doi.org/10.1016/j.envpol.2022.119600] [PMID: 35691442]

[47]   Duan Y, Tarafdar A, Kumar V, *et al.* Sustainable biorefinery approaches towards circular economy for conversion of biowaste to value added materials and future perspectives. Fuel 2022; 325124846.
[http://dx.doi.org/10.1016/j.fuel.2022.124846]

[48]   Kumar V, Sharma N, Umesh M, *et al.* Emerging challenges for the agro-industrial food waste utilization: A review on food waste biorefinery. Bioresour Technol 2022; 362127790.
[http://dx.doi.org/10.1016/j.biortech.2022.127790] [PMID: 35973569]

[49]   Awasthi SK, Sarsaiya S, Kumar V, *et al.* Processing of municipal solid waste resources for a circular economy in China: An overview. Fuel 2022; 317123478.
[http://dx.doi.org/10.1016/j.fuel.2022.123478]

[50]   Kumar V, Sharma N, Maitra SS. In vitro and *in vivo* toxicity assessment of nanoparticles. Int Nano Lett 2017; 7(4): 243-56.
[http://dx.doi.org/10.1007/s40089-017-0221-3]

[51]   Vinay K, Neha S, Maitra S. Protein and Peptide Nanoparticles: Preparation and Surface Modification, in Functionalized Nanomaterials I. CRC Press 2020; pp. 191-204.

[52]   Vallinayagam S, *et al.* Recent developments in magnetic nanoparticles and nano-composites for wastewater treatment. J Environ Chem Eng 2021; 9(6)106553.
[http://dx.doi.org/10.1016/j.jece.2021.106553]

[53]   Kumar V, Sharma N, Lakkaboyana SK, Maitra SS. Silver nanoparticles in poultry health: Applications and toxicokinetic effects, in Silver Nanomaterials for Agri-Food Applications. Elsevier 2021; pp. 685-704.
[http://dx.doi.org/10.1016/B978-0-12-823528-7.00005-6]

[54]   Egbosiuba T C. Biochar and bio-oil fuel properties from nickel nanoparticles assisted pyrolysis of cassava peel 2022/08/01/ 2022.
[http://dx.doi.org/10.1016/j.heliyon.2022.e10114]

CHAPTER 7

Citrus Waste Valorization for Value Added Product Production

Lucky Duhan¹, Deepika Kumari¹ and Ritu Pasrija¹,*

¹ Department of Biochemistry, Maharshi Dayanand University, Rohtak, Haryana-121002, India

Abstract: With the growing population, resource production and utilization, including citrus fruit consumption, have amplified tremendously. Citrus foods include sweet orange, sweet blood orange, tangerine, grapefruit, lemon, lime, and Seville orange. Industrial processing of citrus fruits is done to produce various end products like juice concentrates, jams, jellies, sweets, candies, marmalades, and ice creams, which simultaneously produce tons of peels and waste as well. Like all industrial waste dumping, the negligent discard of citrus waste has legal repercussions. Therefore, the global treatment seems to be a virtuous option, which results in improved earnings, thereby ultimately reducing the reprocessing expenditure.

Conversely, despite the low cost, citrus waste management and valorization still have not reached a virtue that makes it an ideal candidate. Valorization technically refers to the process of industrial recycling or waste composting into commercially valuable products. To fix the citrus wast essential to understand the various ways to recycle and manage the left-over better. This requires research and knowledge of different techniques involved in the commercial utilization of citrus waste for the production of various components, counting-essential oils, flavonoids, pectin, enzymes, ethanol and methane *etc.,* along with the applications of these bioactive components in various ventures. This study summarizes the bioactive components obtained from citrus foods and their possible industrial utilization.

Keywords: Biofuel, Citrus waste, D-limonene, Dietary fibre (DF), Essential oil, Enzymes, Flavonoids, Hydro-distillation (HD), Industrial processing, Microwave-assisted Steam Distillation (MSD), Microwave Hydro-diffusion and gravity method (MHG), Nutritious supplement, Organic Acids, Pectin, Pharmaceutics and Cosmetics, Single Cell Protein (SCP), Supercritical Fluid Extraction (SFE), Subcritical Water Extraction method (SWE), Ultrasonic-Accelerated Extraction method (UAE), Valorization.

* **Corresponding author Ritu Pasrija:** Department of Biochemistry, Maharshi Dayanand University, Rohtak, Haryana, India; E-mail: ritupasrija@yahoo.com

Vinay Kumar, Sivarama Krishna Lakkaboyana & Neha Sharma (Eds.)
All rights reserved-© 2023 Bentham Science Publishers

INTRODUCTION

Citrus fruits are affiliated with the Rutaceae family. The term 'citrus fruits' refers to several varieties, including-sweet orange, sweet blood orange, tangerine, grapefruit, lemon, lime, bitter/Seville orange, *etc.* These fruits typically have a sour and sweet flavor in varied ratios. They are rich in juice, have attractive colors, and are known for their taste and health benefits. The harvest is done mainly for juice, which comprises 45% weight, consumed either fresh or in refined form. The waste is produced as pulp and seeds, which are generally discarded Fig. (1) & Table 1, describes the worldwide citrus fruit production. The juice obtained can be used for making concentrates, jams, jellies, sweets, candies, marmalades, ice creams, *etc* [1].

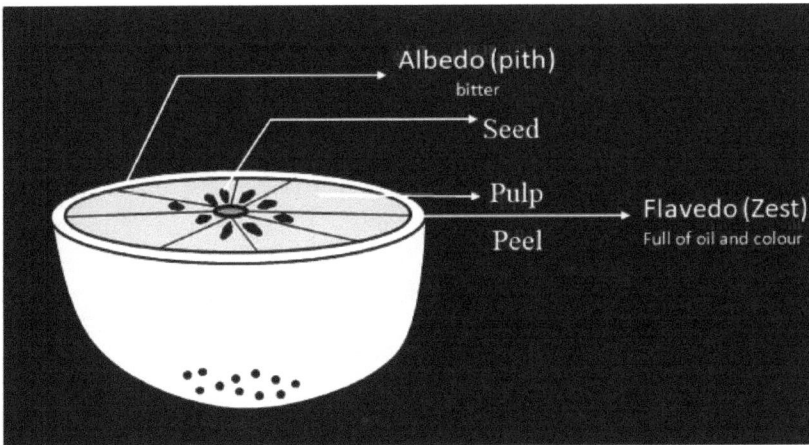

Fig. (1). Diagrammatic composition of citrus fruits

Table 1. Total production of citrus in the world in thousand tons from 2010-2016 [2].

	2010	2011	2012	2013	2014	2015	2016 Preliminary
World	**117,441**	**123,824**	**123,002**	**128,611**	**131,707.7**	**130,947.0**	**124,246.0**
Northern Hemisphere	88,058.5	91,905.4	93,412.8	99,820.1	103,317.4	102,059.5	97,848.9
India	8,855.8	6,875.0	6,955.0	9,235.0	10,401.1	9,216.2	9,755.8
USA	10,193.9	10,919.5	10,813.0	10,301.0	8,751.0	8,208.0	7,829.0
Mediterranean Region	22,355.7	22,689.5	21,945.4	23,195.0	24,541.1	23,825.4	25,216.0
Cyprus	113.3	128.7	112.4	106.4	106.5	118.7	114.4
Greece	1,127.7	1,078.1	1,097.1	1,123.6	958.2	1,049.6	1,041.5
Italy	3,779.3	3 537.0	2,883.9	2,678.7	2,661.6	2,808.5	3,150.2
Spain	6,076.4	5,720.4	5,553.8	6,685.7	7,041.6	6,100.5	6,882.0

	2010	2011	2012	2013	2014	2015	2016 Preliminary
Algeria	788.1	1,106.8	1,087.8	1,204.9	1,271.0	1,289.9	1,372.4
Egypt	3,518.2	3,724.9	3,975.0	4,096.9	4,402.2	4,646.6	4,930.4
Morocco	1,345.5	1 636.3	1,867.0	1,452.1	2,213.6	1,899.4	2,018.9
Tunisia	300.2	325.7	337.8	309.3	326.4	329.9	331.4
Israel	531.9	556.5	467.5	525.9	512.5	534.6	476.0
Lebanon	245.4	230.9	250.8	239.0	239.5	228.7	206.2
Turkey	3,570.0	3,611.6	3,472.9	3,678.6	3,781.4	3,803.3	3,652.1
Portugal	243.2	277.4	258.1	287.3	304.0	296.1	307.9
Japan	850.2	983.4	892.7	937.3	1,273.9	1,103.4	1,143.3
Mexico	6,753.4	7,031.1	6,603.2	7,467.8	7,655.2	7,291.7	6,634.0
China	23,974.9	28,939.9	31,830.4	34,261.7	36,467.0	38,153.9	32,705.9
Indonesia	2,028.9	1,818.9	1,611.8	1,654.7	1,926.6	1,625.9	1,574.8
Pakistan	2,150.0	1,982.2	2,036.0	2,008.8	2,010.4	1,915.8	1,907.4
Thailand	1,089.6	1,030.5	995.5	966.8	1,202.4	1,106.1	1,102.1
Vietnam	1 129.5	955.6	958.3	971.6	1,056.2	985.6	998.7
Others Northern Hemisphere	3,632.1	3,846.2	3,928.3	3,894.0	3,918.5	3,996.6	3,979.4
Southern Hemisphere	29,382.8	31,918.8	29,589.4	28,791.0	28,390.2	28,887.5	26,397.1
Argentina	2,559.4	3,613.4	2,895.8	2,433.7	2,164.2	2,753.3	2,800.7
Bolivia	319.5	324.1	330.2	332.2	337.7	356.8	371.4
Brazil	20,721.1	22,018.8	20,258.5	19,734.7	19,073.9	18,921.6	16,555.1
Paraguay	404.2	403.7	416.9	417.2	429.8	431.0	431.4
Uruguay	315.0	270.2	329.9	234.7	287.7	251.3	270.6
Venezuela	484.6	563.3	474.3	516.7	500.1	460.8	333.9
Chile	287.1	299.5	301.0	303.8	275.0	286.9	282.2
Australia	522.8	423.1	512.6	584.2	487.2	466.6	584.6
South Africa	1,997.0	2,169.2	2,133.6	2,169.9	2,169.9	2,662.6	2,409.2
Others Southern Hemisphere	918.9	948.6	988.6	1,069.4	1,213.2	1,195.4	1,246.0

In citrus fruits-based industries, although the prices paid to the farmers are lesser, transport, packaging, and marketing along with damaged fruits further add up to the prices. Besides that, the processing of fruits ends up rendering loads of discard and waste, like peels, pulp, seeds, rind, *etc.*. The common alternative for the waste produced is dumping to landfills which results in bad stink and disease expand. However, citrus waste has antimicrobial compounds, organic matter, low pH, and high-water contents, which may contaminate soil and water in the environment

[2]. Apparently, citrus waste adversely affects water quality and aquatic life, and its seepage into the earth is not recommended for underground water. The dried citrus peels can also be utilized in agronomic practices like organic compost and as fodders for animals. Although, there is no profitable gain from the thermal production of fodders, the first being energy utilization, the second is rich in sugars merely and deficient in proteins, making it commercially unviable [2].

Therefore, industrious foresight stresses on rationalization of citrus waste as a resource. This achieves many necessities, including-rural economic realization, agro-industrial byproduct valorization comprising waste disposal, as well as obtaining value-added natural food additives for commercial reimbursements suitable for the food industry, cosmetics, and pharmaceutical industries. However, effective extraction procedures require constant industrial research and inputs. The citrus peel waste (CPW) is rich in various components like essential oils, flavonoids, citric acid, pectin, enzymes, phenolics, *etc.* The details of the composition have been compiled in Table **2**. Thus, it is concluded that although CPW is not suitable to be discarded as such, being rich in sugars, cellulose, hemicellulose, and essential oils makes its processing promising for biofuel production and other supplies which are suitable for food, cosmetic and pharmaceutical industries (Fig. **2**).

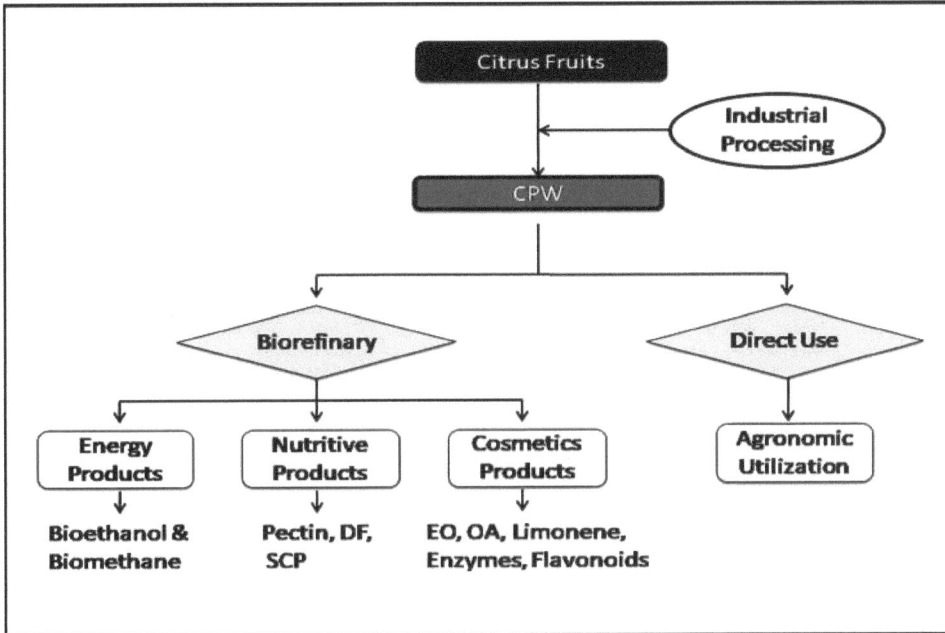

Fig. (2). Scheme flow chart indicating CPW processing and products.

The recovery of various bi-products requires little efforts and suitable techniques to exploit the potential of CPW and is discussed in the next section.

Table 2. Composition of CPW from some citrus fruits (% dry weight). Adapted from [3]

Composition	Composition of CPW (% dry weight)		
	Orange peel	Kinnow peel	Mandarin peel
Cellulose	37.1	10.10	22.5
Hemicellulose	11.0	4.28	6
Pectin	23.0	22.6	16.0
Sugars	9.6	31.58	10.1
Lignin	7.5	0.56	8.6
Fat	4.0	-	1.6

RECOVERY OF VARIOUS COMPONENTS FROM CPW

Biofuel production

Fossil fuels are considered the primary energy source. But they come with their own disadvantages, including limited availability, the release of greenhouse gases and pollution, *etc*. Thus, there is a constant demand to look for renewable energy sources which maintain carbon neutrality. Biofuels or blending of either ethanol or butanol is a possible alternative to reduce dependence on fossil fuels. Industrial synthesis of biofuels is costly. Therefore, biofuels produced from corn, sugar cane, wheat skins, and husks seem better alternatives. In this regard, CPW has also garnered attention for biofuel production due to its availability, abundance as fermentable sugars, and minimal lignin concentrations [4]. The bioethanol produced from CPW has great thermal proficiency and power, which can be expressed in terms of octane numbers and vaporization. Besides this, its blending (3-20% in different countries) would reduce the NO_2 and CO_2 emissions compared to that of conventional unblended fuels. Many countries have already adapted blending of biofuels. These countries include the US, Brazil, Japan, Thailand, and India.

Although molasses is also utilized for bioethanol production, still the demand for sufficient raw material is far from being met. Thus, CPW is seen an alternative addition for meeting requirements due to its ample availability, less dependence on crops, economics, absolute zero ground clearance for farming, and muchneeded disposal. The CPW utilization for biofuel production broadly requires three different major steps, as shown in Fig. (**3**).

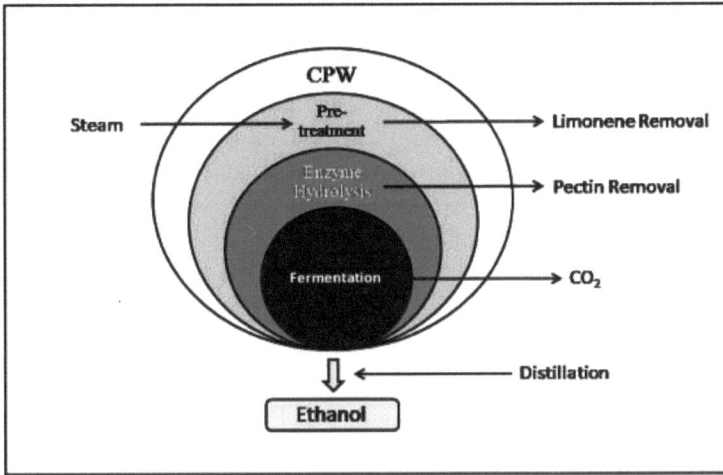

Fig. (3). Process flow chart for the extraction of bioethanol from CPW.

Pre-treatment: The first is actually a pre-treatment step and starts with peels crushing and powdering to lower the particle dimensions (in mm), increase the surface area and biodegradability. This step enhances the efficiency of subsequent steps involving the removal of D-limonene and pectin. The essential oil limonene levels in CPW vary from 0.08-0.15%, and at these levels, it acts as a microbial inhibitor, including yeast, considering its MIC which is close to 0.05% (v/w). Thus, limonene levels essentially need to be less than 0.05% for successful yeast-mediated fermentation [4]. Limonene presumably inhibits yeast growth by reducing its K^+ and oxygen consumption. The removal of limonene involves activated charcoal and raw cotton-based removal columns, with hexane as a solvent, and together increases the removal from hydrolysate to almost 90%. Further, limonene removal also significantly increases ethanol yield. Furthermore, D-limonene recovery adds to the economy of citrus waste valorization (discussed later).

The other component in CPW, called pectin, escalates the thickness of the fermentation liquid and creates difficulty in biofuel production. The pectin cannot be extracted during bioethanol formation, so either pectinases treatment is required, which hydrolyzes it to galacturonic acid and is targeted in the second step.

Hydrolysis: Pectinases, along with polysaccharides and hemicellulose, can be acid hydrolyzed involving combined hydrolysis with acid and enzymes (cellulase, pectolytic) for about 10 min. at a temperature of about 116°C. However, if only hydrochloric acid (HCl) is employed at 4% or less (w/v), then above 160°C temperature serves the purpose. Some studies show the utilization of sulfuric acid (0.5%; pH 2.8) as well.

However, it is seen that acid-alone treatment leaves the majority of pectins intact but increases their sensitivity for enzymatic treatment. The acid treatment step would break the compound structure of cell-wall; liberate, solubilize, and hydrolyze cellulose and hemicellulose to avoid its crystallinity in a cost-effective manner, followed by the action of microbes/yeast at 42°C in anaerobic digestion fermenters for ethanol production [5]. It is important to mention that hydrolysis, without pretreatment, gives a low fuel yield. This set is followed by the removal of contaminants and sterilization by heat, ultra-violet radiation, *etc.* D-limonene and volatile compounds are removed by steam stripping and water impregnation (popping). Additionally, soluble sugars attached to polysaccharides are released as well, reaching 25-32%. Enzymatic saccharification (β-glucosidase and cellulase) can be done additionally to increase sugar levels.

After D-limonene, pectin, nanocellulose, and hemicellulose removal, the CPW is ready to undergo fermentation in the third step to obtain bioethanol.

Fermentation: The 5-C and 6-C sugars obtained from cellulose and hemicellulose can be effortlessly fermented to obtain bio-ethanol. Fermentation can be done in either batches or continuous flow, depending on the composition and microbes used for fermentation, *etc.* As expected, continuous flow gives better results than batch cultures. However, industrial production is preferable with fed-batch, as it encompasses abilities of both, including maximum feasible cells with longer life span and yield collection.

Fermentation requires enzymatic hydrolysis *via* the action of cellulases and xylanases. It can be done either in two steps: Hydrolysis followed by fermentation called Separate Hydrolysis and Fermentation (SHF) or in the same reactor termed Simultaneous Saccharification and Fermentation (SSF). Enzymes involved in SSF, such as cellulase and xylanase, tend to get feedback inhibited by released glucose and xylose. Additionally, conditions (temperature, pH) should meet the requirements of both saccharification and fermentation.

It is important to mention that hydrolysis for sugar release is optimal at 323K and fermentation at 308K. Besides this, if glucose and ethanol are removed continuously, then a high yield can be obtained with lesser reactor volume and shorter retention time. The lysate enters the fermenter for fermentation, followed by distillation and leaving xylose. *Saccharomyces cerevisiae* (baker's yeast) is the most common yeast used for fermentation, although co-culture with *Candida* sp. (*C. parapsilosis*) gives better results. The ethanol concentration generally reaches 40-42 g/L in monoculture, which further increases in co-culture, as *S. cerevisiae* lacks enzymes for 5-C sugars like xylose and pentose. Besides this, ethanol accumulation inhibits *Ethanol-tolerant S. cerevisiae* can be a better alternative instead, and many labs are working to develop these tolerant varieties of yeast. Microbes Immobilization can be employed to reuse them for fermentation, as it has many advantages, including-reduced fermentation period, higher cell density and forbearance against inhibitors, and more straightforward downstream handling. Other microbes- *Candida brassicae, Pichia stipitis, Pachysolen tannophilus, Mucor indicus, Kluyveromyces marxianus, Zymomonas mobilis,* and *Escherichia coli* are used for fermentation at times. These microbes (besides *S. cerevisiae*) have fermentation temperatures close to enzymatic hydrolysis and ferment pentose sugars, but ethanol yield is lower. Therefore, co-culturing befits better than monoculture. Additionally, depending on geographical locations and suitable harvest, the waste of different varieties of citrus fruits available also varies, and so does their yield, which is compiled in Table **3**.

Table 3. Summary of bioethanol production from various CPW

CPW	Pre-treatment	Hydrolysis	Fermentation	Ethanol Yield	References
Orange peel	Acid steam explosion	Enzyme assisted	SHF	4.85 g/L h^{-1}	[6]
Mixed citrus CPW	Popping	Enzyme assisted	SHF	14.4–29.5 g/L	[7]
Orange	Steam explosion	None	SHF	4.1 g/100 mL	[8]
Orange	Steam	Acid	SHF	21% g g^{-1}	[9]
Sweet lime	Steam explosion	Enzyme assisted	SHF	18%	[10]
Valencia orange	Steam explosion	None	SHF	4.05 (% w/v)	[2]
Lemon	Steam explosion	Enzyme assisted	SHF	60 L per 1000 kg	[11]
Orange	Acidic heat treatment	None	SSF	6.0029%	[12]

*SSF- Simultaneous saccharification and fermentation; **SHF**-Separate hydrolysis and fermentation

However, more research is needed to develop tolerant microbes and those which can utilize a variety of sugars, which is possible by genetic engineering. D-limonene recovery during bio-ethanol production is also lucrative to be pursued further.

Essential Oil And D-limonene

The term essential oil (EO) refers to plant-derived hydrophobic liquid/oil containing the essence of the plant fragrance and has nothing to do with the indispensable nature of the compounds. The EO is found in sacs or glands present at different parts of the peel and flavedo (Fig. **1**) of citrus fruits and usually comes out during juice extraction due to the crushing of oil sacs in citrus fruits (2). The EO extracted from various CPW includes 17 compounds, including limonene [13]. D-limonene is the main derivative of the EO, and the amount of D-limonene (terpenic) is found to be about 32-98% in the case of orange. D-limonene is a prominent worthy bi-product found in EO extracted from different citrus fruit wastes. It costs close to $10,000 per ton (t). The EO and its component D-limonene from citrus plants are used in the manufacturing of aromatic soaps, perfumes, flavoring agents in drinks and food. As earlier mentioned, limonene extracted from citrus fruits has inhibitory activity against microorganisms as it disrupts the cellular structure of the microbial cell and inhibition of microbial respiration [14]. The orange oil contains many essential oils as a secondary metabolite, essentially a mixture of many volatile terpenic compounds and a characteristic citrus aroma imparting terpenoids (oxygenated terpene derivatives with hydroxyl and carbonyl groups). The percentage composition of essential oils extracted from various citrus fruit wastes is compiled in Table **4**.

Table 4. Composition of Essential Oils (EO) extracted from CPW Source [21, 22, 23].

EO composition	Essential Oil content in various CPW (in %)			
	Sweet orange	Bitter orange	Bergamot	Lemon
Limonene	89.9	90.3	72.9	59.1
Linalool	0.3	1.5	10.2	0.2
α–pinene	2.3	1.5	1.4	5.2
Sabinene	1.0	-	-	0.9
β-pinene	0.1	0.2	0.1	5.2
Myrcene	1.0	-	-	0.9
Valencene	0.3	0	0	0.8
Geranial	0.1	-	-	2.1
Butylacetate	-	0	5	-
Carvone	0.6	-	-	0.1
Methanol	0.4	0.3	0.3	-
Isopropanol	0.9	0.8	0.3	-
γ-terpinene	-	-	-	9.7
3-heptanone	0.4	1.2	0.9	-
α-terpineol	0.45	1.1	0	0

To overcome these common shortcomings in traditional physical techniques, some new extraction methods have been devised involving Supercritical Fluid Extraction (SFE), Microwave Hydro-diffusion and gravity method (MHG), Microwave-assisted Steam Distillation (MSD), Ultrasonic-Accelerated Extraction method (UAE), and Subcritical Water Extraction method (SWE) [20]. These alternatives show promising results in extracting maximum EO yield with low energy utilization and are discussed in the next section.

Extraction methods for EO/limonene: The extraction of EO from citrus waste can either be done by simple removal or pretreated removal. But, mostly, pre-treatments are done for better yields; usually, thermo-physical or thermo-chemical methods are applied for EO extraction [15].

The thermo-physical methods involve steam distillation, cold pressing, or steam explosion of citrus materials. By exercising these processes, the EO is physically detached from the water phase [16]. Although, traditionally, cold pressing of CPW has been more popular and employed for EO extraction. Cold pressing involves the mechanical pressing of CPW, and oils are released in the form of water emulsion. Further, the EO is extracted from water emulsion with the help of centrifugation [2]. Further distillation of this extracted EO ensures the recovery of D-limonene from other components. Steam distillation method is commonly used for obtaining an industrial yield of limonene (50%). In this method, steaming or boiling of CPW is done, and the EO is extracted by evaporation [17]. Hydro-distillation (HD) is another method exploited for the separation of EO and involves the mixture evaporation at a low boiling point [18]. In a comparative study involving HD, cold pressing and microwave-accelerated distillation of lemon peels released EO content of 0.21%, 0.05%, and 0.24%, respectively [19]. However, HD requires more extraction time and can cause disruption of terpenic substances, especially D-limonene [20]. The probability of chemical modifications of the compounds in traditional physical methods is possible but unwarranted.

Supercritical Fluid Extraction (SFE) Technique involves CO_2 or any other gas above the critical pressure and temperature of the solvent. Its principal benefits includes-a faster speed extraction, high selectivity as well as low consumption of solvent [24]. However, the equipment utilized in this extraction is expensive and does not permit tapping its full potential.

Microwave Hydro-diffusion and gravity (MHG) Method is an alternative to the HD method because this helps in reducing the processing time, as well as lower energy needs compared to the HD method. A 15 min processing time in MHG yields almost the same yield as 180 min in the HD method, without any chemical composition disruptions [17].

Microwave Assisted Steam Distillation (MSD) or **Ultrasonic-Accelerated Extraction (UWE)** Process is an alternative to simple steam distillation because MSD helps in saving the processing time, and energy needs, with a considerable increase in the yields. It involves micro or ultrasound waves (higher than 20 kHz) for EO extraction, and these waves effectively disrupt the cell walls by accelerating heat transfer, which quickens the extraction of EO compounds. The strong microwaves bring high-speed tearing of various membranes of citrus fruit without disrupting the chemical composition of EO and its various components [25]. It also offers the benefits of high reproducibility and low solvent consumption *etc*. By utilizing ultrasound waves, it is possible to extract phenolic compounds as well [26].

Subcritical Water Extraction method (SWE) Involves the use of water as a solvent at optimum pressure and temperature, and thus derived its name. It is a good alternative to solid sample extraction because of the good solubility of organic compounds [27].

Flavonoids

Flavonoids are polyphenolic secondary metabolites present in high quantities in the CPW. These impart varied colors and a bitter flavor to the peels (Fig. **1**) and waste, which protects plants and fruits from herbivorous animals, as well as from ultraviolet radiations [14]. Apart from vitamin C, dietary fibers, carotenoids, and folate, citrus fruits are also rich in flavonoids, and more than 60 of them identified so far have been partitioned into six classes- flavones, flavanones, flavonols, isoflavones, anthocyanidins and flavanols (or catechins). Some common ones are hesperetin, quercetin, naringin, diosmin, tangeretin, naringenin *etc*. Their structures are species-specific, and thus they are used as adulteration markers in the juice industry [28]. These compounds also confer antimicrobial action and defend plants from pathogens' strikes. Flavonoids display various biological properties of being an anti-oxidant, anti-carcinogenic, anti-inflammatory, anti-aging, cardio-protective, anti-lipoperoxidation activity, and chemo-protective agent., and thus termed as 'neutraceutical compounds' [18]. The results shared by Tripoli *et al.,* reflect the antiviral activity of flavonoids against Rhino and Poliomyelitis viruses [29].

Since, the antioxidant activity of flavonoids from various fruits and vegetables is already well known, the correlation between flavonoids extracted from citrus fruits and their antioxidant activity does not come as a surprise. However, not all flavonoids display radical scavenging activity; rather, some specific hydroxyl group-bearing flavonoids at particular positions exhibit this activity. Ghasemi *et al.,* reported the inhibitory concentration (IC_{50}) index of radical scavenging activity of phenols and flavonoids extracted from 13 different citrus fruits [30]. Studies suggest that nobiletin and tangerine flavonoids extracted from citrus fruits may also act as alternatives in the treatment of Alzheimer's and Parkinson's disease due to their neuroprotective effects [31]. Rutin, diosmin, and hesperidin show good activity in vascular syndromes and thrombotic disease treatment [18].

A major reason for the underutilization of flavonoids from CPW is a dearth of extraction procedures. The limited extraction of flavonoids involves both physical and chemical methods. The finer steps include techniques analogous to those described for the extraction of EO/limonene, *i.e.,* SFE, UAE, MAE, SWE, *etc* [20]. Although, HPLC and C-18 columns with methanol and acetonitrile solvent are successfully used for extraction, followed by LC-MS for confirmation for research purpose [32]. Thus, looking at the wide range of beneficial properties displayed by them, further research is required to improve practices and methods for their extraction scale-up.

Organic Acids

The most commonly obtained organic acids for industrial usage are citric acid (CA; ($C_6H_8O_7$), succinic acid, acetic acid, *etc*. W. Scheele (1784) first isolated the CA from the lemon juice in the form of calcium citrate after treatment with sulfuric acid [33]. CA can be produced either by chemical methods or by fermentation techniques. However, the production of CA through fermentation is more frequently employed than in chemical methods [34]. Fermentation for CA method is accomplished through *Aspergillus niger* cultures. Besides being biodegradable in nature, economical, and less toxic, CA also demonstrates varied applications in buffering, cleaning, and dispersing agent [24]. Several other sources can be used for organic acid production [79 - 89]. About 1.7 million t of CA is produced annually, and the annual growth rate touches ~5% globally [35]. Its usage ranges from an anticoagulant blood preservative, antioxidant in fats and oils; flavoring agents (due to acidic property), cleaning agents, and water softener, and suggest the unlimited economic potential still waiting to be tapped.

Extraction of Organic acids: The extraction of organic acids from CPW is achieved through two fermentation processes: one is Solid State Fermentation (SSF), and the other is Submerged Fermentation (SmF). Both methods consume CPW as a substrate, with various microorganisms like fungi, yeasts, bacteria, *etc*. Torrado *et al.,* reported the SSF extraction from the sweet orange peel by using *A. niger* CECT 2090 strain. They observed the yield of CA about 193mg/g dry orange peel without the addition of any sugar nutrient [36]. Jayaprakash *et al.,* improvised the SSF method with the help of the shake culture method and achieved a higher yield of CA of about 38.16%, 33.60%, 27.84%, and 27.60% from four types of different citrus waste. This study suggested 3% methanol addition as an enhancer to achieve a higher yield [37]. Another study reports the highest production of CA, about 11.36 g/L with orange peel and about 9.6 g/L amount while using lemon peel, with *A. niger* strain isolated from decaying oranges and carrots [38]. Kuforiji *et al.,* utilized SSF and recorded the yield of

57.6% and 55.4% by using two fungal strains of *A. niger* NRRL 567 and 328, respectively, but the yield decreased in both samples with the addition of 1-3% methanol [39]. Satheesh kumar *et al.,* studied CA production from orange peel using *A. niger I* and *A. niger II* showing the yield of about 6.0% and 9.8%, respectively [34].

Succinic acid (SA) ($C_4H_6O_4$) is a dicarboxylic acid and can be extracted from citrus waste. SA is utilized for making polymers, detergents, solvents, pigments, fragrances, *etc*. The most preferred method to produce SA is the catalytic hydrogenation, paraffin oxidation, and electrolytic reduction of maleic acid or anhydride and then conversion into SA. However, SA can also be produced through fermentation by using microorganisms [24]. As SA production through fermentation involves waste material, it is economical and eco-friendly and is a better alternative to the maleic anhydride method. The production of SA from citrus waste through microbial fermentation is well documented in various studies. The successful conversion is done through either *Mannheimia succinici-producens, Anaerobiospirillum succiniciproducens, Basfia succiniciproducens,* or *Actinobacillus succinogenes*. Li *et al.* reported the production of SA from orange peel involving hydrolysis of cellulose, followed by fermentation mediated by the bacterium *Fibrobacter succinogenes* S85. The highest succinate titer obtained is 1.9 g/L [40]. The higher yield of SA (22.4 g/L is achieved when the fed-batch fermentation process is applied [13]. In a separate study, Patsalou *et al.,*. performed acid pre-treatment as well as combined dilute acid hydrolysis and enzyme pre-treatment (with cellulases and β-glucosidases), which improved the yields of SA to 0.70 g_{sa}/g_{tsc} (sa:succinic acid; tsc: total sugar consumed) at 109-116°C [41]. However, the cost factor due to the use of enzymes can be skipped by compromising the yields.

Nutritious Supplement

In addition to EO and organic acids, a carbohydrate called pectin (a component of the cell walls), composed of poly-galacturonic acid with varying groups of methyl esters, is also obtained from CPW [42]. It is commercially available as a white powder or colorless liquid. The structure of pectin is shown in Fig. (**4**), along with other by-products. It is extracted from the orange peels' albedo region. It is a form of soluble dietary fiber and its content varies from 16% in mandarin to 23% in oranges. Pectin exhibits various applications like gels, thickeners, bulking agents, emulsifiers, and stabilizers in pharmaceuticals, as well as in food processing industries [4]. Pectin also shows various applications in food packaging industries, as it is an edible as well as a biodegradable product in nature. Reports claim its anti-colon and anti-prostate cancer properties, along with its use as medicine for diarrhea control and throat sores, although scientific evidence is still anticipated.

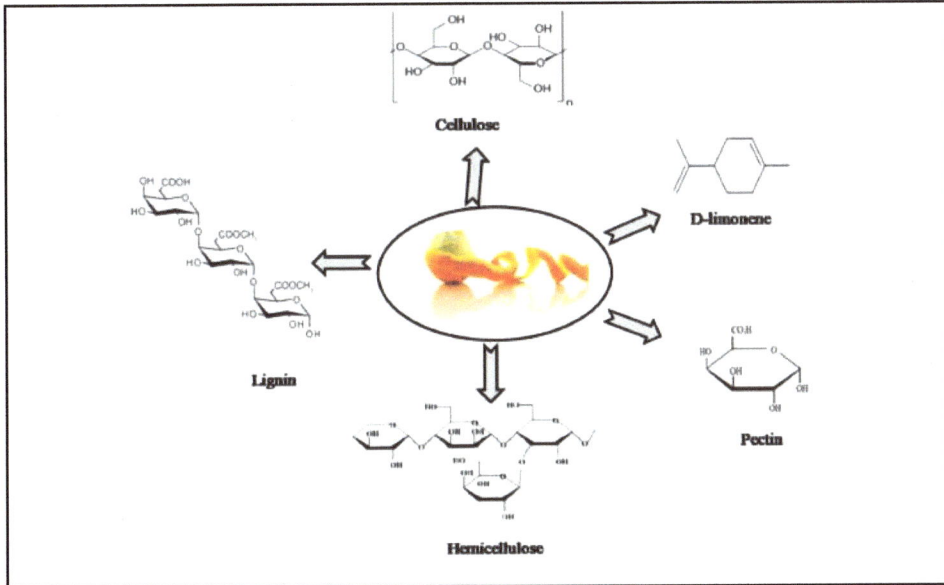

Fig. (4).   Structure of different compounds present in CPW (Modified and adapted from [10].

Extraction of Pectin from CPW: The most common strategy for industrial extraction of pectin from CPW involves water bathing or drying as a pre-treatment, followed by acid extraction of pectin [43]. It proceeds with washed citrus peel being subjected to dilute HCl acid (pH 2) hydrolysis for 2-4 hrs, filtering, and is followed by precipitation with ethanol or isopropyl alcohol. It then undergoes centrifugation or cheesecloth extraction for the recovery of precipitated pectin, and this provides a yield of about 14.3% and 10.6%, respectively, from fresh sweet orange peels [43]. Latest reports suggest even a 30% recovery of pectin through the bio-refinery process [13]. In the water bath heating extraction method, nitric acid use on sweet lemon peels show pectin % yields close to 30.2, 37.6 and 46.4%, at different temperatures of 60°, 70° and 80°C respectively (1.5 pH); while citric acid use gives little lower yields of 41.4, 45.6 and 47.8% at temperatures 60°, 70° and 80°C, respectively [44]. Other methods of pectin extraction involve- ultrasonic-assisted methods and Soxhlet methods, but have not been tried at industrial levels.

Dietary fibre (DF)

The DF is neither readily digestible nor adsorbed in the human small intestine

[45], but it has a significant function in alleviating various digestive disorders like constipation, diabetes, diverticulosis, and obesity [46]. There are chiefly two types of DF in nature: soluble dietary fibres (SDF) and insoluble dietary fibres (IDF). SDF includes pectin, gums, mucilage, β-glucan, fructans, hemicelluloses, *etc.* They are not easily digestible. In the large intestine, its fermentation into short-chain fatty acids (SCFA) is triggered by bacteria [47]. The IDF comprises cellulose, hemicelluloses, and lignin (Fig. **4**), which are compiled in Table **5**. They completely remain indigestible and just move through the digestive tract without any fermentation [48]. The SCFA production in the intestine is crucial for maintaining the local microbiota, which affects our general well-being and digestion as well as has hypocholesterolemic effects [49]. The SDF generated from citrus peels is effective in balancing blood cholesterol as well [50]. The popularity of nutrition-rich DF is rising tremendously due to its multiple health effects and no adverse results. Thus, CPW is industrially processed for the production of DF powder [51]. Most nutritionists recommend almost 20-30% of SDF of our total dietary needs. Citrus fruits and their waste can serve as an alternative for the production of high-quality or cheaper DF, with rich antioxidant bioactive compounds [52]. Hence the bulk quantity of DF, rich in SDF, extracted from orange juice residues is a better option than any other cereals [53].

Table 5. Composition of fibre (% dry weight) of orange and lemon waste [52].

Compounds	Orange peel	Orange pulp	Lemon peel	Lemon pulp
Pectin	23.02 ± 2.12	12.07 ± 1.12	13.00 ± 1.06	22.53 ± 1.95
Cellulose	37.08 ± 3.1	24.52 ± 2.0	23.06 ± 2.11	36.22 ± 3.24
Hemicellulose	11.04 ± 1.05	7.57 ± 0.66	8.09 ± 0.81	11.05 ± 1.09
Lignin	7.52 ± 0.59	7.51 ± 0.62	7.56 ± 0.54	7.55 ± 0.66

Extraction of DF from CPW: Rezzadori *et al.,* suggested an effective method for the valorization of orange wastes for DF production. It primarily involves EO removal in the first step, and the remaining waste can be utilized for DF production. This study recorded approximately 5200 t of DF yield (along with EO) from 8,000 t orange pulps [54]. A similar study on DF abstraction from dried orange peels, with both SDF and IDF, demonstrated yields close to 20.7g/100g and 47.9g/100g, respectively [55]. Citrus peel-extracted fibers act as inhibitors of lipid oxidation in meat products, and thus its usage improves animal protein shelf life effectively [56]. The DF extracted from orange seed could be used to make crackers, which are very helpful in improving health conditions because of the presence of high fiber and flavonoid contents [57]. Therefore, DF production from citrus peel is another good option for industries to use citrus waste appropriately.

Single Cell Protein (SCP)

SCP is the biomass or protein generally extracted from unicellular microorganisms and, being proteinaceous, is a suitable alternative in the diet of humans and animals [58]. The SCP production is triggered mainly by the microorganisms like algae, yeast, fungi, and bacteria [59]. The lipids and vitamin requirements are also fulfilled from SCP, as well as the concentrated nucleic acids in SCP are higher than in conventional proteins [60]. Therefore, SCP can effectively fulfill the increasing demand for nutritious food for the increasing world population. High amounts of cellulose, and hemicelluloses, with lower lignin content in the citrus by-products, make them a good substrate for SCP production. Orange or citrus by-products act as an interesting nutrient source for microorganisms [61].

The utilization of citrus waste in SCP production can definitely minimize environmental pollution to some extent [62]. As citrus waste has very limited protein content, efficient SCP production from citrus waste by various techniques is valuable. The fermentative processes for the production of SCP from citrus waste involve both filamentous as well as unicellular microbes [24]. Some of the studies involving SCP extraction techniques and yield of SCP are discussed here.

Extraction of SCP from CPW: In the process of the production of SCP through solid fermentation, the protein content increases synergistically with simultaneously increased degradation of pectin and crude fibers. Deveci and Ozyurt calculated the crude protein content in orange and lemon fruit peels, which is about 27.7% and 34.22%, respectively, using *A. niger* and approx 27.8% and 19.9%, respectively, with *Penicillium roqueforti* driven fermentation [63].

The SCP obtained from CPW is affected by pH changes and dilution rates (D'). At pH 5 and D' of 0.078, pre-treated peels with acid and *Geotrichum candidum* produce SCP close to0.53 g/hr/Kg, while at pH 4 and the same D' value, the production of SCP drops to 0.40 g/hr/Kg. In addition to this, solid residue having 3.5-4.8% crude protein content is also produced [64]. In another study SCP production from orange peels in submerged fermentation (SmF) by *S. cerevisiae* generated 30.5% crude protein/100g of peels. Interestingly, protein yield increased by a remarkable 60.31% when glucose was added to the process as compared to the fermentation devoid of glucose [65].

Enzymes

An extremely valuable aspect of citrus peel valorization is enzymes extraction, particularly pectinase. The pectinases hydrolyze pectin-containing plant products. The pectinases have got applications in fruit juice extraction (juice clearing), wastewater treatment, paper bleaching, and fermentation of tea and coffee, as well as in feeds and alcoholic beverages (wine clearing) [66]. Citrus peels, mainly dried ones, have got adequate quantity of pectin and thus can be used as a substrate for microbial fermentation-driven pectinolytic enzyme production [67]. The pectinolytic enzymes include- proto-pectinase, polygalacturonase, lyase, and pectin esterase [66]. Apart from them, various other enzymes can also be produced by citrus peel and involve cellulases, xylanases, amylases, endoglucanase, and β-glucosidase, *etc*. Table **6** is a collation of pectinases and some other enzymes produced from citrus peel, extraction strategies, and their yields.

Table 6. Summary of enzyme production from CPW.

CPW	Fermentation	Microbes	Enzyme name	Enzyme yield (activity)	References
Orange	SmF	*A. niger*	Pectinase	117.1 ± 3.4 mM/mL/min	[67]
Orange	SSF	*C. tropicalis* and *S. cerevisiae*	Endo-glucanase	5.5 IU/mL	[68]
			β-glucosidase	7.7 IU/mL	
			Xylanase	2.3 IU/mL	
Citrus limetta	SSF	*C. tropicalis* and *S. cerevisiae*	Endoglucanase	9.03 IU/mL	[68]
			β-glucosidase	7.4 IU/mL	
			Xylanase	10.3 IU/mL	
Orange	SSF	*E. javanicum*	β-glucosidase	49.64 U/g	[69]
			Endoglucanase	46.80 U/g	
			Pectinase	51.87 U/g	
Dried citrus waste	SSF	*A. niger*	Pectinase	5.38 ± 0.2 IU/mL	[70]
			Cellulase	1.17 ± 0.1 FPU	
Orange	SmF	*A. niger*	Pectinase	6888 IU/g	[71]
Lemon	SSF	*A. niger*	Pectinase	2181 U/L	[73]
Dried citrus peel	-	*A. niger*	Pectinase	0.929 M/mL/min	[74]
Sweet lime	SmF	*B. subtilis*	Pectinase	24.18 units/mL	[75]
Orange	SmF	*A. niger*	Pectinase	110.63 μM/mL/min	[76]

(Table 6) cont.....

CPW	Fermentation	Microbes	Enzyme name	Enzyme yield (activity)	References
Orange	SmF	*A. giganteus*	Pectinase	470 U/Ml	[77]
Waste products	Co-culturing	*A. niger* GJ-2 and *S. cerevisiae* J-1	Poly-galacturonase	512.7 U/mL	[78]

Extraction of enzymes from CPW: The two main fermentation strategies to extract the various enzymes from CPW involve either submerged fermentation (SmF) or Solid State Fermentation (SSF). Both fermentation approaches use either one or more than one species of bacteria or yeast. Studies related to fermentation techniques using diverse microbes have been presented in (Table **6**). Enzyme activity from diverse enzymes cannot be compared due to the use of different substrates and conditions. In a study, *Citrus limetta* peels with *C. tropicalis* and *S. cerevisiae* strains in SSF for 72 hrs showed endoglucanase, β-glucosidase, and xylanase activities with a yield of 9.03 IU/mL, 7.4 IU/mL, and 10.3 IU/mL respectively. In SSF of orange peel endo-glucanase, β-glucosidase and xylanase activities were 5.5 IU/mL, 7.7 IU/mL, and 2.3 IU/mL, respectively [68]. The pectinase production from the dried orange peels with *A. niger* mediated fermentation produced 117.1±3.4μM/mL/min at 30°C with 5% substrate concentration [67]. *Citrus limetta* peels with *C. tropicalis* and *S. cerevisiae* strains in SSF for 72 hrs demonstrated activities for β-glucosidase and xylanase with a yield of 9.03 IU/mL, 7.4 IU/mL and 10.3 IU/mL, respectively. In SSF of orange peel, endo-glucanase, β-glucosidase, and xylanase activities were recorded as 5.5 IU/mL, 7.7 IU/mL, and 2.3 IU/mL, respectively [68]. However, *Eupenicillium javanicum* mediated fermentation shows activities for pectinase, endoglucanase, β-glucosidase, and xylanase with 51.87 U/g, 46.80 U/g, 49.64 U/g, and 106.42 U/g, respectively [69]. Pectinase and cellulase from the pre-treated pectin-rich liquid fraction of CPW were recorded for 5.38 ± 0.2 IU/mL and 1.17 ± 0.1 FPU (filter paper unit), respectively [70]. Another study regarding pectinase production from the orange peel extract and *A. niger* in SmF reported an activity of 6888 IU/g with peptone as a nitrogen source [71]. Ismail studied the production of pectinase, cellulase, and xylanase enzymes in SmF of the orange peels by using different fungal strains. Among these, *A. niger* A-20 produced the highest amount of active enzymes after 5 days in culture [72]. However, pectinase extraction from lemon peels using SSF and *A. niger* A-20 demonstrated 2181 U/L recovery [73]. Dhillon *et al.,* studied the production of pectinase from dried citrus peel by using *A. niger* and maximum enzyme activity of about 0.929 M/mL/min with 15% substrate [74].

Pectinase extraction from *C. limetta* peel powder (3%) using SmF with the help of marine *Bacillus subtilis* is also reported. The highest activity of pectinase obtained is found to be approximately 24.18 U/mL after 28 hrs at pH 5.0 [75]. Ali *et al.*, tried the pectinase extraction from dried peels in SmF and *A. niger*. The maximum activity of pectinase (110.63 µM/mL/min) was achieved at 4% substrate concentration [76]. The maximum pectin lyase activity obtained was 470 U/mL, when *A. giganteus* was grown on orange waste in SmF processing [77]. Amid these random studies, some reports unveiled increased production of enzymes when two or more microbial strains were co-cultured in fermentation. The polygalacturonase enzyme activity was 512.7 U/mL when *A. niger* GJ-2 and *S. cerevisiae* J-1 were co-cultured. It was almost double as compared to *A. niger* GJ-2 culture alone [78]. Thus, it is summarized that different fruit wastes can be further optimized for better yields and profits.

CHALLENGES AND FUTURE PERSPECTIVES

Citrus fruits are cultivated in different parts of the continents and have variable harvesting time periods in separate regions. Improved transport ensures their accessibility throughout the year for human consumption and various industries. The total CPW produced is more than 40 million tonnes across the globe. The CPW is generally considered an undesirable waste, but its valorizations can produce various valuable and beneficial by-products with little expenditure. This makes citrus fruits and their waste pulp a viable and economical valorization option. The Insoluble citrus fibre comprises bioactive compounds, carotene, notably flavonoids and polyphenols, which surges its demand in the processed food industry. It has many health benefits, such as coronary heart disease, hypertension, diabetes, and risk of obesity, stroke, and various gastrointestinal (GI) disease. Fiberstar, which has its head office in Wisconsin, is one such biotech company that manufactures and sells Natural line of dietary fibres derived from citrus fruit to improve quality, health, and costs in food, personal care, and industrial markets. The citrus fibre produced has the exceptional water-holding capacity and is reportedly sold to over 69 countries. Being a Vegetarian/vegan/plant-based product is in demand, and also enquired for bakery, beverages, dairy, dressings, dairy alternatives, frozen foods, processed meats, pet foods, plant-based meats, and sauces-based products. Besides them, major key players in the global citrus fiber market are DuPont De Nemours, Edge Ingredients, FGF Trapani S.R.L, Naturex SA, Quadra Chemicals, Yantai Andre Pactin Co., AMC group, Herbafood Ingredients GmbH, Citrus Extracts, CP Kelco, Cargill INC, Carolina Ingredients, CEAMSA, Florida Food Products, Ingredients Nature, JRS Silvateam Ingredients S.r.l, Lucid Collids, Nans Products, and Golden Health Technology, *etc*. This long list suggests the emerging trend in citrus-valorized products.

CPW conversion into energy sources, including biofuels, SCP, pectin, enzymes, and pharmaceutical products like antioxidants, are indeed lucrative options. Besides these, cellulose and hemicellulose extraction is not much explored either. Thus, still many unexplored options and challenges exist in CPW valorizations. First obstacle is the selection of appropriate citrus waste for the targeted production. Different citrus fruits waste contains diverse quantities of products, and their composition in the fruits vary. This would affect the yield and profitability of the whole process. Secondly, even after target-based appropriate selection is complete, the choice of the suitable extraction techniques and their optimization conditions is a critical factor in the economics of valorization. Sometimes, additional purification steps ensure maximum yields but may not turn out cost-effective. Evaluation of bio-accessibility and bioavailability of the targeted compound and its interaction with other components can chemically modify and affect the yield. Thus, various aspects need to be thoroughly analysed in finalising the SOP (standard operating protocols) for every possible bioactive compound.

Another prominent hindrance is the lack of genetically modified microorganisms for the CPW fermentation, and only a few microbes can utilize the pentose sugars in fermentation. This can be alleviated by bringing new modifications /improvements in microorganisms for easy and less time-consuming fermentations, with better yield of bioactive compounds. More and more novel green methods and techniques for by-product extraction need to be developed. The greener methods would reduce wastewater production as well as environmental pollution. Right now, these are costly and not worth the yield in commercial or industrial sight.

Finally, for a much possible boost in CPW utilizations in the immediate future, advanced research is necessary, with emphasis on economies, markets, bio-accessibility, biosafety aspects, *etc*.

LIST OF ABRREVIATIONS

CPW   Citrus Products Waste

DF   Dietary Fibers

SCP   Single Cell Proteins

EO   Essential Oils

OA   Organic Acids

REFERENCES

[1]   Fernandez-Fernandez AM, Dellacassa E, Medrano-Fernandez A, Castillo MD. Citrus waste recovery for sustainable nutrition and health.Food Wastes and By-products. Wiley 2020; pp. 193-222.
       [http://dx.doi.org/10.1002/9781119534167.ch7]

[2] Sharma K, Mahato N, Lee YR. Extraction, characterization and biological activity of citrus flavonoids. Rev Chem Eng 2019; 35(2): 265-84.
[http://dx.doi.org/10.1515/revce-2017-0027]

[3] Oberoi HS, Vadlani PV, Nanjundaswamy A, *et al.* Enhanced ethanol production from Kinnow mandarin (*Citrus reticulata*) waste *via* a statistically optimized simultaneous saccharification and fermentation process. Bioresour Technol 2011; 102(2): 1593-601.
[http://dx.doi.org/10.1016/j.biortech.2010.08.111] [PMID: 20863699]

[4] John I, Muthukumar K, Arunagiri A. A review on the potential of citrus waste for D -Limonene, pectin, and bioethanol production. Int J Green Energy 2017; 14(7): 599-612.
[http://dx.doi.org/10.1080/15435075.2017.1307753]

[5] Patsalou M, Samanides CG, Protopapa E, Stavrinou S, Vyrides I, Koutinas M. A citrus peel waste biorefinery for ethanol and methane production. Molecules 2019; 24(13): 2451.
[http://dx.doi.org/10.3390/molecules24132451] [PMID: 31277372]

[6] Santi G, Crognale S, D'Annibale A, *et al.* Orange peel pretreatment in a novel lab-scale direct steam-injection apparatus for ethanol production. Biomass Bioenergy 2014; 61: 146-56.
[http://dx.doi.org/10.1016/j.biombioe.2013.12.007]

[7] Choi IS, Kim JH, Wi SG, Kim KH, Bae HJ. Bioethanol production from mandarin (*Citrus unshiu*) peel waste using popping pretreatment. Appl Energy 2013; 102: 204-10.
[http://dx.doi.org/10.1016/j.apenergy.2012.03.066]

[8] Joshi SM, Waghmare JS, Sonawane KD, Waghmare SR. Bio-ethanol and bio-butanol production from orange peel waste. Biofuels 2015; 6(1-2): 55-61.
[http://dx.doi.org/10.1080/17597269.2015.1045276]

[9] Tsukamoto J, Durán N, Tasic L. Nanocellulose and bioethanol production from orange waste using isolated microorganisms. J Braz Chem Soc 2013; 24(9): 1537-43.
[http://dx.doi.org/10.5935/0103-5053.20130195]

[10] John I, Yaragarla P, Muthaiah P, Ponnusamy K, Appusamy A. Statistical optimization of acid catalyzed steam pretreatment of citrus peel waste for bioethanol production. Resource-Efficient Technologies 2017; 3(4): 429-33.
[http://dx.doi.org/10.1016/j.reffit.2017.04.001]

[11] Boluda-Aguilar M, López-Gómez A. Production of bioethanol by fermentation of lemon (*Citrus limon* L.) peel wastes pretreated with steam explosion. Ind Crops Prod 2013; 41: 188-97.
[http://dx.doi.org/10.1016/j.indcrop.2012.04.031]

[12] Chahande A, Gedam V, Raut P, Moharkar Y. Pretreatment and production of bioethanol from Citrus reticulata fruit waste with baker's yeast by solid-state and submerged fermentation. 2018.
[http://dx.doi.org/10.1007/978-981-10-5349-8_13]

[13] Patsalou M, Chrysargyris A, Tzortzakis N, Koutinas M. A biorefinery for conversion of citrus peel waste into essential oils, pectin, fertilizer and succinic acid *via* different fermentation strategies. Waste Manag 2020; 113: 469-77.
[http://dx.doi.org/10.1016/j.wasman.2020.06.020] [PMID: 32604008]

[14] Sharma K, Mahato N, Cho MH, Lee YR. Converting citrus wastes into value-added products: Economic and environmently friendly approaches. Nutrition 2017; 34: 29-46.
[http://dx.doi.org/10.1016/j.nut.2016.09.006] [PMID: 28063510]

[15] Choi IS, Lee YG, Khanal SK, Park BJ, Bae HJ. A low-energy, cost-effective approach to fruit and citrus peel waste processing for bioethanol production. Appl Energy 2015; 140: 65-74.
[http://dx.doi.org/10.1016/j.apenergy.2014.11.070]

[16] Sawamura M, *et al.* Citrus essential oils: Flavor and fragrance 2010.
[http://dx.doi.org/10.1002/9780470613160]

[17]   Bousbia N, Vian MA, Ferhat MA, Meklati BY, Chemat F. A new process for extraction of essential oil from Citrus peels: Microwave hydrodiffusion and gravity. J Food Eng 2009; 90(3): 409-13.
[http://dx.doi.org/10.1016/j.jfoodeng.2008.06.034]

[18]   Zema DA, Calabrò PS, Folino A, Tamburino V, Zappia G, Zimbone SM. Valorisation of citrus processing waste: A review. Waste Manag 2018; 80: 252-73.
[http://dx.doi.org/10.1016/j.wasman.2018.09.024] [PMID: 30455006]

[19]   Ferhat MA, Meklati BY, Chemat F. Comparison of different isolation methods of essential oil fromCitrus fruits: cold pressing, hydrodistillation and microwave 'dry' distillation. Flavour Fragrance J 2007; 22(6): 494-504.
[http://dx.doi.org/10.1002/ffj.1829]

[20]   Negro V, Mancini G, Ruggeri B, Fino D. Citrus waste as feedstock for bio-based products recovery: Review on limonene case study and energy valorization. Bioresour Technol 2016; 214: 806-15.
[http://dx.doi.org/10.1016/j.biortech.2016.05.006] [PMID: 27237574]

[21]   Espina L, Somolinos M, Lorán S, Conchello P, García D, Pagán R. Chemical composition of commercial citrus fruit essential oils and evaluation of their antimicrobial activity acting alone or in combined processes. Food Control 2011; 22(6): 896-902.
[http://dx.doi.org/10.1016/j.foodcont.2010.11.021]

[22]   Badee AZM, Helmy SA, Morsy NFS. Utilisation of orange peel in the production of α-terpineol by Penicillium digitatum (NRRL 1202). Food Chem 2011; 126(3): 849-54.
[http://dx.doi.org/10.1016/j.foodchem.2010.11.046]

[23]   Moufida S, Marzouk B. Biochemical characterization of blood orange, sweet orange, lemon, bergamot and bitter orange. Phytochemistry 2003; 62(8): 1283-9.
[http://dx.doi.org/10.1016/S0031-9422(02)00631-3] [PMID: 12648552]

[24]   Mamma D, Christakopoulos P. Biotransformation of citrus by-products into value added products. Waste Biomass Valoriz 2014; 5(4): 529-49.
[http://dx.doi.org/10.1007/s12649-013-9250-y]

[25]   Mahato N, Sharma K, Koteswararao R, Sinha M, Baral E, Cho MH. Citrus essential oils: Extraction, authentication and application in food preservation. Crit Rev Food Sci Nutr 2019; 59(4): 611-25.
[http://dx.doi.org/10.1080/10408398.2017.1384716] [PMID: 28956626]

[26]   Ma YQ, Chen JC, Liu DH, Ye XQ. Simultaneous extraction of phenolic compounds of citrus peel extracts: Effect of ultrasound. Ultrason Sonochem 2009; 16(1): 57-62.
[http://dx.doi.org/10.1016/j.ultsonch.2008.04.012] [PMID: 18556233]

[27]   Gámiz-Gracia L, Luque de Castro MD. Continuous subcritical water extraction of medicinal plant essential oil: comparison with conventional techniques. Talanta 2000; 51(6): 1179-85.
[http://dx.doi.org/10.1016/S0039-9140(00)00294-0] [PMID: 18967949]

[28]   Spinelli FR, Dutra SV, Carnieli G, Leonardelli S, Drehmer AP, Vanderlinde R. Detection of addition of apple juice in purple grape juice. Food Control 2016; 69: 1-4.
[http://dx.doi.org/10.1016/j.foodcont.2016.04.005]

[29]   Tripoli E, Guardia ML, Giammanco S, Majo DD, Giammanco M. Citrus flavonoids: Molecular structure, biological activity and nutritional properties: A review. Food Chem 2007; 104(2): 466-79.
[http://dx.doi.org/10.1016/j.foodchem.2006.11.054]

[30]   Ghasemi K, Ghasemi Y, Ebrahimzadeh MA. Antioxidant activity, phenol and flavonoid contents of 13 citrus species peels and tissues. Pak J Pharm Sci 2009; 22(3): 277-81.
[PMID: 19553174]

[31]   Braidy N, Behzad S, Habtemariam S, et al. Neuroprotective effects of citrus fruit-derived flavonoids, nobiletin and tangeretin in Alzheimer's and Parkinson's disease. CNS Neurol Disord Drug Targets 2017; 16(4): 387-97.
[http://dx.doi.org/10.2174/1871527316666170328113309] [PMID: 28474543]

[32]   Delourdesmatabilbao M, Andreslacueva C, Jauregui O, Lamuelaraventós R. Determination of flavonoids in a Citrus fruit extract by LC–DAD and LC–MS. Food Chem 2007; 101(4): 1742-7. [http://dx.doi.org/10.1016/j.foodchem.2006.01.032]

[33]   P. Dhanke, 'A review on citric acid production and its applications', Int. J. Curr. Adv. Res., Oct. 2018, doi: 10.24327/ijcar.2017.5883.0825.

[34]   Satheeshkumar S, Sivagurunathan P, Muthulakshmi K, Uma C. Utilization of fruit waste for the production of citric acid by using *Aspergillus Niger*. J Drug Deliv Ther 2019; 9(4-A): 9-14. [http://dx.doi.org/10.22270/jddt.v9i4-A.3487]

[35]   Dhillon GS, Brar SK, Verma M, Tyagi RD. Utilization of different agro-industrial wastes for sustainable bioproduction of citric acid by *Aspergillus niger*. Biochem Eng J 2011; 54(2): 83-92. [http://dx.doi.org/10.1016/j.bej.2011.02.002]

[36]   Torrado AM, Cortés S, Salgado JM, *et al*. Citric acid production from orange peel wastes by solid-state fermentation. Braz J Microbiol 2011; 42(1): 394-409. [http://dx.doi.org/10.1590/S1517-83822011000100049] [PMID: 24031646]

[37]   Jayaprakash J, Babu R, Packiam M, Roch V. Eco-friendly utilization of citrus peels for citric acid production by *Aspergillus niger*. Int J Recent Sci Res 2019; (Jul): [http://dx.doi.org/10.24327/IJRSR]

[38]   Vidya P. Optimization and utilisation of various fruit peel as substrate for citric acid production by Aspergillus niger isolated from orange and carrot 2019.

[39]   Kuforiji O, Kuboye A, Odunfa S. Orange and pineapple wastes as potential substrates for citric acid production. Int J Plant Biol 2010; 1(1)e4. [http://dx.doi.org/10.4081/pb.2010.e4]

[40]   Li Q, Siles JA, Thompson IP. Succinic acid production from orange peel and wheat straw by batch fermentations of *Fibrobacter succinogenes* S85. Appl Microbiol Biotechnol 2010; 88(3): 671-8. [http://dx.doi.org/10.1007/s00253-010-2726-9] [PMID: 20645087]

[41]   Patsalou M, Menikea KK, Makri E, Vasquez MI, Drouza C, Koutinas M. Development of a citrus peel-based biorefinery strategy for the production of succinic acid. J Clean Prod 2017; 166: 706-16. [http://dx.doi.org/10.1016/j.jclepro.2017.08.039]

[42]   Liu Y, Shi J, Langrish T. Water-based extraction of pectin from flavedo and albedo of orange peels. Chem Eng J 2006; 120(3): 203-9. [http://dx.doi.org/10.1016/j.cej.2006.02.015]

[43]   Abebe Alamineh E. Extraction of pectin from orange peels and characterizing its physical and chemical properties. American Journal of Applied Chemistry 2018; 6(2): 51. [http://dx.doi.org/10.11648/j.ajac.20180602.13]

[44]   Wahengbam E, *et al*. Extraction of pectin from citrus fruit peel and its utilization in preparation of jelly 2014.

[45]   Romero-Lopez MR, Osorio-Diaz P, Bello-Perez LA, Tovar J, Bernardino-Nicanor A. Fiber concentrate from orange (*Citrus sinensis* L.) bagase: characterization and application as bakery product ingredient. Int J Mol Sci 2011; 12(4): 2174-86. [http://dx.doi.org/10.3390/ijms12042174] [PMID: 21731434]

[46]   Villanueva-Suárez MJ, Redondo-Cuenca A, Rodríguez-Sevilla MD, de las Heras Martínez M. Characterization of nonstarch polysaccharides content from different edible organs of some vegetables, determined by GC and HPLC: comparative study. J Agric Food Chem 2003; 51(20): 5950-5. [http://dx.doi.org/10.1021/jf021010h] [PMID: 13129300]

[47]   Soliman GA. Dietary fiber, atherosclerosis, and cardiovascular disease. Nutrients 2019; 11(5): 1155. [http://dx.doi.org/10.3390/nu11051155] [PMID: 31126110]

[48]   Putnik P, Bursać Kovačević D, Režek Jambrak A, *et al.* Innovative "Green" and novel strategies for the extraction of bioactive added value compounds from citrus wastes-A review. Molecules 2017; 22(5): 680.
[http://dx.doi.org/10.3390/molecules22050680] [PMID: 28448474]

[49]   Sun NX, Tong LT, Liang TT, *et al.* Effect of oat and tartary buckwheat – Based food on cholesterol – lowering and gut microbiota in hypercholesterolemic hamsters. J Oleo Sci 2019; 68(3): 251-9.
[http://dx.doi.org/10.5650/jos.ess18221] [PMID: 30760672]

[50]   Berk Z. Nutritional and health-promoting aspects of citrus consumption.Citrus Fruit ProcessingChapter 13Berk ZSan DiegoAcademic Press. 2016; pp. 261-79.
[http://dx.doi.org/10.1016/B978-0-12-803133-9.00013-8]

[51]   Elleuch M, Bedigian D, Roiseux O, Besbes S, Blecker C, Attia H. Dietary fibre and fibre-rich by-products of food processing: Characterisation, technological functionality and commercial applications: A review. Food Chem 2011; 124(2): 411-21.
[http://dx.doi.org/10.1016/j.foodchem.2010.06.077]

[52]   Marín FR, Soler-Rivas C, Benavente-García O, Castillo J, Pérez-Alvarez JA. By-products from different citrus processes as a source of customized functional fibres. Food Chem 2007; 100(2): 736-41.
[http://dx.doi.org/10.1016/j.foodchem.2005.04.040]

[53]   Grigelmo-Miguel N, Martín-Belloso O. Characterization of dietary fiber from orange juice extraction. Food Res Int 1998; 31(5): 355-61.
[http://dx.doi.org/10.1016/S0963-9969(98)00087-8]

[54]   Rezzadori K, Benedetti S, Amante ER. Proposals for the residues recovery: Orange waste as raw material for new products. Food Bioprod Process 2012; 90(4): 606-14.
[http://dx.doi.org/10.1016/j.fbp.2012.06.002]

[55]   R. C. Bortoluzzi and C. Marangoni, 'Characterization of dietery fibre from orange juice extraction', Rev. Bras. Prod. Agroindustriais, p. 6, 2006.

[56]   Fernandez-Ginés JM, Fernandez-Lopez J, Sayas-Barbera E, Sendra E, Perez-Alvarez JA. Effect of storage conditions on quality characteristics of Bologna sausages made with citrus fiber. J Food Sci 2003; 68(2): 710-4.
[http://dx.doi.org/10.1111/j.1365-2621.2003.tb05737.x]

[57]   Yilmaz E, Karaman E. Functional crackers: incorporation of the dietary fibers extracted from citrus seeds. J Food Sci Technol 2017; 54(10): 3208-17.
[http://dx.doi.org/10.1007/s13197-017-2763-9] [PMID: 28974806]

[58]   Nangul A, Bhatia R. Microorganisms: A marvelous source of single cell proteins. J Microbiol Biotechnol Food Sci 2013; 3(1): 15-8.

[59]   Sharif M, Zafar MH, Aqib AI, Saeed M, Farag MR, Alagawany M. Single cell protein: Sources, mechanism of production, nutritional value and its uses in aquaculture nutrition. Aquaculture 2021; 531735885.
[http://dx.doi.org/10.1016/j.aquaculture.2020.735885]

[60]   Najafpour GD. Single-Cell Protein.Biochemical Engineering and BiotechnologyChapter 142nd edNajafpour GDAmsterdamElsevier. 2015; pp. 417-34.
[http://dx.doi.org/10.1016/B978-0-444-63357-6.00014-6]

[61]   Kantifedaki A, Kachrimanidou V, Mallouchos A, Papanikolaou S, Koutinas AA. Orange processing waste valorisation for the production of bio-based pigments using the fungal strains Monascus purpureus and Penicillium purpurogenum. J Clean Prod 2018; 185: 882-90.
[http://dx.doi.org/10.1016/j.jclepro.2018.03.032]

[62]   Azam S, Khan Z, Bashir A, Khan I, Ali J. Production of single cell protein from orange peels using *Aspergillus niger* and *Saccharomyces cerevisiae*. Glob J Biotechnol Biochem 2014; 9: 14-8.

[http://dx.doi.org/10.5829/idosi.gjbb.2014.9.1.82314]

[63]   Deveci E, Ozyurt M. Evaluation of citrus wastes for production of single cell protein. Adv Food Sci 2004; 26: 130-4.

[64]   Lo Curto R, Tripodo MM, Leuzzi U, Giuffrè D, Vaccarino C. Flavonoids recovery and SCP production from orange peel. Bioresour Technol 1992; 42(2): 83-7.
[http://dx.doi.org/10.1016/0960-8524(92)90065-6]

[65]   Mondal AK, Sengupta S, Bhowal J, Bhattacharya D. Utilization of fruit wastes in producing single cell protein. Int J Sci 2012; 1(5): 430-8.

[66]   Jayani RS, Saxena S, Gupta R. Microbial pectinolytic enzymes: A review. Process Biochem 2005; 40(9): 2931-44.
[http://dx.doi.org/10.1016/j.procbio.2005.03.026]

[67]   Ahmed I, Zia MA, Hussain MA, Akram Z, Naveed MT, Nowrouzi A. Bioprocessing of citrus waste peel for induced pectinase production by *Aspergillus niger*; its purification and characterization. Journal of Radiation Research and Applied Sciences 2016; 9(2): 148-54.
[http://dx.doi.org/10.1016/j.jrras.2015.11.003]

[68]   Shariq M, Sohail M. *Citrus limetta* peels: a promising substrate for the production of multienzyme preparation from a yeast consortium. Bioresour Bioprocess 2019; 6(1): 43.
[http://dx.doi.org/10.1186/s40643-019-0278-0]

[69]   Tao N, Shi W, Liu Y, Huang S. Production of feed enzymes from citrus processing waste by solid-state fermentation with *Eupenicillium javanicum*. Int J Food Sci Technol 2011; 46(5): 1073-9.
[http://dx.doi.org/10.1111/j.1365-2621.2011.02587.x]

[70]   Mathias DJ, Kumar S, Rangarajan V. An investigation on citrus peel as the lignocellulosic feedstock for optimal reducing sugar synthesis with an additional scope for the production of hydrolytic enzymes from the aqueous extract waste. Biocatal Agric Biotechnol 2019; 20101259.
[http://dx.doi.org/10.1016/j.bcab.2019.101259]

[71]   Rangarajan V, Rajasekharan M, Ravichandran R, Sriganesh K. Int J Biotechnol Biochem 2010; 6: 973-2691.

[72]   Ismail AMS. Utilization of orange peels for the production of multienzyme complexes by some fungal strains. Process Biochem 1996; 31(7): 645-50.
[http://dx.doi.org/10.1016/S0032-9592(96)00012-X]

[73]   Ruiz HA, Rodríguez-Jasso RM, Rodríguez R, Contreras-Esquivel JC, Aguilar CN. Pectinase production from lemon peel pomace as support and carbon source in solid-state fermentation column-tray bioreactor. Biochem Eng J 2012; 65: 90-5.
[http://dx.doi.org/10.1016/j.bej.2012.03.007]

[74]   Dhillon SS, Gill RK, Gill SS, Singh M. Studies on the utilization of citrus peel for pectinase production using fungus *Aspergillus niger*. Int J Environ Stud 2004; 61(2): 199-210.
[http://dx.doi.org/10.1080/00207230320001433346]

[75]   Joshi M, Nerurkar M, Adivarekar R. Use of *Citrus limetta* peels for pectinase production by marine *Bacillus subtilis*. Innov Rom Food Biotechnol 2013; 9.

[76]   Ali J, Jaffery SA, Assad Q, Hussain A, Abid H, Gul F. Optimization of pectinase enzyme production using sour oranges peel (*Citrus aurantium* L.) as substrate. Pak J Biochem Mol Biol Pak 2010; 43(3): 126-30.

[77]   Pedrolli DB, Carmona EC. Pectin lyase from *Aspergillus giganteus*: Comparative study of productivity of submerged fermentation on citrus pectin and orange waste. Appl Biochem Microbiol 2009; 45(6): 610-6.
[http://dx.doi.org/10.1134/S0003683809060064] [PMID: 20067152]

[78]   Zhou JM, Ge XY, Zhang WG. Improvement of polygalacturonase production at high temperature by

mixed culture of *Aspergillus niger* and *Saccharomyces cerevisiae*. Bioresour Technol 2011; 102(21): 10085-8.
[http://dx.doi.org/10.1016/j.biortech.2011.08.077] [PMID: 21908185]

[79]   Awasthi MK, Kumar V, Yadav V, *et al.* Current state of the art biotechnological strategies for conversion of watermelon wastes residues to biopolymers production: A review. Chemosphere 2022; 290133310.
[http://dx.doi.org/10.1016/j.chemosphere.2021.133310] [PMID: 34919909]

[80]   Liu H, Kumar V, Jia L, *et al.* Biopolymer poly-hydroxyalkanoates (PHA) production from apple industrial waste residues: A review. Chemosphere 2021; 284131427.
[http://dx.doi.org/10.1016/j.chemosphere.2021.131427] [PMID: 34323796]

[81]   Awasthi SK, Kumar M, Kumar V, *et al.* A comprehensive review on recent advancements in biodegradation and sustainable management of biopolymers. Environ Pollut 2022; 307119600.
[http://dx.doi.org/10.1016/j.envpol.2022.119600] [PMID: 35691442]

[82]   Duan Y, Tarafdar A, Kumar V, *et al.* Sustainable biorefinery approaches towards circular economy for conversion of biowaste to value added materials and future perspectives. Fuel 2022; 325124846.
[http://dx.doi.org/10.1016/j.fuel.2022.124846]

[83]   Kumar V, Sharma N, Umesh M, *et al.* Emerging challenges for the agro-industrial food waste utilization: A review on food waste biorefinery. Bioresour Technol 2022; 362127790.
[http://dx.doi.org/10.1016/j.biortech.2022.127790] [PMID: 35973569]

[84]   Awasthi SK, Sarsaiya S, Kumar V, *et al.* Processing of municipal solid waste resources for a circular economy in China: An overview. Fuel 2022; 317123478.
[http://dx.doi.org/10.1016/j.fuel.2022.123478]

[85]   Kumar V, Sharma N, Maitra SS. *In vitro* and *in vivo* toxicity assessment of nanoparticles. Int Nano Lett 2017; 7(4): 243-56.
[http://dx.doi.org/10.1007/s40089-017-0221-3]

[86]   Vinay K, Neha S, Maitra S. Protein and Peptide Nanoparticles: Preparation and Surface Modification, in Functionalized Nanomaterials I. CRC Press 2020; pp. 191-204.

[87]   Vallinayagam S, *et al.* Recent developments in magnetic nanoparticles and nano-composites for wastewater treatment. J Environ Chem Eng 2021; 9(6)106553.
[http://dx.doi.org/10.1016/j.jece.2021.106553]

[88]   Kumar V, Sharma N, Lakkaboyana SK, Maitra SS. Silver nanoparticles in poultry health: Applications and toxicokinetic effects, in Silver Nanomaterials for Agri-Food Applications. Elsevier 2021; pp. 685-704.
[http://dx.doi.org/10.1016/B978-0-12-823528-7.00005-6]

[89]   Egbosiuba T C. Biochar and bio-oil fuel properties from nickel nanoparticles assisted pyrolysis of cassava peel 2022.
[http://dx.doi.org/10.1016/j.heliyon.2022.e10114]

CHAPTER 8

Valorization of Waste Plastics to Produce Fuels and Chemicals

Varsha Sharma[1,*]

[1] *Central Pollution Control Board (CPCB), Waste Management Division-I, East Arjun Nagar, Vishwas Nagar Extension, Vishwas Nagar, Shahdara, Delhi, 110032, India*

Abstract: The increase in the use of plastic products caused the major worldwide disposal problem of plastic solid waste (PSW). Plastics are becoming appropriate materials of interest for everyone due to their attractive applications in households, packaging, healthcare, and industries owing to their durability and versatile functionality at affordable prices. Statistics show that a large number of waste plastics are dumped in landfills, and only a tiny amount of plastic is recycled for making valuable materials *e.g.*, shampoo bottles, film, sheets, trash bags, kitchen-wares and packing materials. About 26,000 tonnes of plastic waste is generated in India every day, of which 40% remains uncollected and littered leading to adverse impacts on human health and the environment. Further, the incineration of plastic wastes emits many harmful gases such as nitrous oxide, sulfur oxides, dust clouds, dioxins and other toxins that pollute the atmosphere. To reduce waste plastics generation in the environment, the Indian government has implemented the Plastic Waste Management Rules, 2016 and its amendments, which explain ways for collection and management of plastic waste, its recycling, and utilization. Plastic wastes can be valorized to produce fuels using techniques such as thermal degradation, catalytic cracking, and gasification. This chapter is focused on waste plastic handling approaches, and novel routes to convert plastic wastes into energy and other valuable chemicals. This approach may compensate for high-energy demands and plastic waste management.

Keywords: Biodegradable, Catalytic, Chemicals, Conversion, Disposal, Degradation, Environment, Energy, Fuel, Hydrogen, Management, Plastic, Polymer, Production, Pyrolysis, Recycling, Revenue, Sustainable, Valorization, Waste.

INTRODUCTION

In recent years, the disposal of waste plastics has become a major worldwide environmental problem. Plastics are made from natural materials such as cellulose

---

* **Corresponding author Varsha Sharma:** Central Pollution Control Board (CPCB), Waste Management Division-I, East Arjun Nagar, Vishwas Nagar Extension, Vishwas Nagar, Shahdara, Delhi, 110032 India; E-mail: varshasharma277@gmail.com

Vinay Kumar, Sivarama Krishna Lakkaboyana & Neha Sharma (Eds.)

coal, natural gas and crude oil through polymerization and polycondensation processes. Further, physical properties of plastics i.e, lightweight, durability, versatility and relatively low cost make them a suitable candidate for applications in materials such as concrete, glass, metals, wood, natural fibers, and paper. Plastic production has increased by 3-4% annually since the 1990s and its consumption is projected to increase dramatically in developing countries due to economic expansion [1]. Nowadays, it is reported that only 9 -12% of global plastic waste is recycled and incinerated, while up to 79% is discarded into landfills or the natural environment, indicating that there is a great need for exploring innovative recycling methods to dispose of plastic wastes [2]. Over the past seventy years, the plastic industry has witnessed drastic growth, in the production of synthetic polymers represented by polyethylene (PE), polypropylene (PP), polystyrene (PS), polyethylene terephthalate (PET), polyvinyl alcohol (PVA) and polyvinyl chloride (PVC). Plastic packaging is the largest application by weight, but plastics are also used widely in the textile, consumer goods, transport, and construction sectors. Further, the disposal of plastics has become a major environmental and economic issue. Moreover, inadequate waste plastic management causes serious environmental impacts, such as their accumulation in the oceans leading to marine debris [3]. The harmful effects of plastic are shown in Fig. (**1**). Plastics are typically organic polymers of high molecular mass and often contain other substances. The burning of plastics releases toxic gases like dioxins, furans, hydrogen chloride, airborne particles, and carbon dioxide into the atmosphere which contribute to climate change and air pollution. Burning of plastic wastes increases the risk of heart disease, aggravates respiratory ailments such as asthma and emphysema and causes rashes, nausea, or headaches, and damages the nervous system [4]. Recycling plastic and conversion of waste into energy are the best possible solutions for the management of plastic waste. Due to the high cost and poor biodegradability, it is undesirable to dispose of plastics in a landfill. Comparatively, plastic recycling based on pelletizing and molding to low-grade plastics has attracted the interest of many scientists worldwide, but recycled plastic possesses poor mechanical strength and color properties, hence having low market values and restricted applications [5]. The recycling of virgin plastic material can be done 2-3 times only because after every recycling, the plastic material deteriorates due to thermal pressure resulting in a reduced lifespan. Further, waste-to-energy technologies enable converting waste plastics into heat, hydrocarbon fuels and chemicals, therefore reducing the number of plastics to be landfilled [6]. Plastic Solid Waste (PSW) recycling processes could be allocated to four major categories, re-extrusion (primary), mechanical (secondary), chemical (tertiary), and energy recovery (quaternary). Each method provides a unique set of advantages that make it particularly beneficial for specific locations, applications or requirements [7]. The re-extrusion (primary) process involves re-introducing scrap plastics into valuable products. Mechanical recycling (Secondary) involves various operations that aim to recover plastics *via* mechanical processes (grinding, washing, separating, drying, re-granulating, and compounding), thus producing recyclates that can be converted into plastic products, substituting virgin plastics. Chemical recycling (tertiary), that is, the conversion of waste plastics into feedstock or fuel has been recognized as an ideal approach and could significantly reduce the net cost of disposal. The energy recovery (quaternary) process involves complete or partial oxidation of the material, producing heat, power, gaseous fuels, oils, and chars. These by-products must be disposed of. Among these processes, the chemical recycling process is useful in the production of fuel. Chemical recycling processes are like those employed in the petrochemical industry *e.g.*, Pyrolysis, liquid gas hydrogenation, viscosity breaking, steam or catalytic cracking and the use of plastic solid waste as a reducing agent in furnaces. These are suitable methods for producing different fuels from plastic solid waste. Developed countries like japan, Germany and the United States have successfully implemented plastic in the fuel conversion process in their countries [8]. This review is helpful to convert waste plastic into value-added chemicals.

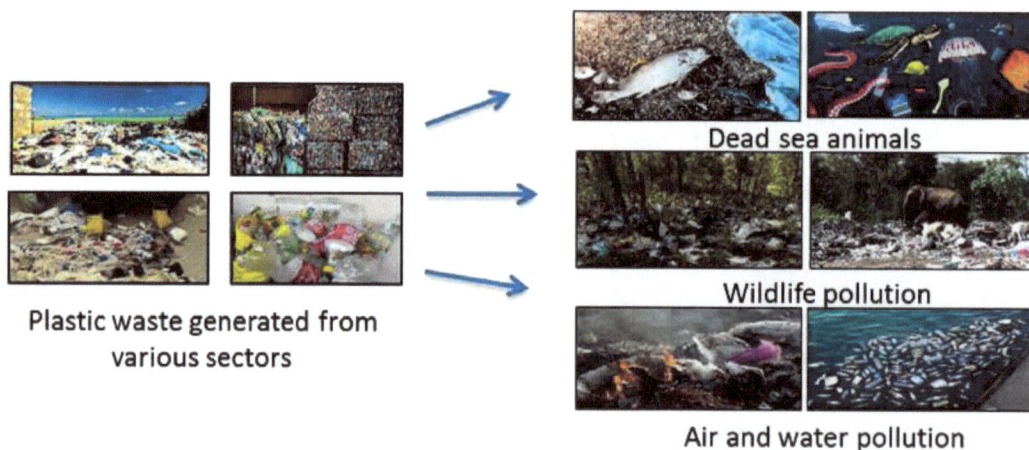

Fig. (1). Waste plastic and its harmful effect.

GLOBAL SCENARIO OF WASTE PLASTICS PRODUCTION AND ITS MANAGEMENT

On an average, the production of plastic globally crosses 150 million tonnes per year. Plastic waste is also a growing concern and is present in all the world's ocean basins, including around remote islands, the poles and in the deep seas. Since 1950, close to half of all plastic has ended up in landfill or dumped in the wild and only 9% of used plastic has been recycled. Every year, it is estimated that 4 to 12 million metric tons of plastic waste ends up in the oceans. The amount of plastic used has indeed grown constantly over the past 30 years, reaching over 300 million metric tons in 2017. This growth is set to continue, driven in large part by the demands of the Chinese and Indian middle classes; the demand may double by 2050. Over 90% of raw plastic is produced from fossil fuels (oil or natural gas) [9]. The plastic is sold to plastic manufacturers to make objects, mostly by injection, blow molding or heat forming. It is assessed that approximately 70% of plastic packaging products are converted into plastic waste in a short span. Approximately 9.4 million (tonnes per annum) plastic wastes are generated in the country, which amounts to 26,000 (tonnes per day). Of this, about 60% is recycled, most of it by the informal sector. While the recycling rate in India is higher than the global average of 20%, there is still over 9,400 tonnes of plastic waste which is either landfilled or ends up polluting streams or groundwater resources [10]. Furthermore, laws and guidelines enforced by the government across the country for plastic waste disposal management are necessary. As per the Annual Report information for year 2017-18, 9.4 million tonnes of plastic waste was estimated by the Central Pollution Control Board (CPCB) and out of this, approximately 5.6 Million tonnes per annum of plastic waste is recycled and 3.8 Million tonnes per annum plastic waste is left uncollected or littered. Currently, compostable plastics (i.e. 100% bio-based) have been introduced which is a good alternative to petro-based plastic carry bags. The Government of India notified Plastic Waste Management (PWM) Rules, on 18th March, 2016, superseding Plastic Waste (Management & Handling) Rules, 2011. These rules were further amended and named as 'Plastic Waste Management (Amendment) Rules, 2018 (Ministry of Housing & Urban Affairs Government of India). In addition, as per provisions under Rule 4 (h) of PWM Rules, 2016, the producer or sellers of compostable plastic carry bags are required to apply for a certificate and should obtain this certificate from the Central Pollution Control Board (CPCB) before marketing or selling of products. Recycled plastics are more harmful to the environment than virgin products due to the mixing of additives, colors, stabilizers, halogenated flame-retardants, and so on. Virgin plastic is mostly made in North America (18%), Europe (19%), and Asia (50%, with China accounting for 29%). The U.S is one of the largest plastic wastes generating countries. The U.S. exports most of its plastic waste to countries including China, Hong Kong, Ecuador, Colombia, Indonesia, Vietnam, and Thailand, where it is processed for energy generation or disposed of in landfills. As shown in Fig. (**2**), the current per capita consumption of plastics in the U.S. is 109 kg and in China and India, it is 38 and 11 kg respectively. Plastic waste will add an economic value to materials to create an extra source of revenue for stakeholders as shown in Fig. (**3**).

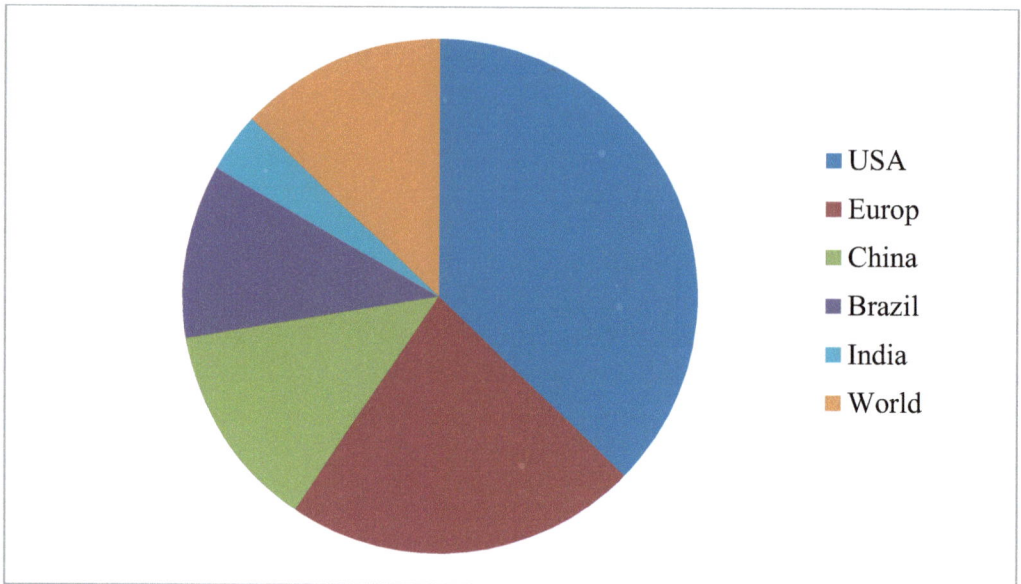

Fig. (2). Per capita plastic products consumption (Kg/person).

WASTE PLASTIC TO FUEL CONVERSION TECHNOLOGIES

The global recycled plastic and plastic waste to oil market accounted for $47.29 million in 2019 and are expected to reach $112.96 million by 2027 growing at a CAGR of 11.5% during the forecast period [13]. Geographically, Europe is the largest market for recycled plastic and plastic waste to oil. Various types of waste plastics i.e., polypropylene (PP), polystyrene (PS), polyethylene (PE), polyethylene terephthalate (PET) and polyvinyl chloride (PVC) can be converted into gasoline, diesel, synthetic gases, and kerosene through technologies such as catalytic de-polymerization gasification and pyrolysis p. Adopting plastic to the fuel procedure provides significant economic benefits to the region and creates abundant job opportunities. The major product of plastics-to-fuel is diesel. The produced diesel has an exceptionally low content of sulfur and due to its low content, it reduces the negative impact on the environment. The plastic to oil process is one of the best-emerging technologies to convert plastic into synthetic crude oil and other petroleum products [14]. The plastic to oil (fuel) process not only aids in converting plastic to synthetic crude oil (fuel) but also enables eco-friendly disposal of plastic waste. Pyrolysis and Gasification produce fuels or electricity and provide a more flexible way of storing energy than incineration. These processes also generate incredibly low sulfur and nitrogen oxide as compared to incineration and are less harmful to the environment. Further, plastic to fuel conversion through various processes is summarized in Table **1**. The U.S. plastic-to-fuel market is dominated by major players, such as Plastic2Oil, Agilyx

Corporation, Vadxx Energy, Green Envirotec Holdings LLC and RES polyflow [15]. List of companies which are converting waste plastic into fuels is summarized in Table **2**.

Table 1. Plastic to fuel conversion through various processes.

Type of Plastic	Process/System used for conversion	Oil/Fuel produced	Ref.
Polypropylene (PP)	Catalytic Pyrolysis	Diesel	[19]
High-density polyethylene (HDPE)	Integrated continuous stirred tank reactor – reactive distillation column (CSTRRDC) pyrolysis system	Diesel	[20]
HDEP	Pyrolysis reactor	Diesel	[21]
HDEP	Laboratory scale externally heated fixed bed pyrolysis batch reactor (Thermal Pyrolysis	Mixture of oil, gas and small amount of char	[22]
Waste plastic from electrical and electronic equipment (WEEE)	Catalytic Pyrolysis	Aromatic oil	[23]
HDEP	Laboratory-scale, continuous fluidized bed reactor	Pyrolysis oil, gas & Char	[24]
Low density polyethylene (LDPE)	Catalytic microwave degradation	Jet fuel	[25]
High-density polyethylene (HDPE-wax) and polypropylene (PP-wax).	Catalytic cracking	Gasoline	[26]
polyethylene (PE), polypropylene (PP), polystyrene (PS) and polyethylene terephthalate (PET)	Catalytic Pyrolysis	Aromatic hydrocarbons	[27]
LDPE	Catalytic cracking	Liquid Fuel	[28]
LDPE	Catalytic Pyrolysis	Jet fuel	[29]
Waste Plastic	Gasification system	Catalytic hydrogen production	[30]
PP	Catalytic Pyrolysis	Liquid fuel	[31]
PP		Liquid Hydrocarbons	[32]
mix plastics waste i.e PET, HDPE, LDPE, PVC, PP and PS		Waste plastic oil	[33]
HDPE	Catalytic co-pyrolysis	petroleum-like hydrocarbons	[34]
LDPE	Catalytic Pyrolysis	Waste plastic oil	[35]
PET	Pyrolysis in fixed bed reactor	Waste plastic oil	[36]

(Table 1) cont.....

Type of Plastic	Process/System used for conversion	Oil/Fuel produced	Ref.
HDPE, PP and PS	Continuous and batch pyrolysis reactors	liquid fuel oil	[37]
PP	Integrated pyrolysis	Liquid fuel	[38]

Table 2. Companies globally converting waste plastics to fuel and oil.

Name of Company	Location/Country
Nexus Fuels	Atlanta, Georgia
Plastic Energy	London, UK
MK Aromatics Limited	Tamilnadu, India
Vadxx Energy LLC	Cleveland, Ohio
Res Polyflow	Akron, Ohio, USA
Plastic2Oil Inc. (JBI Inc.)	Niagara Falls, New York
Agile Process Chemicals LLP	Maharashtra, India
Northwood Exploration Israel Ltd	Israel
Agilyx, Inc.	Tigard, Oregon
Niutech	China
Clean Blue Technologies Inc	North Carolina, US
Polycycl Private Limited	Haryana, India
Anhui Oursun Environmental Technologies	China
Blest	Japan
Cynar Plc	Ireland
ECO – Int'l Marketing	Korea
Klean Industries, Inc.	Vancouver, BC Canada
P-Fuel, Ltd.	Australia
Plastic Advanced Recycling Corp.	Illinois, China
PlastOil	Switzerland
Promeco/Cimelia	Italy/Singapore
T- Technology	Poland, Spain, Italy
Green Fuels AG	Germany
Sapporo Plastic Recycling	Japan
Rudra Environmental Solutions	Pune, India
Paterson Energy Private Limited	India

To solve the problem of waste plastic management, researchers have developed processes to convert waste plastic into hydrogen which could also be used as fuel.

Pyrolysis of waste plastics has been suggested as a fossil fuel-free method to produce hydrogen. However, this still requires elevated temperatures of 500–800 °C. A continuous hydrogen production from waste plastic (WP) by fluidized-bed gasification (FBG) and plastic waste pyrolysis-reforming strategy has been developed [16]. Hydrogen is produced *via* catalytic thermo-chemical conversion of plastic waste [17] and catalytic pyrolysis with activated carbon and MgO [18].

WASTE PLASTICS TO FUEL CONVERSION TECHNOLOGIES

Pyrolysis and Thermal Decomposition

One of the first extensive technologies for pyrolysis appeared in 1978 with the name of PYROPLEQ. Pyrolysis consists of thermal decomposition under an inert gas like nitrogen and a process conducted in the absence of oxygen. The plastic materials are introduced into a reactor where they decompose at 400°C-600°C; the major product of the process is an oily mixture of liquid hydrocarbons obtained through the condensation of the decomposed vapors. The evaporated oil may also be further cracked with a catalyst. The use of catalysts enables to decrease the cracking temperature. In addition, they increase the reaction, selectivity and quality. The most used catalysts in the literature for plastic waste pyrolysis include silica alumina, zeolites (beta, USY, ZSM-5, REY, clinoptilolite, *etc.*), and MCM-41 [39]. The typical scheme of pyrolysis for the production of fuel from waste plastic is reported in Fig. (**3**).

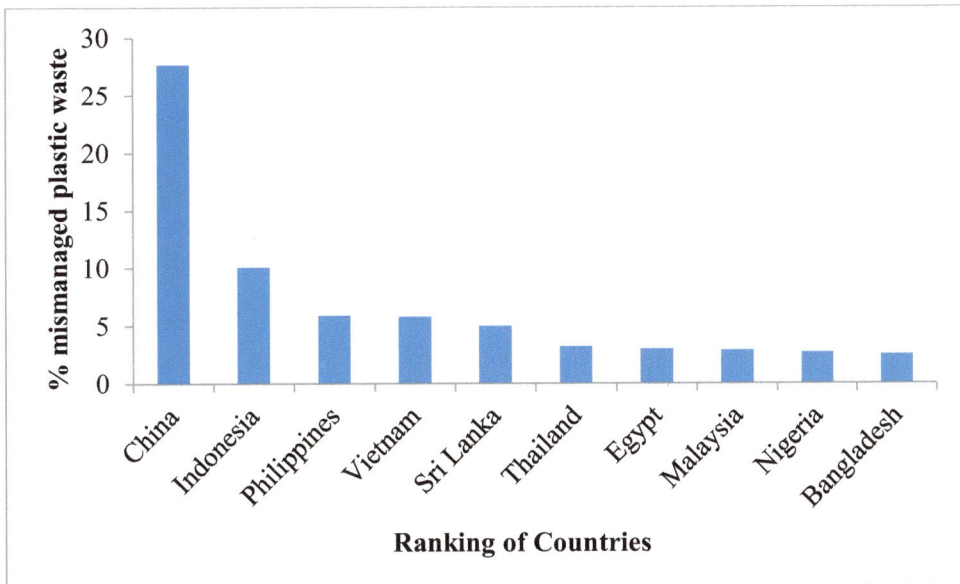

Fig. (3). Top ten countries mismanaged plastic waste. Source: [12].

Catalytic Degradation

In this method, waste plastic is melted and cracked in the absence of oxygen and at extremely high temperatures, the resulting gases were cooled by condensation and the resulting crude oil was recovered. A suitable catalyst is used to conduct the cracking reaction. The reuse of catalysts and the use of effective catalysts in lesser quantities can optimize this option [40]. This process can be developed into a cost-effective commercial polymer recycling process for solving the acute environmental problem of the disposal of plastic waste. It also offers the higher cracking ability of plastics, and a lower concentration of solid residues in the product. Compared to thermal degradation, which usually occurs *via* a free radical mechanism, catalytic degradation occurs by carbocations, which consist of hydrocarbon ions carrying a single positive charge. The catalysts allow the reduction of the processing temperature that leads to a decrease in energy consumption; at the same time, they improve the output quality and the corresponding yield (gas & fuel).

Gasification

In gasification, plastic waste is reacted with gasifying agent (*e.g.*, steam, oxygen and air) at high temperature around 500–1300 °C. The process oxidizes the hydrocarbon feedstock to generate the endothermic de-polymerization heat and produce (primary) a gaseous mixture of carbon monoxide and hydrogen, with minor percentages of gaseous hydrocarbons. In this process, syngas is produced which may be used as an energy source for combustion processes [41]. In recent years, many research studies are focusing on a fluidized bed, co-gasification and two-stage gasification to overcome the related technological issues. If the process does not occur with help of an oxidizing agent, it is called indirect gasification. Fig. (**4**) presents the process of pyrolysis.

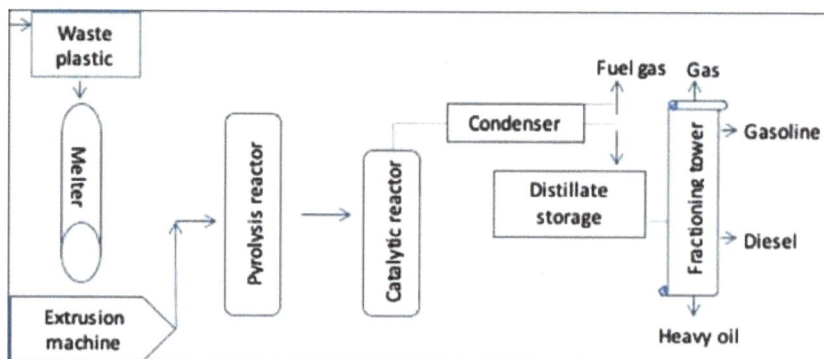

Fig. (4). Pyrolysis process.

WASTE PLASTIC TO CHEMICALS AND OTHER VALUE-ADDED PRODUCTS

Nanyang Technological University, Singapore has reported their work in Advance Science about the discovery of a method that breaks down non-biodegradable plastics into formic acid a chemical with multiple uses using sunlight and a catalyst that does not contain heavy metals. Formic acid is a naturally occurring preservative and antibacterial agent, which can be used for energy generation by power plants and in hydrogen fuel cell vehicles [42]. In another research, waste plastic polyethylene terephthalate (PET) bottles have been utilized to prepare copper-1,4-benzenedicarboxylate metal for the removal of methylene blue from wastewater [43]. Effective conversion of polyethylene waste (plastic bags *etc*.) into high value-added porous carbon materials has been reported *via* a simple mechano-mixed method and subsequent carbonization with a flame retard agent, magnesium carbonate pentahydrate (MCHP). The introduction of MCHP not only provides an in-situ hard template (MgO), but also stabilizes the polyethylene during thermal treatment. It provides a promising approach for the disposal of waste plastics and opens new applications in various energy fields [44]. Synthesis and characterization of sulfonated activated carbon as a catalyst for bio-jet fuel production from biomass and waste plastics have been reported [45]. Cubic phase TaC nanomaterials have been synthesized by a solid-state reaction of waste Polyvinyl Chloride, Ta_2O_5, and metallic Lithium (Li) in a stainless-steel autoclave at 700-1000°C. TaC nanoparticles with different compositions have been obtained at different reaction temperatures. This method provides a novel way to resource utilization of waste plastics [46]. Several other biochemical and valorization methods are available to produce value-added products [47 - 57]. In addition, Terephthalonitrile (TPN) has been directly produced from polyethylene terephthalate (PET) plastic *via* catalytic fast pyrolysis with ammonia [58]. Further, an overview of chemical additives present in plastics has been summarized by researchers [59].

Practices Adopted Worldwide for Energy Recovery from Plastic Waste

European nations, like Switzerland and Norway, recycle most of their plastic waste into fuel and use it for energy recovery. Both M/s Green Fuels AG, Germany, and M/s Sapporo Plastic Recycling ("SPR"), Japan, are producing fuel oil from plastic waste and have established a fully commercial plant capable of recycling over fifty tonnes of mixed plastic waste per day. In India, M/s Polycyclic Private Limited Company, Haryana; M/s Rudra Environmental Solutions, Pune and M/S Paterson Energy Private Limited, Chennai (Mathura Plant) are converting waste plastic into fuel.

CONCLUSION AND FUTURE PROSPECTUS

Due to the depletion of fossil fuel sources such as coal, natural gas and crude oil, there is an urgent need to search for new alternatives such as plastic waste for the production of fuel. Plastics have a high energy content that can be converted into electricity, synthetic gas, fuels, and other products of chemistry. Recovering this abundant energy also reduces waste sent to landfills and complements plastics recycling. The current global market scenario for the conversion of plastic into fuel was reviewed in this paper. Various techniques of converting waste plastics into fuel including thermal degradation, pyrolysis, and gasification were highlighted. Pyrolysis is a promising technology for polymeric waste disposal. Thus, an effort was made to solve the problem of plastic waste disposal and optimize the use of plastic waste for the production of fuel and useful chemical. It is required from businesses, governments, and the international community to solve the global plastic waste pollution problem. By converting plastics into fuel, it is possible to reduce large plastic wastes and fulfill the demand for fuel. Further, plastic to fuel technologies will surely provide a strong platform for a sustainable, clean and green future.

ACKNOWLEDGEMENTS

The author, V. S. would like to thank Central Pollution Control Board (Under the Ministry of Environment, Forest and Climate Change), Delhi for the stipend of Research Associate.

REFERENCES

[1] Xue Y, Johnston P, Bai X. Effect of catalyst contact mode and gas atmosphere during catalytic pyrolysis of waste plastics. Energy Convers Manage 2017; 142: 441-51.
[http://dx.doi.org/10.1016/j.enconman.2017.03.071]

[2] Ru J, Huo Y, Yang Y. Microbial degradation and valorization of plastic wastes. Front Microbiol 2020; 11(442): 442.
[http://dx.doi.org/10.3389/fmicb.2020.00442] [PMID: 32373075]

[3] Barbarias I, Lopez G, Artetxe M, Arregi A, Bilbao J, Olazar M. Valorisation of different waste plastics by pyrolysis and in-line catalytic steam reforming for hydrogen production. Energy Convers Manage 2018; 156: 575-84.
[http://dx.doi.org/10.1016/j.enconman.2017.11.048]

[4] Verma R, Vinoda KS, Papireddy M, Gowda ANS. Toxic Pollutants from Plastic Waste- A Review. Procedia Environ Sci 2016; 35: 701-8.
[http://dx.doi.org/10.1016/j.proenv.2016.07.069]

[5] Burange AS, Gawande MB, Lam FLY, Jayaram RV, Luque R. Heterogeneously catalyzed strategies for the deconstruction of high density polyethylene: plastic waste valorisation to fuels. Green Chem 2015; 17(1): 146-56.
[http://dx.doi.org/10.1039/C4GC01760A]

[6] Lee U, Han J, Wang M. Evaluation of landfill gas emissions from municipal solid waste landfills for the life-cycle analysis of waste-to-energy pathways. J Clean Prod 2017; 166: 335-42.

[http://dx.doi.org/10.1016/j.jclepro.2017.08.016]

[7]   Al-Salem SM, Lettieri P, Baeyens J. The valorization of plastic solid waste (PSW) by primary to quaternary routes: From re-use to energy and chemicals. Pror Energy Combust Sci 2010; 36(1): 103-29.
        [http://dx.doi.org/10.1016/j.pecs.2009.09.001]

[8]   A. K. Awasthi, M. Shivashankar, S. Majumder, Plastic solid waste utilization technologies: A Review, IOP Conf. Series: Mater. Sci. Eng., vol. 263: 022024, pp. 1-13, doi:10.1088/1757-899X/263/2/022024.

[9]   http://journals.openedition.org/factsreports/5102 Woldemar d'Ambrières, Plastics recycling worldwide: current overview and desirable changes, Field Actions Science Reports [Online], vol. Special Issue 19, pp. 12-21, Oct. 2019.

[10]   http://164.100.228.143:8080/sbm/content/writereaddata/SBM%20Plastic%20Waste%20Book.pdf

[11]   http://ficci.in/spdocument/20872/report-Plastic-infrastructure-2017-ficci.pdf FICCI, 3rd national conference on Sustainable Infrastructure with Plastics, Knowledge paper on Plastic Industry for Infrastructure, pp. 1-48, Feb. 2017,

[12]   https://www.earthday.org/top-20-countries-ranked-by-mass-of-mismanaged-plastic-waste/ Earth day 2018, "End plastic Pollution," April 2018

[13]   https://www.researchandmarkets.com/reports/5116399/recycled-plastic-and-plastic-waste-to-oil-rela0-4766764 Recycled Plastic and Plastic Waste to Oil - Global Market Outlook (2019-2027), June 2020, pp. 1-157,

[14]   Antelava A, Damilos S, Hafeez S, *et al.* Plastic Solid Waste (PSW) in the Context of Life Cycle Assessment (LCA) and Sustainable Management. Environ Manage 2019; 64(2): 230-44.
        [http://dx.doi.org/10.1007/s00267-019-01178-3] [PMID: 31230103]

[15]   https://www.futuremarketinsights.com/reports/us-plastic-to-fuel-market Plastic-to-fuel Market: U.S. Industry analysis and opportunity assessment 2015-2020.

[16]   Dou B, Wang K, Jiang B, *et al.* Fluidized-bed gasification combined continuous sorption-enhanced steam reforming system to continuous hydrogen production from waste plastic Int J Hydrogen Energy Feb. 2016.; 3803-10.

[17]   Sharma SS, Batra VS. Production of hydrogen and carbon nanotubes *via* catalytic thermo-chemical conversion of plastic waste: review. J Chem Technol Biotechnol 2020; 95(1): 11-9.
        [http://dx.doi.org/10.1002/jctb.6193]

[18]   E. Huo, H. Lei, C. Liu, Y. Zhang, L. Xin, Y. Zhao, M. Qian, Q. Zhang, X. Lin, C. Wang, W. Mateo, E. M. Villota, R. Ruan, Jet fuel and hydrogen produced from waste plastics catalytic pyrolysis with activated carbon and MgO, Science of total environment, vol. 727, pp. 138411, July 2020.

[19]   Mangesh VL, Padmanabhan S, Tamizhdurai P, Ramesh A. Experimental investigation to identify the type of waste plastic pyrolysis oil suitable for conversion to diesel engine fuel. J Clean Prod 2020; 246119066.
        [http://dx.doi.org/10.1016/j.jclepro.2019.119066]

[20]   Auxilio AR, Choo WL, Kohli I, Chakravartula Srivatsa S, Bhattacharya S. An experimental study on thermo-catalytic pyrolysis of plastic waste using a continuous pyrolyser. Waste Manag 2017; 67: 143-54.
        [http://dx.doi.org/10.1016/j.wasman.2017.05.011] [PMID: 28532621]

[21]   Kaimal VK, Vijayabalan P. A study on synthesis of energy fuel from waste plastic and assessment of its potential as an alternative fuel for diesel engines. Waste Manag 2016; 51: 91-6.
        [http://dx.doi.org/10.1016/j.wasman.2016.03.003] [PMID: 26969288]

[22]   Khan MZH, Sultana M, Al-Mamun MR, Hasan MR. Pyrolytic Waste Plastic Oil and Its Diesel Blend: Fuel Characterization. J Environ Public Health 2016; 2016: 1-6.
        [http://dx.doi.org/10.1155/2016/7869080] [PMID: 27433168]

[23] Muhammad C, Onwudili JA, Williams PT. Catalytic pyrolysis of waste plastic from electrical and electronic equipment. J Anal Appl Pyrolysis 2015; 113: 332-9.
[http://dx.doi.org/10.1016/j.jaap.2015.02.016]

[24] Xue Y, Zhou S, Brown RC, Kelkar A, Bai X. Fast pyrolysis of biomass and waste plastic in a fluidized bed reactor. Fuel 2015; 156: 40-6.
[http://dx.doi.org/10.1016/j.fuel.2015.04.033]

[25] Zhang X, Lei H, Zhu L, *et al.* From plastics to jet fuel range alkanes *via* combined catalytic conversions. Fuel 2017; 188: 28-38.
[http://dx.doi.org/10.1016/j.fuel.2016.10.015]

[26] Lovás P, Hudec P, Jambor B, Hájeková E, Horňáček M. Catalytic cracking of heavy fractions from the pyrolysis of waste HDPE and PP. Fuel 2017; 203: 244-52.
[http://dx.doi.org/10.1016/j.fuel.2017.04.128]

[27] Xue Y, Johnston P, Bai X. Effect of catalyst contact mode and gas atmosphere during catalytic pyrolysis of waste plastics. Energy Convers Manage 2017; 142: 441-51.
[http://dx.doi.org/10.1016/j.enconman.2017.03.071]

[28] Wong SL, Ngadi N, Abdullah TAT, Inuwa IM. Conversion of low density polyethylene (LDPE) over ZSM-5 zeolite to liquid fuel. Fuel 2017; 192: 71-82.
[http://dx.doi.org/10.1016/j.fuel.2016.12.008]

[29] Zhang Y, Duan D, Lei H, Villota E, Ruan R. Jet fuel production from waste plastics *via* catalytic pyrolysis with activated carbons. Appl Energy 2019; 251113337.
[http://dx.doi.org/10.1016/j.apenergy.2019.113337]

[30] Yang RX, Xu LR, Wu SL, Chuang KH, Wey MY. Ni/SiO$_2$ core–shell catalysts for catalytic hydrogen production from waste plastics-derived syngas. Int J Hydrogen Energy 2017; 42(16): 11239-51.
[http://dx.doi.org/10.1016/j.ijhydene.2017.03.114]

[31] Panda AK, Alotaibi A, Kozhevnikov IV, Shiju NR. Pyrolysis of Plastics to Liquid Fuel Using Sulphated Zirconium Hydroxide Catalyst. Waste Biomass Valoriz 2020; 11(11): 6337-45.
[http://dx.doi.org/10.1007/s12649-019-00841-4]

[32] Gaurh P, Pramanik H. Thermal and catalytic pyrolysis of plastic waste polypropylene for recovery of valuable petroleum range hydrocarbon Int J Res Sci Engineer, 2018; 228-33. CHEMCON Special Issue: March 2018; e-ISSN: 2394-8299.

[33] Damodharan D, Sathiyagnanam AP, Rana D, Kumar BR, Saravanan S. Combined influence of injection timing and EGR on combustion, performance and emissions of DI diesel engine fueled with neat waste plastic oil. Energy Convers Manage 2018; 161: 294-305.
[http://dx.doi.org/10.1016/j.enconman.2018.01.045]

[34] Ryu HW, Kim DH, Jae J, Lam SS, Park ED, Park YK. Recent advances in catalytic co-pyrolysis of biomass and plastic waste for the production of petroleum-like hydrocarbons. Bioresour Technol 2020; 310123473.
[http://dx.doi.org/10.1016/j.biortech.2020.123473] [PMID: 32389430]

[35] Sambandam P, Venu H, Narayanaperumal BK. Effective utilization and evaluation of waste plastic pyrolysis oil in a low heat rejection single cylinder diesel engine. Energy Sources A Recovery Util Environ Effects 2020; (Aug): 1-17.
[http://dx.doi.org/10.1080/15567036.2020.1803453]

[36] Khairil, T. M. I. Riayatsyah, S. Bahri, S. E. Sofyan, J. Jalaluddin, F. Kusumo, A. S. Sillitonga, Y. Padli, M. Jihad, A. H. Shamsuddin, Experimental Study on the Performance of an SI Engine Fueled by Waste Plastic Pyrolysis Oil–Gasoline Blends,Energies, vol. 13: 4196, Aug. 2020.
[http://dx.doi.org/10.3390/en13164196]

[37] Owusu PA, Banadda N, Zziwa A, Seay J, Kiggundu N. Reverse engineering of plastic waste into useful fuel products. J Anal Appl Pyrolysis 2018; 130: 285-93.

[http://dx.doi.org/10.1016/j.jaap.2017.12.020]

[38]   R. Thahir, A. Altway, S. R. Juliastuti, Susianto, Production of liquid fuel from plastic waste using integrated pyrolysis method with refinery distillation bubble cap plate column, Energy and Reports, vol. 5, 70-77, No. 2019.

[39]   Cleetus C, Thomas S, Varghese S. Synthesis of Petroleum-Based Fuel from Waste Plastics and Performance Analysis in a CI Engine. J Energy 2013; 2013: 1-10.
       [http://dx.doi.org/10.1155/2013/608797]

[40]   Tiwari DC, Ahmad E, Singh K. Catalytic degradation of waste plastic into fuel range hydrocarbons. Int J Chem Res 2009; 1: 31-6.

[41]   Saebea D, Ruengrit P, Arpornwichanop A, Patcharavorachot Y. Gasification of plastic waste for synthesis gas production. Energy Rep 2020; 6: 202-7.
       [http://dx.doi.org/10.1016/j.egyr.2019.08.043]

[42]   Gazi S, Ðokić M, Chin KF, Ng PR, Soo HS. Visible Light–Driven Cascade Carbon–Carbon Bond Scission for Organic Transformations and Plastics Recycling. Adv Sci (Weinh) 2019; 6(24)1902020.
       [http://dx.doi.org/10.1002/advs.201902020] [PMID: 31871870]

[43]   Doan VD, Do TL, Ho TMT, Le VT, Nguyen HT. Utilization of waste plastic pet bottles to prepare copper-1,4-benzenedicarboxylate metal-organic framework for methylene blue removal. Sep Sci Technol 2020; 55(3): 444-55.
       [http://dx.doi.org/10.1080/01496395.2019.1577266]

[44]   Y. Lian, M. Ni, Z. Huang, R. Chen, L. Zhou, W. Utetiwabo, W. Yang, Polyethylene waste carbons with a mesoporous network towards highly efficient Supercapacitors, Chemical Engineering Journal, vol. 366, pp. 313-320, June 2019.

[45]   W. Mateo, H. Lei, E. Villota, M. Qian, Y. Zhao, E. Huo, Q. Zhang, X. Lin, C. Wang, Z. Huang, Synthesis and Characterization of Sulfonated Activated Carbon as a Catalyst for Bio-jet Fuel Production from Biomass and Waste Plastics, Bioresource Technology, vol. 297: 122411, Feb. 2020.
       [http://dx.doi.org/10.1016/j.biortech.2019.122411] [PMID: 31767431]

[46]   L. Wang, F. Zhang, W. Dai, Q. Cheng, L. Lu, K. Zhang, M. Lin, M. Shen, D. Wang, One step transformation of waste polyvinyl chloride to tantalum carbide@carbon nanocomposite at low temperature, Journal of American Ceramic Society, vol. 102, pp. 6455–6462, June 2019.

[47]   M. K. Awasthi *et al.*, Current state of the art biotechnological strategies for conversion of watermelon wastes residues to biopolymers production: A review, Chemosphere, p. 133310, 2021.

[48]   H. Liu *et al.*, Biopolymer poly-hydroxyalkanoates (PHA) production from apple industrial waste residues: A review, Chemosphere, vol. 284, p. 131427, 2021.

[49]   S. K. Awasthi *et al.*, A comprehensive review on recent advancements in biodegradation and sustainable management of biopolymers,Environmental Pollution, p. 119600, 2022.

[50]   Y. Duan *et al.*, Sustainable biorefinery approaches towards circular economy for conversion of biowaste to value added materials and future perspectives,Fuel, vol. 325, p. 124846, 2022.

[51]   V. Kumar *et al.*, Emerging challenges for the agro-industrial food waste utilization: A review on food waste biorefinery, Bioresource Technology, p. 127790, 2022.

[52]   S. K. Awasthi et al., Processing of municipal solid waste resources for a circular economy in China: An overview, Fuel, vol. 317, p. 123478, 2022.

[53]   Kumar V, Sharma N, Maitra SS. *In vitro* and *in vivo* toxicity assessment of nanoparticles. Int Nano Lett 2017; 7(4): 243-56.
       [http://dx.doi.org/10.1007/s40089-017-0221-3]

[54]   Vinay K, Neha S, Maitra S. Protein and Peptide Nanoparticles: Preparation and Surface Modification, in Functionalized Nanomaterials I. CRC Press 2020; pp. 191-204.

[55]   Vallinayagam S, Rajendran K, Lakkaboyana SK, Soontarapa K, Remya RR, Sharma VK, Kumar V, Venkateswarlu K, Koduru JR. Recent developments in magnetic nanoparticles and nano-composites for wastewater treatment. Journal of Environmental Chemical Engineering. 2021 Dec 1;9(6):106553.

[56]   Kumar V, Sharma N, Lakkaboyana SK, Maitra SS. Silver nanoparticles in poultry health: Applications and toxicokinetic effects, in Silver Nanomaterials for Agri-Food Applications. Elsevier 2021; pp. 685-704.
         [http://dx.doi.org/10.1016/B978-0-12-823528-7.00005-6]

[57]   Egbosiuba T C. Biochar and bio-oil fuel properties from nickel nanoparticles assisted pyrolysis of cassava peel Heliyon, vol 8, no 8, p e10114, 2022.
         [http://dx.doi.org/10.1016/j.heliyon.2022.e10114]

[58]   L. Xu, L. Y. Zhang, H. Song, Q. Dong, G. H. Dong, X. Kong, Z. Fang, Catalytic fast pyrolysis of polyethylene terephthalate plastic for the selective production of terephthalonitrile under ammonia atmosphere, Waste Management, vol. 92, pp. 97-106, June 2019.

[59]   Hahladakis JN, Iacovidou E. An overview of the challenges and trade-offs in closing the loop of post-consumer plastic waste (PCPW): Focus on recycling. J Hazard Mater 2019; 380120887.
         [http://dx.doi.org/10.1016/j.jhazmat.2019.120887] [PMID: 31330387]

CHAPTER 9

Wood Biomass Valorization for Value-added Chemicals

Vinay Kumar[1], Neha Sharma[2] and **Subhrangsu Sundar Maitra[2,*]**

[1] *Department of Community Medicine, Saveetha Medical College & Hospital, Saveetha Institute of Medical and Technical Sciences (SIMATS), Chennai, Thandalam-602105, India*

[2] *Bioprocess Design Laboraotry, School of Biotechnology, Jawaharlal Nehru University, New Delhi, India*

Abstract: Wood biomass is a vital component in producing various value-added products. It can be used to produce biofuels and chemicals. Agriculture practices produce a lot of lignocellulosic biomass, a waste management concern for years. Most of this lignocellulosic biomass is considered waste. But in recent years, efforts have been made to utilize and valorize this biomass to produce value-added products. The major challenge with lignocellulosic biomass is that it cannot be used in production processes. Therefore, it requires several physical and chemical pretreatments. This chapter discusses various pretreatment technologies involved in valorizing lignocellulosic biomass. In addition, it also discusses lignin pretreatment, saccharification, and microbial biodiesel production.

Keywords: Biodiesel, Lignocellulosic biomass, Pretreatment, Saccharification, Value-added chemicals, Wood biomass, Waste.

INTRODUCTION

Biomass as a feedstock is one of the most copious materials on the earth, and it is considered an essential renewable supply [1]. It has various advantages. For instance, it is a green sustainable feedstock with zero carbon emissions. It saves the planet from the effect of global warming. Several studies have shown that lignocellulosic biomass can be used as a renewable feedstock for better quality of biofuels and biochemicals. On an annual basis, agriculture projects produce many lignocellulosic residues and create an issue of waste management. Lignocellulosic waste biomass can be converted into valuable products such as fuels. The adequate consumption of these wastes can provide a solution to meet the demand for value-added products worldwide.

---

* **Corresponding author Subhrangsu Sundar Maitra:** Bioprocess Engineering Laboratory, School of Biotechnology, Jawaharlal Nehru University, New Delhi, India; Email: ssm2100@mail.jnu.ac.in

Over the last decades, it was observed that vigorous growth of biorefining produces value-added fossil fuel substitutes and biochemicals (like furfurals, organic acids, and alcohol). Bio-based technologies application continuously bring up sustainable bioeconomy growth. As per the European context, the economy needs to have circularity and sustainability at its heart to enable the change from a linear economy to a circular economy [2]. As per the prediction of the International Energy Agency, the bioenergy demand will increase considerably by almost 3-times by 2060 worldwide [3]. Sustainable biomass resources like crops, waste, and algae must be used more effectively for this drastic change. Moreover, to achieve biochemical and bioenergy targets in the future, it is necessary to develop circular and bio-cascading approaches [4]. Biorefinery and bioenergy products have become progressively inter-disciplinary, bridging several chemicals, biological, and physical technologies [5]. Hydrothermal treatment is an important technique for treating recalcitrant biomass into valuable products.

Pretreatment Technologies

Lignocellulosic biomass can be converted into cellulose, hemicellulose and lignin fractions. These products can further be converted into intermediate compounds such as 5-Hydroxymethylfurfural and furfural. The pretreatment technologies to convert lignocellulosic biomass are pyrolysis, hydro-liquefaction, gasification, catalytic hydrolysis, and solvolysis [1]. Among various pretreatment technologies, pyrolysis and gasification are considered essential. In pyrolysis, the thermal decomposition of cellulose, lignin and hemicellulose takes place in the absence of oxygen and the presence of a heterogeneous catalyst. Fast pyrolysis is much more efficient in lignocellulosic biomass hydrolysis than slow pyrolysis and hydrolysis. It produces green aromatics, phenolic compounds, furfural, hydroxy-methylfurfural and levoglucosenone. The pyrolysis process involves liquid evaporation and mass transfer of vapors through solid, solid-phase chemical, and liquid-phase reactions. The pyrolysis products are divided into gases, biochar, and pyrolysis oil. For the first and foremost product, the composition of crude bio-oil and pyrolysis oil mainly depends on the source and type of lignocellulosic biomass. Gasification produces gaseous fuel by burning biomass in a medium such as steam, oxygen and air to produce a mixture of gases at high temperatures, *i.e.*, 500-1500°C and pressure, *i.e.*, 30-40 bars [1]. Fig. (**1**) presents a lignocellulosic biomass pretreatment method.

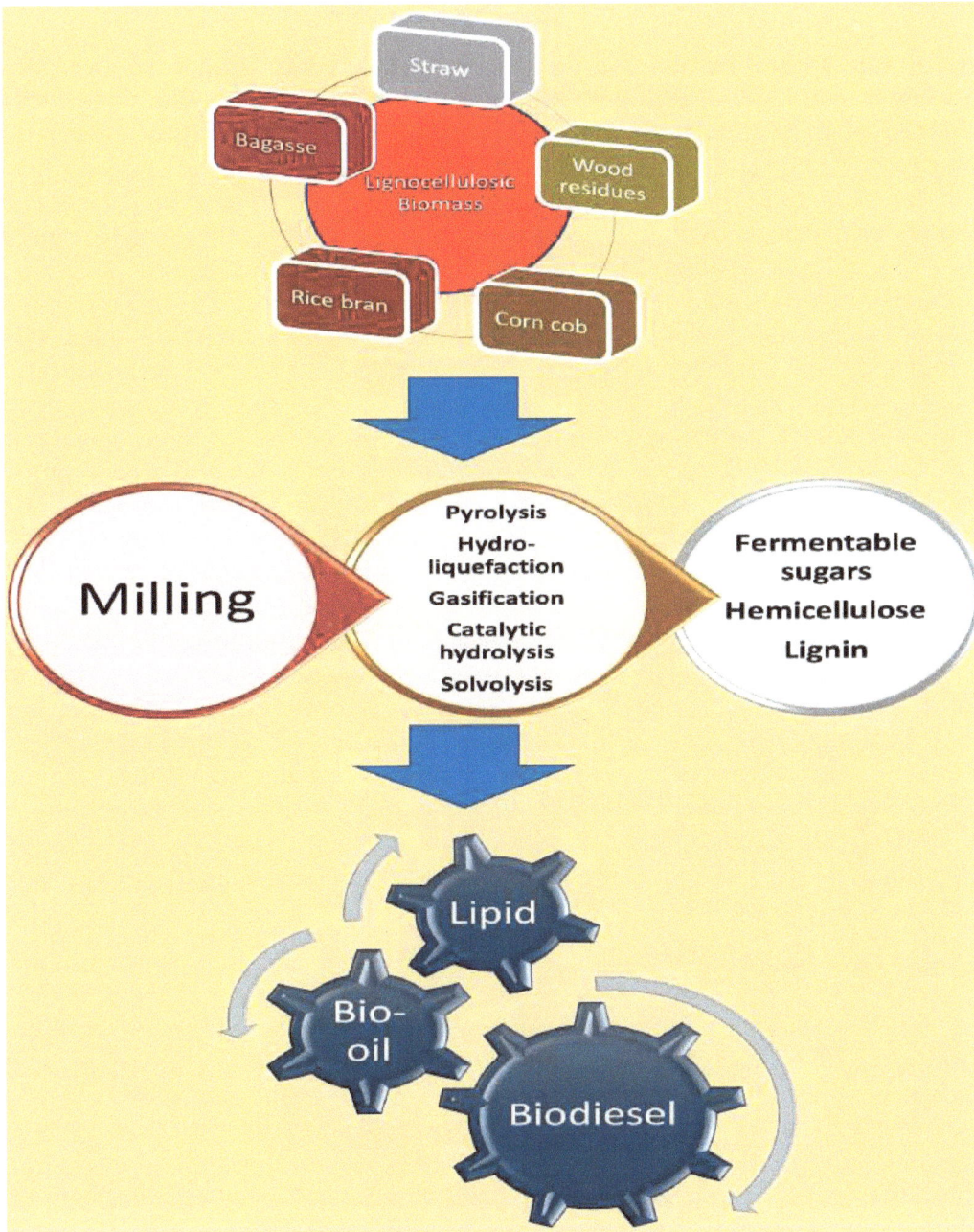

Fig. (1). A schematic route for lignocellulosic biomass treatment methods.

Although biomass valorization and waste management can use biological techniques to treat a broad range of biodegradable wastes like municipal, kitchen, agriculture, and industrial waste [6], even so, both techniques have their drawbacks. Biological treatment is time-consuming, and breaking recalcitrant substrate will be less effective, although hydrothermal treatment generally requires high energy input and solvents or catalysts [4]. So, the combination of both the treatment that is biological and hydrothermal can thus be favorable for biomass valorization by combining the advantages of both treatments to conquer the recalcitrant and diversified nature of biomass for a specific platform to produce biofuels or biochemicals. Fig. (**1**) presents a scheme of biomass treatment from a combination of hydrothermal and biological treatments. Compared to biological treatment, reaction environment and substrate type are less susceptible to hydrothermal treatment. But it is appropriate for biological treatment as either pretreatment or post-treatment. Using biological procedures for the raw materials, pretreatment is promising to enhance the performances of subsequent hydrothermal treatment. By carefully monitoring the conditions, biological treatment efficiency is heavily influenced by the performance and survival of the microorganisms [4]. When biological treatment is applied at the subsequent stage, the selection of appropriate hydrothermal technique is most important because of the following reasons [1]: It may produce various inhibitors like furfural in biological processes [2]. Due to an extra effect, this technology is expensive in terms of economy and energy [3]. The enzymes and microorganisms' activity in hydrothermal pretreatment is hindered by solvents [4]. Several reviews have discussed the advances in biomass valorization through biological and hydrothermal techniques [7, 8]. Pretreatment can also be performed using organic solvents, ionic liquids (ILs), deep eutectic solvents (DESs) and anaerobic digestion [9 - 13]. Thermochemical pretreatment and anaerobic digestion can be used to pretest lignocellulosic biomass [1, 15].

Lignin Pretreatment

The most abundant aromatic compound is lignin, *i.e.*, 30% of organic carbon on earth. The U.S. Energy Independence and Security act of 2007 is expected to produce up to 21 billion gallons of biofuels from lignocellulose biomass by 2022. Sixty-two million tons of lignin were generated as a byproduct. The pulp and paper industry produces annually approximately 50 million tons of waste lignin [16]. Different kinds of Lignin are produced, for instance, organosolv lignin, soda lignin, Kraft lignin, residual lignin, and lignosulfonate, depending on the pulping process. Most of the lignin is used for ignition to produce extra energy and heat, although cellulose, a carbohydrate polymer, has been successfully used for biofuel fermentation [16]. Recently, to produce value-added products, yearly, less than 2% of lignin is used [17]. Many lignin conversion technologies based on biological and thermochemical processes have been found [18]. Lignin depolymerization and upgrading of degradation products are the two forms of lignin valorization. It has been found that some microorganisms can perform biological depolymerization of lignin into monomers. But because of the complex nature and insolubility of lignin structure, microorganisms can only degrade a fraction of lignin compounds [18, 19]. So, currently, the degradation of lignin is based on thermochemical processes. In the process, lignin is converted into its monomers, and produced monomers can be used to generate value-added products [16]. Lignocellulosic biomass pretreatment gives lignin, a byproduct from paper and pulp industries, and biofuel production. Accessibility of Hemi cellulosic and cellulosic fractions is enhanced during the pretreatment process because of the breaking down of the cell wall. Simultaneously, lignin fractions are removed as insoluble residues [16]. Complete lignin degradation requires microbial pathway degradation enzymes operated in the microbial population. Another way to depolymerize lignin can be thermochemical degradation, with industrial characteristics like high yield and the release of monomer [16]. Hydrothermal technology is one of the effective pretreatment technologies. Based on variations in the pressure and temperature, hydrothermal treatment can be classified into [1] water treatment - the temperature between 25-100°C and 0.1 MPa pressure is ambient liquid [2]; supercritical water treatment - temperature more than 374°C and pressure more than 22 MPa; and [3] subcritical water treatment - pressure from saturated steam pressure to 22 MPa where temperature varies between 100 to 374°C to maintain water in the liquid form [4].

Saccharification

Saccharification of lignocellulosic biomass produces sugar monomers such as glucose and xylose with a combination of hydrothermal treatment and enzymatic hydrolysis methods. In the combinatorial process, the enzymatic hydrolysis success of the enzyme on cellulose is suppressed by the presence of Xylo oligomers. Moreover, cellulose absorbs lignin, which makes it less accessible to cellulase [20]. During the enzymatic hydrolysis of cellulose, the crystalline index of cellulose is related to the initial rate of hydrolysis because of the rigid crystalline structure of cellulose that is very highly resistant to hydrolysis [21]. Lignocellulosic biomass is composed of cellulose and hemicellulose connected with lignin. Lignin blocks the enzyme accessibility for cellulose degradation [22].

Microbial Biodiesel Production

Biodiesel is a renewable fuel produced from various sources, such as microbial lipids, animal fats, plant oils, and waste oils [23, 24]. There are various disadvantages to using the above feedstocks. These include the requirement of a huge area to grow the crops, high prices of oils, and low product yields [25]. In biodiesel production, feedstock accounts for about 75% of the total production cost [26]. Therefore, the sustainable and economical feedstock is the demand in the current scenario. In this contest, waste resources can serve the purpose of production cost reduction [27]. Oleaginous microorganisms can potentially use waste residues such as lignocellulosic biomass for bioconversion to biodiesel [28]. These microbes produce byproducts such as glycerol, which can be used further for lipid production by microbes [29, 30].

Feedstock plays an essential role in biodiesel production costs. Therefore, a careful selection of the feedstock is a mandatory step. Vegetable oil is considered the most common feedstock for biodiesel production. It was observed that vegetable oil competes with food and its production cost is very high. Therefore, it cannot be considered a sustainable feedstock. In contrast, if a non-edible feedstock should be selected, it should be available throughout the year to produce biodiesel.

Oleaginous microbes can effectively utilize lignocellulosic biomass to produce biodiesel. Oleaginous microbes can accumulate lipids higher than 20% of the cell's dry weight under stress [31]. Under high carbon and low nitrogen conditions, these cells can accumulate up to 70% of lipids. Transesterification of the lipid converts it into biodiesel [32]. The biodiesel production process includes culturing oleaginous microbes, cell harvesting, drying, lipid extraction, and, ultimately, transesterification [33]. There are various advantages to using oleaginous microbes for biodiesel production. These include non-competence with edible feedstocks, shorter culture duration timing, and no effect of seasonal variations [34]. The extracted lipid has a comparable composition of biodiesel to that obtained from other sources. These microbes can convert various wastes into biodiesel [35]. Lignocellulosic biomass can serve as an inexpensive feedstock. But lower yield is a major bottleneck in the process. Moreover, inhibitory compounds such as furfural may inhibit microbial growth [36]. Recently, microalgae have been considered a candidate to produce high-yield lipids for biodiesel. The biodiesel produced is many times higher than other carbon feedstocks [37]. They do not compete with edible food crops and do not require arable land for cultivation as their plant counterparts. Lipid content in microalgae may reach up to 90% in some instances [38]. Table **1** presents various lipid-producing oleaginous microbes along with the lignocellulosic biomass residues.

Table 1. Microorganisms producing biodiesel from the lignocellulosic biomass.

Microbes	Feedstock	Lipid production	References
Schizochytrium mangrove	Food waste hydrolysate	3.52 g/L	[39]
Nannochloropsis sp.	Effluent from palm oil	3.2 g/L	[40]
Chlorella protothecoides	Sugarcane bagasse	5.8 g/L	[41]
Auxenochlorella protothecoides	Spruce biomass	5.3g/L	[42]
Schizochytrium sp.	Sugarcane bagasse hydrolysate	45.15%	[43]
Mucor Circinelloides Q531	Mulberry branches	28.8%	[44]
Mucor circinelloides	Hydrolyzed whey permeates	32%	[45]
Cutaneotrichosporon dermatis	Hydrolysate from corn stover	56%	[46]
Pichia kudriavzevii	Hydrolysate from rice husk	0.009%	[47]
Naganishia albida	Hydrolyzed biowaste	20%	[48]
Rhodotorula paludigenum	Hydrolysate from corncob	58%	[49]
Rhodotorula glutinis	Hydrolysate from cassava	51%	[50]
Candida phangngensis PT1-17	Hydrolysate from switchgrass	9.8g/L	[51]
Cutaneotrichosporon cutaneum	Hydrolysate from corn stover	5g/L	[52]
Rhodococcus opacus	Lignocellulosic pretreatment effluent	26.9%	[53]

The major hurdle in biodiesel production from microalgae is the high production cost, which is about 5-8 USD per liter [32]. Lipid content, biomass productivity, the scale of production and oil recovery costs are the significant factors determining microalgal-based biodiesel [54]. Inexpensive carbon feedstock is the primary requirement for higher and economic biodiesel production. In this context, lignocellulosic waste is a suitable candidate, which can serve as a cost-effective feedstock. It has been observed that molds can also accumulate lipids as biomass and store up to 80% of their biomass composition. It was observed that compared to yeast, their biomass contains more unsaturated fatty acids [55]. The advantage of using a mold is that it can be cultivated in traditional reactors, which reduces the production cost. Several yeast species are considered oleaginous and can be used for biodiesel production [56]. Moreover, they have the potential for large-scale production and can be genetically manipulated for improvement [57]. In starvation and stress conditions, they can accumulate up to 70% lipids [26]. They are mainly composed of triglycerides, and their growth rate is relatively higher than microalgae [58]. Several studies have used yeast to produce lipids using lignocellulosic biomass [59]. It has been observed that bacteria are advantageous in several properties. They have a high growth rate and can be easily manipulated. Bacteria generally do not accumulate higher lipids except a few, such as *Mycobacterium, Rhodococcus and Streptomyces* [60]. Actinomycetes are known to store up to 70% of their dry cell weight [61]. Bacteria, such as *E. coli.*, are known to accumulate up to 2.5 g/L of lipids after genetic manipulation [62]. Bacteria are also known to utilize lignocellulosic biomass to produce lipids. *Rhodococcus opacus* is known to accumulate up to 80% of triglycerides of dry cell biomass when grown on cob waste as carbon feedstock [63]. Lignocellulose hydrolysate produced lipids by engineered bacterial strains [64] a unique approach, where algae-bacteria consortium was used for microbial oil production [65].

Lignocellulosic Biomass As Carbon Feedstock

Lignocellulosic biomasses are produced through photosynthesis and are one of the most important alternatives to fossil fuel resources [66]. Lignocellulosic biomass can be categorized into four types: forest residues, agricultural residues, weeds, and herbaceous grass [24]. From a commercial biodiesel perspective, the carbon feedstock incurred more than half of the total production cost. Lignocellulosic biomass is one of the cheaper and more sustainable materials that can serve the purpose of reducing the production cost. Biodiesel production from lignocellulosic biomass is the fastest and cheaper [67]. The utilization of lignin from lignocellulosic biomass to produce value-added products such as biodiesel and polymers are a significant challenge [68]. This is due to the recalcitrant nature of lignin. The recalcitrancy of lignin can be removed using several pretreatment techniques, which include methods such as physical, chemical, and biological. Milling is the first step before biomass pretreatment, costing about 5% of the total cost. But pretreatment of the biomass results in the production of inhibitory compounds such as 5-hydroxymethylfurfural and furfural, which are known to inhibit microbial growth [69]. In addition, many other feedstocks are available which can be converted into produce value-added products [70 - 80]. It was observed that some of the oleaginous microbes could degrade furfural and 5-hydroxymethylfurfural to produce alcohols, acetic acid, and formic acid.

CONCLUSION

Wood biomass produced from various agriculture practices can be pretreated to produce various products including biofuels. The technologies that can be used include pyrolysis, hydro-liquefaction, gasification, catalytic hydrolysis and solvolysis. Gasification and pyrolysis are considered advanced and recent technologies. Wood biomass such as straw, corn cob, bagasse, rice bran, *etc*. can be processed, milled, and pretreated to produce fermentable sugars. The fermentable sugars produced can be used by microorganisms to produce value-added products such as lipids, bio-oil, and biodiesel.

ACKNOWLEDGEMENTS

The authors are thankful to the funding from Jawaharlal Nehru University for the support of this study.

REFERENCES

[1]   Khemthong P, Yimsukanan C, Narkkun T, *et al.* Advances in catalytic production of value-added biochemicals and biofuels *via* furfural platform derived lignocellulosic biomass. Biomass Bioenergy 2021; 148106033..
[http://dx.doi.org/10.1016/j.biombioe.2021.106033]

[2]   Strategy UB. A sustainable bioeconomy for Europe: strengthening the connection between economy, society and the environment 2018.

[3]   IEA I. Energy technology perspectives 2017. Catalysing Energy Technology Transformations. 2017

[4]   Song B, Lin R, Lam CH, Wu H, Tsui TH, Yu Y. Recent advances and challenges of inter-disciplinary biomass valorization by integrating hydrothermal and biological techniques. Renew Sustain Energy Rev 2021; 135110370.
[http://dx.doi.org/10.1016/j.rser.2020.110370]

[5]   Song B, Buendia-Kandia F, Yu Y, Dufour A, Wu H. Importance of lignin removal in enhancing biomass hydrolysis in hot-compressed water. Bioresour Technol 2019; 288121522.
[http://dx.doi.org/10.1016/j.biortech.2019.121522] [PMID: 31130346]

[6]   Tsui TH, Wong JWC. A critical review: emerging bioeconomy and waste-to-energy technologies for sustainable municipal solid waste management. Waste Disposal & Sustainable Energy 2019; 1(3): 151-67.
[http://dx.doi.org/10.1007/s42768-019-00013-z]

[7]   Luo L, Kaur G, Wong JWC. A mini-review on the metabolic pathways of food waste two-phase anaerobic digestion system. Waste Manag Res 2019; 37(4): 333-46.
[http://dx.doi.org/10.1177/0734242X18819954] [PMID: 30696377]

[8]   Holzhäuser FJ, Creusen G, Moos G, Dahmen M, König A, Artz J, Palkovits S, Palkovits R. Electrochemical cross-coupling of biogenic di-acids for sustainable fuel production. Green chemistry. 2019;21(9):2334-44.

[9]   Tan YT, Chua ASM, Ngoh GC. Deep eutectic solvent for lignocellulosic biomass fractionation and the subsequent conversion to bio-based products – A review. Bioresour Technol 2020; 297122522.
[http://dx.doi.org/10.1016/j.biortech.2019.122522] [PMID: 31818720]

[10]   Usmani Z, Sharma M, Gupta P, Karpichev Y, Gathergood N, Bhat R. Ionic liquid based pretreatment of lignocellulosic biomass for enhanced bioconversion. Bioresource technology. 2020;304:123003.

[11]   Ferreira JA, Taherzadeh MJ. Improving the economy of lignocellulose-based biorefineries with organosolv pretreatment. Bioresour Technol 2020; 299122695.
[http://dx.doi.org/10.1016/j.biortech.2019.122695] [PMID: 31918973]

[12]   Abraham A, Mathew AK, Park H, Choi O, Sindhu R, Parameswaran B. Pretreatment strategies for enhanced biogas production from lignocellulosic biomass. Bioresource technology. 2020;301:122725

[13]   Millati R, Wikandari R, Ariyanto T, Putri RU, Taherzadeh MJ. Pretreatment technologies for anaerobic digestion of lignocelluloses and toxic feedstocks. Bioresour Technol 2020; 304122998.
[http://dx.doi.org/10.1016/j.biortech.2020.122998] [PMID: 32107151]

[14] Pecchi M, Baratieri M. Coupling anaerobic digestion with gasification, pyrolysis or hydrothermal carbonization: A review. Renew Sustain Energy Rev 2019; 105: 462-75.
[http://dx.doi.org/10.1016/j.rser.2019.02.003]

[15] Deng C, Lin R, Kang X, Wu B, O'Shea R, Murphy JD. Improving gaseous biofuel yield from seaweed through a cascading circular bioenergy system integrating anaerobic digestion and pyrolysis. Renew Sustain Energy Rev 2020; 128109895.
[http://dx.doi.org/10.1016/j.rser.2020.109895]

[16] Nguyen LT, Phan DP, Sarwar A, Tran MH, Lee OK, Lee EY. Valorization of industrial lignin to value-added chemicals by chemical depolymerization and biological conversion. Ind Crops Prod 2021; 161113219.
[http://dx.doi.org/10.1016/j.indcrop.2020.113219]

[17] Cao L, Iris K, Liu Y, Ruan X, Tsang DC, Hunt AJ. Lignin valorization for the production of renewable chemicals: State-of-the-art review and future prospects. Bioresource technology. 2018;269:465-75.

[18] Xu Z, Lei P, Zhai R, Wen Z, Jin M. Recent advances in lignin valorization with bacterial cultures: microorganisms, metabolic pathways, and bio-products. Biotechnol Biofuels 2019; 12(1): 32.
[http://dx.doi.org/10.1186/s13068-019-1376-0] [PMID: 30815030]

[19] Cao Y, Chen SS, Zhang S, Ok YS, Matsagar BM, Wu KC-W. Advances in lignin valorization towards bio-based chemicals and fuels: Lignin biorefinery. Bioresource technology. 2019;291:121878.

[20] Lai C, Tu M, Xia C, Shi Z, Sun S, Yong Q. Lignin alkylation enhances enzymatic hydrolysis of lignocellulosic biomass. Energy & Fuels. 2017;31(11):12317-26

[21] Li C, Knierim B, Manisseri C, Arora R, Scheller HV, Auer M. Comparison of dilute acid and ionic liquid pretreatment of switchgrass: biomass recalcitrance, delignification and enzymatic saccharification. Bioresource technology. 2010;101(13):4900-6.

[22] Yoo CG, Meng X, Pu Y, Ragauskas AJ. The critical role of lignin in lignocellulosic biomass conversion and recent pretreatment strategies: A comprehensive review. Bioresour Technol 2020; 301122784.
[http://dx.doi.org/10.1016/j.biortech.2020.122784] [PMID: 31980318]

[23] Banerjee R, Chintagunta AD, Ray S. Laccase mediated delignification of pineapple leaf waste: an ecofriendly sustainable attempt towards valorization. BMC Chem 2019; 13(1): 58.
[http://dx.doi.org/10.1186/s13065-019-0576-9] [PMID: 31384806]

[24] Kumar SJ, Avanthi A, Chintagunta AD, Gupta A, Banerjee R. Oleaginous lipid: a drive to synthesize and utilize as biodiesel Practices and Perspectives in Sustainable BioenergY. Springer 2020; pp. 105-29.

[25] Martin NR, Kelley P, Klaski R, Bosco A, Moore B, Traviss N. Characterization and comparison of oxidative potential of real-world biodiesel and petroleum diesel particulate matter emitted from a nonroad heavy duty diesel engine. Sci Total Environ 2019; 655: 908-14.
[http://dx.doi.org/10.1016/j.scitotenv.2018.11.292] [PMID: 30481717]

[26] Subramaniam R, Dufreche S, Zappi M, Bajpai R. Microbial lipids from renewable resources: production and characterization. J Ind Microbiol Biotechnol 2010; 37(12): 1271-87.
[http://dx.doi.org/10.1007/s10295-010-0884-5] [PMID: 21086103]

[27] Jeevan Kumar SP, Banerjee R. Enhanced lipid extraction from oleaginous yeast biomass using ultrasound assisted extraction: A greener and scalable process. Ultrason Sonochem 2019; 52: 25-32.
[http://dx.doi.org/10.1016/j.ultsonch.2018.08.003] [PMID: 30563792]

[28] Jeevan Kumar SP, Sampath Kumar NS, Chintagunta AD. Bioethanol production from cereal crops and lignocelluloses rich agro-residues: prospects and challenges. SN Applied Sciences 2020; 2(10): 1673.
[http://dx.doi.org/10.1007/s42452-020-03471-x]

[29] Garlapati VK, Chandel AK, Kumar SJ, Sharma S, Sevda S, Ingle AP. Circular economy aspects of

lignin: Towards a lignocellulose biorefinery. Renewable and Sustainable Energy Reviews. 2020;130:109977

[30] Chandel AK, Garlapati VK, Jeevan Kumar SP, Hans M, Singh AK, Kumar S. The role of renewable chemicals and biofuels in building a bioeconomy. Biofuels Bioprod Biorefin 2020; 14(4): 830-44.
[http://dx.doi.org/10.1002/bbb.2104]

[31] Meng X, Yang J, Xu X, Zhang L, Nie Q, Xian M. Biodiesel production from oleaginous microorganisms. Renew Energy 2009; 34(1): 1-5.
[http://dx.doi.org/10.1016/j.renene.2008.04.014]

[32] Pinzi S, Leiva D, López-García I, Redel-Macías MD, Dorado MP. Latest trends in feedstocks for biodiesel production. Biofuels Bioprod Biorefin 2014; 8(1): 126-43.
[http://dx.doi.org/10.1002/bbb.1435]

[33] Chintagunta AD, Zuccaro G, Kumar M, *et al.* Biodiesel Production From Lignocellulosic Biomass Using Oleaginous Microbes: Prospects for Integrated Biofuel Production. Front Microbiol 2021; 12658284.
[http://dx.doi.org/10.3389/fmicb.2021.658284] [PMID: 34475852]

[34] Gujjala LK, Kumar SJ, Talukdar B, Dash A, Kumar S, Sherpa KC. Biodiesel from oleaginous microbes: opportunities and challenges. Biofuels. 2019;10(1):45-59.

[35] Cho HU, Park JM. Biodiesel production by various oleaginous microorganisms from organic wastes. Bioresour Technol 2018; 256: 502-8.
[http://dx.doi.org/10.1016/j.biortech.2018.02.010] [PMID: 29478783]

[36] Jin M, Slininger PJ, Dien BS, Waghmode S, Moser BR, Orjuela A. Microbial lipid-based lignocellulosic biorefinery: feasibility and challenges. Trends Biotechnol. 2015;33(1):43-54.

[37] Yousuf A, Sannino F, Pirozzi D. Lignocellulosic biomass to liquid biofuels. Academic Press 2019.

[38] Pantazaki AA, Papaneophytou CP, Pritsa AG, Liakopoulou-Kyriakides M, Kyriakidis DA. Production of polyhydroxyalkanoates from whey by Thermus thermophilus HB8. Process Biochem 2009; 44(8): 847-53.
[http://dx.doi.org/10.1016/j.procbio.2009.04.002]

[39] Papanikolaou S, Aggelis G. Lipids of oleaginous yeasts. Part I: Biochemistry of single cell oil production. Eur J Lipid Sci Technol 2011; 113(8): 1031-51.
[http://dx.doi.org/10.1002/ejlt.201100014]

[40] Cheirsilp B, Thawechai T, Prasertsan P. Immobilized oleaginous microalgae for production of lipid and phytoremediation of secondary effluent from palm oil mill in fluidized bed photobioreactor. Bioresour Technol 2017; 241: 787-94.
[http://dx.doi.org/10.1016/j.biortech.2017.06.016] [PMID: 28628983]

[41] Mu J, Li S, Chen D, Xu H, Han F, Feng B. Enhanced biomass and oil production from sugarcane bagasse hydrolysate (SBH) by heterotrophic oleaginous microalga Chlorella protothecoides. Bioresource Technology. 2015;185:99-105.

[42] Patel A, Matsakas L, Rova U, Christakopoulos P. Heterotrophic cultivation of *Auxenochlorella protothecoides* using forest biomass as a feedstock for sustainable biodiesel production. Biotechnol Biofuels 2018; 11(1): 169.
[http://dx.doi.org/10.1186/s13068-018-1173-1] [PMID: 29946359]

[43] Nguyen HC, Su CH, Yu YK, Huong DTM. Sugarcane bagasse as a novel carbon source for heterotrophic cultivation of oleaginous microalga Schizochytrium sp. Ind Crops Prod 2018; 121: 99-105.
[http://dx.doi.org/10.1016/j.indcrop.2018.05.005]

[44] Qiao W, Tao J, Luo Y, Tang T, Miao J, Yang Q. Microbial oil production from solid-state fermentation by a newly isolated oleaginous fungus, *Mucor circinelloides* Q531 from mulberry branches. R Soc Open Sci 2018; 5(11)180551.

[http://dx.doi.org/10.1098/rsos.180551] [PMID: 30564386]

[45]   Chan LG, Dias FF, Saarni A, Cohen J, Block D, Taha AY. Scaling up the bioconversion of cheese whey permeate into fungal oil by Mucor circinelloides. Journal of the American Oil Chemists' Society. 2020;97(7):703-16.

[46]   Yu Y, Xu Z, Chen S, Jin M. Microbial lipid production from dilute acid and dilute alkali pretreated corn stover *via* Trichosporon dermatis. Bioresour Technol 2020; 295122253.
[http://dx.doi.org/10.1016/j.biortech.2019.122253] [PMID: 31630000]

[47]   Ananthi V, Prakash GS, Chang SW, Ravindran B, Nguyen DD, Vo D-VN. Enhanced microbial biodiesel production from lignocellulosic hydrolysates using yeast isolates. Fuel. 2019;256:115932

[48]   Sathiyamoorthi E, Kumar P, Kim BS. Lipid production by Cryptococcus albidus using biowastes hydrolysed by indigenous microbes. Bioprocess Biosyst Eng 2019; 42(5): 687-96.
[http://dx.doi.org/10.1007/s00449-019-02073-1] [PMID: 30661102]

[49]   Chaiyaso T, Manowattana A, Techapun C, Watanabe M. Efficient bioconversion of enzymatic corncob hydrolysate into biomass and lipids by oleaginous yeast *Rhodosporidium paludigenum* KM281510. Prep Biochem Biotechnol 2019; 49(6): 545-56.
[http://dx.doi.org/10.1080/10826068.2019.1591985] [PMID: 30929597]

[50]   Liu L, Chen J, Lim PE, Wei D. Enhanced single cell oil production by mixed culture of Chlorella pyrenoidosa and Rhodotorula glutinis using cassava bagasse hydrolysate as carbon source. Bioresour Technol 2018; 255: 140-8.
[http://dx.doi.org/10.1016/j.biortech.2018.01.114] [PMID: 29414159]

[51]   Quarterman JC, Slininger PJ, Hector RE, Dien BS. Engineering Candida phangngensis—an oleaginous yeast from the Yarrowia clade—for enhanced detoxification of lignocellulose-derived inhibitors and lipid overproduction. FEMS Yeast Res 2018; 18(8)foy102.
[http://dx.doi.org/10.1093/femsyr/foy102] [PMID: 30247683]

[52]   Wang J, Gao Q, Zhang H, Bao J. Inhibitor degradation and lipid accumulation potentials of oleaginous yeast Trichosporon cutaneum using lignocellulose feedstock. Bioresour Technol 2016; 218: 892-901.
[http://dx.doi.org/10.1016/j.biortech.2016.06.130] [PMID: 27441826]

[53]   Goswami L, Tejas Namboodiri MM, Vinoth Kumar R, Pakshirajan K, Pugazhenthi G. Biodiesel production potential of oleaginous Rhodococcus opacus grown on biomass gasification wastewater. Renew Energy 2017; 105: 400-6.
[http://dx.doi.org/10.1016/j.renene.2016.12.044]

[54]   Balat M. Potential alternatives to edible oils for biodiesel production – A review of current work. Energy Convers Manage 2011; 52(2): 1479-92.
[http://dx.doi.org/10.1016/j.enconman.2010.10.011]

[55]   Papanikolaou S, Aggelis G. Sources of microbial oils with emphasis to Mortierella (Umbelopsis) isabellina fungus. World J Microbiol Biotechnol 2019; 35(4): 63.
[http://dx.doi.org/10.1007/s11274-019-2631-z] [PMID: 30923965]

[56]   Garay LA, Sitepu IR, Cajka T, Chandra I, Shi S, Lin T. Eighteen new oleaginous yeast species. Journal of industrial microbiology & biotechnology. 2016;43(7):887-900

[57]   Athenaki M, Gardeli C, Diamantopoulou P, Tchakouteu SS, Sarris D, Philippoussis A. Lipids from yeasts and fungi: physiology, production and analytical considerations. J Appl Microbiol. 2018;124(2):336-67.

[58]   Ratanaporn L, Phajongjit K. Kinetic growth of the isolated oleaginous yeast for microbial lipid production. Afr J Biotechnol 2011; 10(63): 13867-77.
[http://dx.doi.org/10.5897/AJB10.2162]

[59]   Sreeharsha RV, Mohan SV. Obscure yet Promising Oleaginous Yeasts for Fuel and Chemical Production. Trends Biotechnol 2020; 38(8): 873-87.
[http://dx.doi.org/10.1016/j.tibtech.2020.02.004] [PMID: 32673589]

[60] Zuccaro G, Pirozzi D, Yousuf A. ignocellulosic biomass to biodiesel.Lignocellulosic Biomass to Liquid Biofuels.Chapter 4Yousuf A., Pirozzi D., Sannino F.Academic Press 2020; pp. 127-67.
[http://dx.doi.org/10.1016/B978-0-12-815936-1.00004-6]

[61] Kalscheuer R, Stölting T, Steinbüchel A. Microdiesel: Escherichia coli engineered for fuel production. Microbiology (Reading) 2006; 152(9): 2529-36.
[http://dx.doi.org/10.1099/mic.0.29028-0] [PMID: 16946248]

[62] Lu X, Vora H, Khosla C. Overproduction of free fatty acids in E. coli: Implications for biodiesel production. Metab Eng 2008; 10(6): 333-9.
[http://dx.doi.org/10.1016/j.ymben.2008.08.006] [PMID: 18812230]

[63] H A, A S. Triacylglycerols in prokaryotic microorganisms. Appl Microbiol Biotechnol 2002; 60(4): 367-76.
[http://dx.doi.org/10.1007/s00253-002-1135-0] [PMID: 12466875]

[64] Kumar KK, Deeba F, Sauraj , Negi YS, Gaur NA. Harnessing pongamia shell hydrolysate for triacylglycerol agglomeration by novel oleaginous yeast *Rhodotorula pacifica* INDKK. Biotechnol Biofuels 2020; 13(1): 175.
[http://dx.doi.org/10.1186/s13068-020-01814-9] [PMID: 33088345]

[65] Zhang B, Li W, Guo Y, Zhang Z, Shi W, Cui F. Microalgal-bacterial consortia: From interspecies interactions to biotechnological applications. Renewable and Sustainable Energy Reviews. 2020;118:109563.

[66] Shaheena S, Chintagunta AD, Dirisala VR, Sampath Kumar NS. Extraction of bioactive compounds from Psidium guajava and their application in dentistry. AMB Express 2019; 9(1): 208.
[http://dx.doi.org/10.1186/s13568-019-0935-x] [PMID: 31884522]

[67] Huber G. Breaking the chemical and engineering barriers to lignocellulosic biofuels: hydrocarbon biorefineries. Am Chem Soc. 2007; pp. 1-177.

[68] Zhou Z, Liu D, Zhao X. Conversion of lignocellulose to biofuels and chemicals *via* sugar platform: An updated review on chemistry and mechanisms of acid hydrolysis of lignocellulose. Renew Sustain Energy Rev 2021; 146111169.
[http://dx.doi.org/10.1016/j.rser.2021.111169]

[69] Valdés G, Mendonça RT, Aggelis G. Lignocellulosic biomass as a substrate for oleaginous microorganisms: a review. Appl Sci (Basel) 2020; 10(21): 7698.
[http://dx.doi.org/10.3390/app10217698]

[70] Awasthi MK, Kumar V, Yadav V, Sarsaiya S, Awasthi SK, Sindhu R, Binod P, Kumar V, Pandey A, Zhang Z. Current state of the art biotechnological strategies for conversion of watermelon wastes residues to biopolymers production: A review. Chemosphere. 2022 Mar 1;290:133310.

[71] Liu H, Kumar V, Jia L, Sarsaiya S, Kumar D, Juneja A, Zhang Z, Sindhu R, Binod P, Bhatia SK, Awasthi MK. Biopolymer poly-hydroxyalkanoates (PHA) production from apple industrial waste residues: A review. Chemosphere. 2021 Dec 1;284:131427.

[72] Awasthi SK, Kumar M, Kumar V, Sarsaiya S, Anerao P, Ghosh P, Singh L, Liu H, Zhang Z, Awasthi MK. A comprehensive review on recent advancements in biodegradation and sustainable management of biopolymers. Environmental Pollution. 2022 Aug 15;307:119600.

[73] Duan Y, Tarafdar A, Kumar V, Ganeshan P, Rajendran K, Giri BS, Gomez-Garcia R, Li H, Zhang Z, Sindhu R, Binod P. Sustainable biorefinery approaches towards circular economy for conversion of biowaste to value added materials and future perspectives. Fuel. 2022 Oct 1;325:124846.

[74] Kumar V, Sharma N, Umesh M, Selvaraj M, Al-Shehri BM, Chakraborty P, Duhan L, Sharma S, Pasrija R, Awasthi MK, Lakkaboyana SR. Emerging challenges for the agro-industrial food waste utilization: A review on food waste biorefinery. Bioresource Technology. 2022 Aug 13:127790.

[75] Awasthi SK, Sarsaiya S, Kumar V, Chaturvedi P, Sindhu R, Binod P, Zhang Z, Pandey A, Awasthi

MK. Processing of municipal solid waste resources for a circular economy in China: An overview. Fuel. 2022 Jun 1;317:123478.

[76] Kumar V, Sharma N, Maitra SS. *In vitro* and *in vivo* toxicity assessment of nanoparticles. Int Nano Lett 2017; 7(4): 243-56.
[http://dx.doi.org/10.1007/s40089-017-0221-3]

[77] Vinay K, Neha S, Maitra S. Protein and Peptide Nanoparticles: Preparation and Surface Modification, in Functionalized Nanomaterials I.. CRC Press 2020; pp. 191-204.

[78] Vallinayagam S, Rajendran K, Lakkaboyana SK, Soontarapa K, Remya RR, Sharma VK, Kumar V, Venkateswarlu K, Koduru JR. Recent developments in magnetic nanoparticles and nano-composites for wastewater treatment. Journal of Environmental Chemical Engineering. 2021 Dec 1;9(6):106553.

[79] Kumar V, Sharma N, Lakkaboyana SK, Maitra SS. Silver nanoparticles in poultry health: Applications and toxicokinetic effects, in Silver Nanomaterials for Agri-Food Applications.. Elsevier 2021; pp. 685-704.
[http://dx.doi.org/10.1016/B978-0-12-823528-7.00005-6]

[80] Egbosiuba T C. Biochar and bio-oil fuel properties from nickel nanoparticles assisted pyrolysis of cassava peel 2022.
[http://dx.doi.org/10.1016/j.heliyon.2022.e10114]

CHAPTER 10

Food Waste Valorization for Bioplastic Production

Mridul Umesh[1,*]**, Suma Sarojini**[1]**, Debasree Dutta Choudhury**[1]**, Adhithya Sankar Santhosh**[1] **and Sapthami Kariyadan**[1]

[1] *Department of Life Sciences, CHRIST (Deemed to be University), Bengaluru-560029, Karnataka, India*

Abstract: The alarming concern over the environment created due to the uncontrolled use of based petrochemical-based synthetic plastic created a research thrust on bioplastics. Bioplastics, in general, refers to the polymers derived from plants, animals, and microorganisms that have close material properties to their synthetic counterparts. Despite having good biodegradability, their commercialization still faces hurdles majorly contributed by the high production cost involved. An integrated strategy of waste valorization with bioplastic production was a sustainable approach toward their cost-effective production and commercialization. Food waste represents a continuous and rapidly available substrate containing high-value nutrients that can be exploited for the production of bioplastics through microbial fermentation and chemical treatment methods. This chapter describes the biotechnological strategies for valorizing food waste into commercially important biopolymeric components like chitosan, polyhydroxyalkanoates, HAp, and cellulose-based polymers. It presents a comprehensive outlook on their chemical nature, production strategy, and application in various fields.

Keywords: Biocompatibility, Biodegradability, Biopolymers, Bioactivity, Biomaterial, Bleaching, Chitosan, Cellulose, Crystallinity, Calcination, Deacetylation, Demineralization, Deproteinization, Dewaxing, Food waste, Hydrolysis, Hydroxyapatite, Polyhydroxyalkanoates, Thermoplastic, Valorization.

INTRODUCTION

The demand for food, fuel, and feed will keep increasing as long as the population of human beings keeps increasing. The big question looming around us is how best we can utilize available resources to the maximum extent in a sustainable manner. The drive towards sustainable development will be significant in the coming years as we have already exploited almost all-natural resources to a large extent, and it's high time that we give back or at least stop the over-exploitation of

---

* **Corresponding author Mridul Umesh** Department of Life Sciences, CHRIST (Deemed to be University), Bengaluru - 560029, Karnataka, India; Email: mridul.umesh@christuniversity.in

Vinay Kumar, Sivarama Krishna Lakkaboyana & Neha Sharma (Eds.)

Mother Earth. In this context, food waste valorization assumes significance. Every year 1300 million tonnes of food is wasted globally. Most of these are dumped into landfills or into water bodies. The dangers of this are twofold. First, we are underutilizing a potential energy source without a sustainable approach. Second, the huge amount of dumped waste pollutes the environment. In water bodies, it can lead to dangers like eutrophication, which can cause havoc to the entire aquatic ecosystem, and groundwater pollution, which can affect water portability.

According to FAO, one-third of the food produced in the world goes to waste. This happens at multiple levels starting from the farm to food reaches the dining table. 30% of cereals and 20% of pulses are lost. Also, almost 8% of the caught fish is thrown back into the sea, mostly in dead or damaged condition, and 20% of meat also goes to waste. The maximum wastage is in the case of vegetables and fruits (45%). This implies that almost half of the global production of vegetables and fruits is wasted in some manner. One should translate this wastage to the enormous losses incurred in the form of resources utilized to produce these food items [1].

The concept of bioplastic manufacture from food waste has multiple advantages. It can lead to the reduced use of synthetic plastic and also prevent food wastage. The more considerable advantages also include reaping maximum benefits from the resources invested in making the food products, starting from the water and fertilizers in the field, labor cost, transportation, and processing costs, etc. If one accounts for all these factors, food waste valorization for biopolymer manufacture becomes highly sustainable. This is because the per capita cost of synthetic plastic production is much less than that incurred for bioplastics. This is very environmentally friendly as it is a renewable and sustainable process in which materials are synthesized from carbon-neutral resources. Bioplastics produced in this manner are primarily biodegradable and compostable [2].

Creating biopolymers from food waste can reduce food wastage and generate more employment in the processing sector. This will, in turn, help boost the local economy too. This is in line with goal number 12 of the "UN2030 agenda for sustainable development" to valorize food waste into commercially important products, thereby increasing employment with fewer resources via the circular economy model [3].

The long periods of industrialization have paved the way for the depletion of many natural resources. So in the 21st century, humankind may have to rely on sustainable methods to satisfy the ever-increasing needs for food, feed, fuel, and other luxuries. Soon a time will come when the natural wholly get completely depleted, and we may have to rely on biological systems more than chemical ones. In this scenario, biopolymer production from food wastes assumes paramount importance. While the policymakers in different countries have to look at this from different angles, it is better to give more thrust on increasing food waste utilization based on local needs. For instance, if this has to be implemented in a predominantly seafood-based economy in a coastal area, studies can be oriented toward fish waste valorization. This is the need of the hour as it simultaneously takes care of multiple aspects- reducing food waste, preventing pollution, full utilization of resources, and increasing employment among local people. This chapter focuses on the production of four major biopolymers from food wastes- chitosan, cellulose, polyhydroxyalkanoates, and hydroxyapatite.

Cellulose is the most abundant biopolymer on planet Earth. Non-toxic and biodegradable properties have led to its use in industries like food, paper, cosmetics, textiles, and pharmaceuticals. In the biomedical field, it is used in drug delivery, scaffolds, implants, etc. A significant amount of cellulose-based waste is generated across the world, most of which remain untapped. Hence ways of extracting cellulose from waste materials can have dual advantages in waste management and sustainability.

Chitosan is one of the naturally occurring polymers which is highly abundant in nature after cellulose. Its physicochemical and unique biological properties like biocompatibility and biodegradability, have given chitosan an important place in many industries, including food, medical, cosmetics, water treatment, metal extraction, etc. Different structural forms like gels, beads, membranes, films, sponges, etc. are made using chitosan and its derivatives. Chitosan can be derived from insects, mollusks, crustaceans, etc. More than 2000 tons of chitosan is produced annually, which is mainly extracted from shrimp and crab shell residues [4]. Since chitosan has a lot of applications and the demand is very high, tapping the best and cheapest sources could be a matter of great interest. To this effect, utilizing the byproducts of crustacean processing can be profitable as they yield high-value compounds like chitosan and its derivatives [5].

Hydroxyapatite (HAp) is yet another essential biopolymer used in bone repair and substitution and as scaffolds in tissue engineering for bone regeneration. It is compatible with bone without causing any toxic or inflammatory responses, and hence it is widely used as a scaffold for animals and plants [6]. HAp can be synthesized chemically or extracted from biological sources. A source of potential interest is fish waste in the form of scales and bones, as these are rich in phosphate, calcium, and carbonate [7].

Polyhydroxyalkanoates (PHA) are biopolymers that have attracted significant attention as they can be substitutes for synthetic plastics. Though their predominant application is in the packaging material industry, they also find use in surgical stitching, implants, and drug delivery systems [8]. More research is being undertaken to find economical ways to commercially produce PHA from various food wastes.

Valorization of food waste should indeed be a global agenda, as we can curb the substantial loss of resources such as land, water, energy, and labor and utilize them to the fullest extent in a sustainable manner. This chapter focuses majorly on various methods of production of four commercially important bioplastics from food waste. It further discusses the application of these biopolymers in diverse sectors.

Chitosan

Chitosan is a linear polyamine produced through the deacetylation of chitin, the second most abundant biopolymer on Earth after cellulose [9]. This biopolymer is a linear polysaccharide consisting of 2-amino-2-deoxy- (1-4)-β-D-glucopyranose residues (D-glucosamine units) [10]. The classification of biopolymers as chitin and chitosan is basically done based on deacetylation percentage. Acetylation degree higher than 50% is an attribute of chitin, whereas lower than 50% deacetylation is a characteristic feature of chitosan [11]. Chitosan, a polyamine, has a reactive amine group responsible for its solubility in acids and its moldability into various forms. Being cationic polymer, chitosan is insoluble in water, organic solvents, and aqueous bases. However, it is soluble in organic acids and diluted acidic solutions [12]. Chemical or enzymatic modification of chitosan into derivatives like alkylated chitosan and carboxymethyl chitosan is highly effective due to the presence of amino and hydroxyl functional groups [13]. Chitosan is highly biocompatible and biodegradable, thereby making it an ideal component in the medical field. Inside the mammalian body, the digestion of chitosan yields oligosaccharides that are channeled to specific biochemical pathways for further degradation or used for the production of glycoproteins [14]. Biodegradability of chitosan is attributed to its molecular weight, degree of deacetylation, and distribution [15].

Extraction of chitosan from food waste basically involves screening of food waste rich in chitin followed by chitin extraction and deacetylation. Chitin is widely distributed in the exoskeleton of marine organisms, especially crustaceans. In crustaceans, chitin along with proteins and calcium carbonate forms the hard exoskeleton [16]. Other food-based chitosan sources include fish and mushroom waste [17, 18].

Extraction of Chitosan From Food Waste

Food waste, especially from fisheries and related industries, is an excellent source for the extraction of chitosan due to its high content of chitin that can be easily deacetylated to obtain chitosan. The seafood industry, on average, generates 106 tons of waste per year that are usually converted into low-value-added products like fertilizers and animal feed [19]. The major sources from which chitosan is commercially extracted include shrimp waste, crab waste, fish scales, and another crustacean exoskeleton. A significant proportion of chitosan is derived from mushroom waste as well. A list of sources used for extracting chitosan on a commercial scale is provided in Table **1**.

Table 1. Chitosan extraction from selected food waste

Source for chitosan extraction	Degree of deacetylation	References
Mixture of shrimp shells	89.06%	[20]
Metapeneaus monoceros shell protein hydrolysate	89.5%	[21]
Agaricus bisporus stalks	87.6%	[18]
Labeo rohita scales	61%	[17]
Shrimp head and scales	95.19%	[22]
Shrimp shell waste	81.5%	[23]
Waste *Aspergillus niger* mycelium from a citric acid production plant	73.6%	[24]
Nephrops norvegicus and *Labeo rohita scales*	45%	[25]
Prawn shells waste	78.40%	[26]
Persian Gulf shrimp waste	89.34%	[27]
Penaeus merguiensis waste	87%	[28]
Parapenaeus Longirostris shrimp shell waste	83.55%	[29]
Penaeus kerathurus waste	12%	[30]
Carcinus mediterraneus shells	17%	[30]
Sepia officinalis bones	5%	[30]
Callinectes sapidus waste	80.8%	[31]
Nototodarus sloanii pen	NA	[32]
Metapenaeus stebbingi shells	92.19%	[33]
Champignon stipe of the mushroom	NA	[34]
Callinectes sapidus shell	71%	[35]
Procambarus clarkii exoskeleton	87%	[36]
White snapper (Lates sp.) scales	84.05%	[37]

Steps in Chitosan Extraction From Food Waste

Extraction of chitosan from food waste, especially shells, and scales, is done through chemical or enzymatic methods [19]. The chemical method of extraction basically involves several rounds of acid and alkali treatment, whereas the biological method relies on the use of enzymes or microbial fermentation [38, 39]. A brief outlook on both extraction methods is represented in Fig. (**1**).

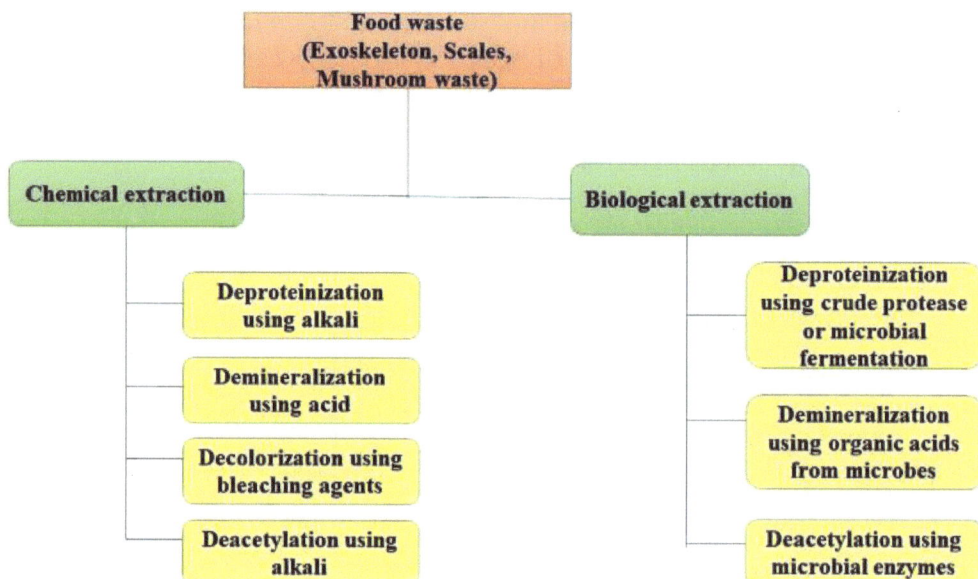

Fig. (1). Schematic representation of chitosan extraction methods from food waste.

Chemical Extraction

Chemical extraction basically involves four major processes- deproteinization, demineralization, pigment removal, and deacetylation.

Chemical deproteinization: Chitin present in the scales and shells of marine food waste sources is linked with proteins. As a first step towards chitosan extraction, the removal of the protein from the waste to recover high-purity chitin to serve as a precursor for chitosan formation is usually done [40]. Deproteinization is done by treating the substrate with a concentrated alkali solution (usually NaOH or KOH). Other alkali solutions like Na_2CO_3, $NaHCO_3$, KOH, K_2CO_3, $Ca(OH)_2$, Na_2SO_3, $NaHSO_3$, $CaHSO_3$, Na_3PO_4, and Na_2S have been applied for deproteinization [41]. This step is usually repeated multiple times. The alkali treatment is done at an elevated temperature (usually above 50°C) for several hours to remove the adhered proteins, depending on the nature of the targeted substrate [42].

Chemical demineralization: This step involves the removal of inorganic fractions of the exoskeleton from marine organisms [43]. Calcium carbonate and calcium phosphate are the major mineral components removed in this step [44]. Acids like HCl, HNO_3, H_2SO_4, CH_3COOH, and HCOOH are used for demineralization [45, 46]. When acids react with minerals, it results in the formation of soluble salts, and CO_2 is released [47, 48]. The soluble salts can be removed by filtration and chitin by washing with deionized water, followed by drying.

Chemical decoloration: Decolorizing is the final stage in the chitin's recovery and involves removing any residual pigments in the extracted chitin depending on the nature of the food waste used [49]. Pigment removal is done by treating with sodium hypochlorite, potassium permanganate, or hydrogen peroxide [50]. Dewatering is done to get dry chitin powder subjected to deacetylation, producing chitosan.

Chemical deacetylation: This is the final step in of converting extracted chitin into chitosan and involves treatment with acids or alkalis. It can be done either through heterogenous or homogenous methods. These methods are important with respect to their treatment temperature, time, and concentration of alkali used[51].

Biological Extraction Method

The drawbacks associated with chemical extraction methods, like alteration of physicochemical properties, toxicity, hazardous nature, and high cost associated with energy consumption, lead to the development of biological methods involving microbes and their enzymes.

Biological deproteinization: Proteolytic enzymes derived from microbes, plants or animals are employed to remove proteins from the food waste instead of chemicals. Crude proteases derived from marine bacteria are reported to be effective in removing proteins from fish and shrimp waste for chitin recovery [21]. Fermentation with endogenous microbes or selected proteolytic strains can also be used as a method for the removal of proteins from the raw materials [52].

Biological demineralization: In biological demineralisation, organic acid is produced by certain bacteria like lactic acid bacteria. It precipitates the calcium carbonate as their corresponding salts that can be separated and further utilized as preservatives [41].

Biological deacetylation: It involves the use of thermostable chitin deacetylases that serve as a catalyst for the breakdown of N-acetamido bonds. These enzymes are derived from fungi like *Mucor rouxii* and *Aspergillus nidulans*. Enzymatic deacetylation is a non-hazardous and controlled deacetylation method [53].

Application Of Chitosan

Chitosan and its derivatives, being natural biopolymers, find application in various sectors due to their biocompatibility, biodegradability, and non-toxicity [19]. Some applications of chitosan and its derivatives are represented in Fig. (**2**). Due to their easily modifiable chemical structure, controlled biodegradable nature, and insignificant toxicity, chitosan derivatives are used for biomedical applications like tissue engineering and drug delivery [54, 55]. Chitosan nanoparticles were used to encapsulate drugs like levofloxacin and vaginal delivery of microbicides in sexually transmitted infections [56, 57]. Chitosan blended with other biopolymers like PLA has been used as an immunomodulator and an antibiotic delivery system [58]. The presence of hydroxyl and amino groups in chitosan makes it an ideal material for the removal of pesticides, dyes, and heavy metals from wastewater [59, 60]. Chitosan derivatives, due to their antimicrobial properties, are used as food preservatives to prevent microbial spoilage of food stuff [61]. Chitosan was reported to have the ability to absorb bradykinin, thus acting as an analgesic [62]. Chitosan, with its reactive amino and hydroxyl group, is capable of removing fatty acids and thus reported to have anticholesterolemic properties [63]. Currently a lot of research is focused on modifying chitosan to increase its solubility without altering its biodegradability and biocompatibility to serve as an ideal candidate for biomedical and environmental applications.

Polyhydroxyalkanoates

Polyhydroxyalkanoates (PHA) are the polymers of hydroxy alkanoic acid monomers linked by ester bonds. These polyesters are produced by microorganisms and stored in the intra-cellular granular bodies as an energy reserve. Their production is carried out when the organisms are under stress due to unbalanced nutrient supply (carbon in excess with limitation of any other essential nutrients). Other than their energy storage application, PHA granules help the microbes to cope up with stress factors like osmotic shock [64]. Out of the many types of PHA more than 80% are detected in different strains of bacteria [65]; hence they can be easily degraded into carbon dioxide and water in both aerobic and anaerobic conditions. The most commonly studied PHA molecules are poly-beta hydroxybutyric acid (PHB) and poly-beta hydroxy valeric acid (PHBV). The degradation of PHA can happen intra cellularly or extracellularly [66]. When the cells are experiencing stress due to a limited carbon source, intracellular PHA degradation takes place [67]. The extracellular PHA degradation is observed in the non-PHA accumulating organisms, in which the extracellular degrading enzymes convert PHA into its water-soluble monomers. These monomers are taken up by the cells due to their small molecular size for further degradation [68]. The degradation results in the production of acetyl-CoA, which enters Krebs' cycle for energy production [69]. PHA polymers have a low melting temperature, low tensile strength, and high break extension, making them suitable candidates to replace many synthetic polymers. Different PHA molecules also differ in their properties from one another. For example, the scl-PHA has a similar degree of crystallinity and melting temperature to that of polypropylene, but PHBs are stiffer and more brittle than polypropylene. Due to the wide property ranges and their biocompatible nature, PHAs can be ideally used for applications in food packaging [70, 71], drug delivery, and other biomedical applications [72 - 74].

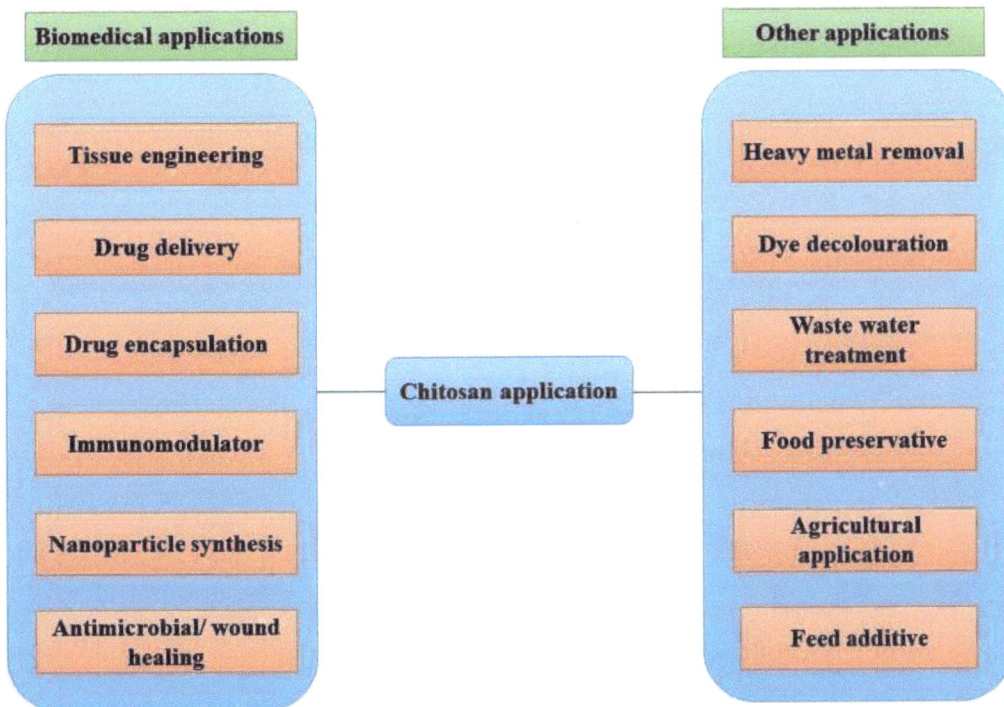

Fig. (2). Applications of chitosan

PHA Fermentation

PHA fermentation from food waste can be carried out using different strategies under optimized conditions. The basic steps involved pretreatment of the substrate, incorporation of the substrate into a media, inoculation of PHA producing bacterial strain, optimal incubation, and PHA recovery. The principle microbial strains which are known to produce PHA are *E. coli, Alcaligenes latus, Cupriavidus necator, Ralstonia eutropha, Pseudomonas putida, Aeromonas hydrophila, Hydrogenophaga pseudoflava, Zobellella denitri-ficans, Thermus thermophilus, Comamonas sp., Rhodobacter sphaeroides, Azotobacter sp., Bacillus sp., Chromobacterium sp., Erwinia sp., Haloferax sp., Methylobacterium sp., Saccharophagus degradans, Cyanobacteria and so on* [75, 76]. The production process can be done through submerged or solid-state fermentation (SSF) [77].

In submerged fermentation, the pretreatment of the food wastes is done in order to break down the complex molecules into simpler molecules (Sugars- glucose or lactose and fatty acids- acetic acid) which can be utilized by microbes to synthesize PHA. Some of the substrates which are used for the production of PHA in a submerged fermentation include agricultural wastes like whey, starch, oils, lignocellulosic materials, legume, and sugar [77]. Submerged fermentation techniques are preferred over solid-state fermentation as this method can control all the factors necessary for high PHA production [77]. The problem associated with submerged fermentation is the high cost of media components [78]. A list of commonly used food waste substrates for PHA production is presented in Table **2**.

Table 2. PHA production from selected food waste residues.

Substrate	Organism used	Type of fermentation	Yield of PHA	Reference
Food waste slurry	*Cupriavidus necator*	submerged	250 mgL^{-1}	[79]
Cane molasses	*Alcaligenes sp.* NCIM 5085	Submerged	2.68 ± 0.15 gL^{-1}	[80]
Kitchen waste	*Cupriavidus necator* CCGUG 52238	Submerged	1.218 gL^{-1}	[81]
Restaurant waste	Recombinant *E. coli pnDTM2*	Submerged	9.2 gL^{-1}	[82]
Waste frying oil	*Cupriavidus necator*	Submerged	1.2 gL^{-1}	[83]
Corn cob	*Bacillus sp.*	Submerged	4.8 ± 0.34 gL^{-1}	[84]
Fish solid waste	*Bacillus subtilis* (KP172548)	Submerged	1.62 gL^{-1}	[85]
Papaya peel	*Bacillus subtilis* NCDC0671	Submerged	4.2 gL^{-1}	[86]
Date molasses	*Lactobacillus acidophilus*	Submerged	43.1 gL^{-1}	[87]
Tapioca industry waste	*Bacillus megaterium* MSBN04	Solid state	8.637 mg g^{-1}	[88]
Ragi bran	*Bacillus thuringiensis IAM 12077*	Submerged	0.32 gL^{-1}	[89]

(Table 2) cont.....

Substrate	Organism used	Type of fermentation	Yield of PHA	Reference
Orange peel	*Bacillus subtilis* NCDC0671	Submerged	5.31 gL^{-1}	[90]
Soy cake with sugar cane molasses	*Ralstonia eutropha*	Solid state	4.9 mg g^{-1}	[91]
Mango peel	*Bacillus thuringiensis* IAM 12077	Submerged	4.03 gL^{-1}	[92]
Sugarcane juice	*Bacillus thuringiensis* B417-5	Submerged	2.768 gL^{-1}	[93]
Groundnut oil cake	*Bacillus endophyticus*	Submerged	24%	[94]
Sugar beet juice	*Alcaligenes latus*	Submerged	4.01 ± 0.95 gL^{-1}	[95]
Pineapple waste	*Cupriavidus necator strain* A-04	Submerged	57.2±1.0 gL^{-1}	[96]
Onion peel	*Bacillus subtilis* JCM 1465	Submerged	5.53 gL^{-1}	[97]
Orange peel peel hydrolysate	*Bacillus subtilis* NCDC0671	Submerged	2.15±0.06 gL^{-1}	[73]
Liquid bean curd waste	*Alcaligenes latus*	Submerged	2.48 gL^{-1}	[98]
Banana peel	*Geobacillus stearothermophilus* R- 35646	Submerged	84.63%	[84]
Distillery waste	*Bacillus endophyticus* MTCC 9021	Submerged	6.45 ± 0.07 gL^{-1}	[99]

The extraction of PHA from the bacterial cell is the most crucial step in PHA recovery. The methods for PHA extractions should be selected carefully so that their interference with the physical and chemical properties of PHA is minimal. The extraction process consists of mainly two steps, pretreatment for cell lysis and recovery of PHA. The PHA recovery is performed right after the pretreatment. Methods routinely used for recovery are enzymatic degradation, using solvents and floatation techniques. The most commonly used method is the recovery of PHA using solvent [100, 101]. PHAs dissolve in solvents of chlorinated hydrocarbons (chloroform, 1,2-dichloroethane) and cyclic carbonates (1,2-propylene carbonate) [102, 103]. From these solvents, PHA can be recovered either by solvent evaporation or by precipitating PHA using an anti-solvent liquid (methanol) in which PHAs are insoluble [104].

In SSF, the microbes are inoculated onto the wet solid substrate bed, which has a lower water content [104]. The SSF allows a significant reduction in the cost of production as it doesn't require complex media components in submerged fermentation. It is also reported that the solid fermented product with PHA can be directly used for the synthesis of biodegradable products, this property reduces the production cost further [78, 105]. For SSF, the microbe producing PHA can be cultured first in appropriate nutrient broth, and the dry cells are harvested after centrifugation. The substrate will be generally ground into a fine powder and will be autoclaved. The autoclaved substrate will be inoculated with minimal amounts of autoclaved distilled water with the previously harvested dry bacteria (e.g. 10ml distilled water with 15 mg dry bacteria) [106]. The PHA recovery in SSF is made by first agitating the fermented media in sterile distilled water; then the cells are separated by filtering [91]. The resulting suspension can be subjected to an extraction and recovery process similar to the PHA extraction from submerged fermentation.

Applications of PHA

Globally, problems associated with the use of synthetic polymers are increasing every day. To solve this problem, the world should shift completely to biopolymers like PHA. PHAs are known to mimic the qualities of petroleum-based plastics, including moldability and thermoplastic properties, which opens up its field of applications in various areas [107].

When PHAs are degraded, it leaves no toxic chemical residues, a property that makes them ideal for biomedical uses [108]. Studies have shown that PHA can be conjugated with enzymes, inorganic materials, and other polymers and can improve their physical nature and biocompatibility. Other biomedical applications of PHA include degradable sutures, nerve guides, orthopedic tools, repair patches, cardiological stents, articular cartilage, repair devices, adhesion barriers, bone plates, osteosynthetic materials, tissue engineering materials, and wound dressings [109]. The surface erosion properties, combined with its biocompatibility, make PHA a potential drug carrier. PHAs are also used as grafts for blood-contacting devices [110]. The thermoplasticity of PHA polymer makes it a fit nominee for industrial applications. It can be used in the packaging and coating sectors of food industries [111]. The PHA copolymers are also used in industries to make flushable, non-woven, synthetic papers, thermoformed articles, and binders. PHAs are proven to be very useful in agricultural sectors; they can be used for the encapsulation of fertilizers and seeds, controlled release of insecticides, and effective nitrogen fixation [112]. Other applications of PHA are cosmetic containers, tissue generation scaffolds, the production of bandages, syringes, and masks [113], and antimicrobial PHA sheets incorporated with herbal extracts [90].

Cellulose

Cellulose is a linear polysaccharide made of D-glucose, bonded by β (1-4) linkage. In plants, the cell wall is comprised of microfibrils formed by cellulose chains that are bound by hydrogen bonds. There are different polymorphs of cellulose; among these, natural cellulose is obtained from polymorph I while the other polymorphs (II, III, IV) are produced by several pre-treatments [114]. Though cellulose can occur in its pure form, it is often accompanied by hemicellulose and lignin constituents when present in nature. About 15-35% of the plant biomass is composed of hemicellulose, which is a heterogeneous polymer of short polysaccharide molecules. It consists of five different sugar monomers, which include xylose, arabinose, galactose, mannose, and glucose [115]. However, lignin is the major component of the plant biomass and consists of phenylpropane units, which vary in substitution of methoxyl groups in aromatic rings [116, 117].

Cellulose is generally considered a hygroscopic matter and is insoluble. It is observed to be soluble only in concentrated acids and swells up in the rest, including water, dilute acids, and in most of the other solvents. The chemical reactions of cellulose are polymorphic in nature; generally, the less ordered amorphous region is more reactive than the ordered crystalline regions. Moreover, amorphous cellulose is more liable to enzymatic hydrolysis than crystalline cellulose. Also, the molecular size of polymers is elucidated in terms of the degree of polymerization, which specifies the average value of the number of monomer units and is generally estimated between the range of 1000-30000 in native cellulose [118]. Furthermore, this natural cellulose can be reformed into microcrystalline cellulose, microfibrillar cellulose, and cellulose whiskers. Characteristics of microcrystalline structures include fine texture and white and odorless crystalline powder, and in micro-fibrillated cellulose, they are known to be resistant to heat as well as pH. However, cellulose whiskers attain excellent properties due to their precise, rigid rod-shaped structure, which ameliorates the mechanical characteristics of various natural as well as synthetic materials and moreover, are known to enhance the crystallinity of the matrix [119].

Steps in Cellulose Extraction

A sequence of pretreatment steps is undertaken prior to the actual extraction of cellulose from food waste. The practice of pretreatment is an essential process to enhance the efficiency of lignocellulosic hydrolysis and to inflate the porosity as well as the concentration of amorphous cellulose by the elimination of lignin and hemicellulose. This can be accomplished by treatment with corrosive chemicals like acids and alkalis at a higher temperature in a specified time [116]. The basic mechanism adopted to extract cellulose from food waste in its purest form is dewaxing, treatment with alkali, and further bleaching. An overall depiction of the extraction process is represented in Fig. (3).

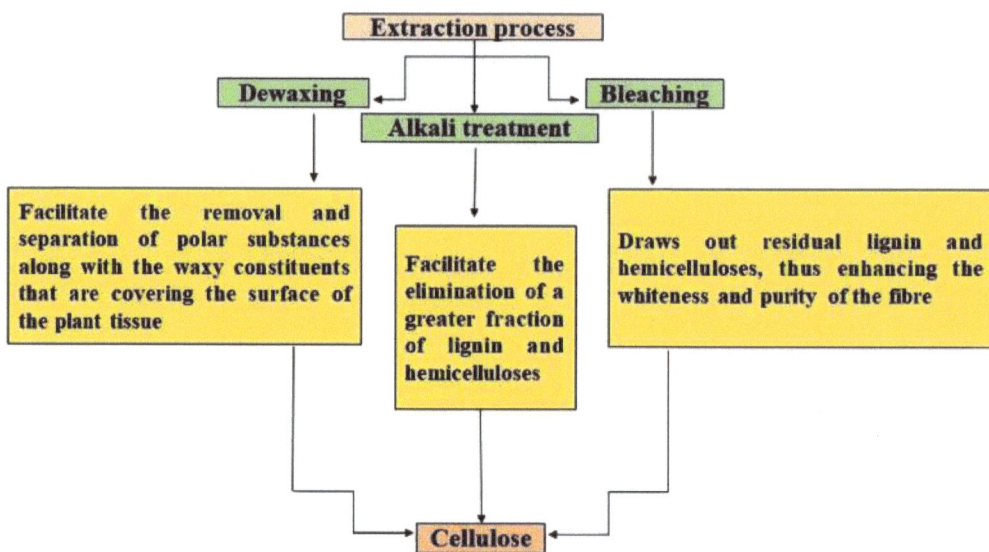

Fig. (3). Cellulose extraction from food waste.

Dewaxing: Epicuticular wax forms the outer surface of the cuticle, which accumulates soluble waxes. The surface action of lye (KOH/ NaOH) initiates the dissolution of epicuticular and cuticular waxes of the outer surface, keeping the cutin matrix and the network of cellulose unaltered. This diffusion of lye enables the breakdown of epidermal and hypodermal cells, which further solubilizes the pectic matter in the middle lamella and thus separates the presence of wax in the sample [120].

Alkali treatment: Sodium hydroxide and sodium carbonate are the major alkaline solutions used to initiate the treatment, which are processed at various concentrations at an applicable temperature and treatment time based upon the type of food waste employed. This step enables the effective removal of a larger proportion of lignin and hemicellulose. It promotes the degradation of the esterification reaction and further triggers the rupture of glycosidic bonds, which consequently alter the structure, followed by degradation of the lignocellulosic biomass [121 - 123].

Bleaching: This process induces the eradication of residual lignin as well as the hemicellulose using sodium chlorite or glacial acetic acid, which would help facilitate obtaining cellulose in its pure whitish form [124]. This treatment allows the breakdown of the phenolic compounds or the molecules that exhibit chromogenic groups in the lignin and also eliminates any by-products produced through the breakdown, intensifying the whiteness of the pulp [125]. Many other solvents, like hydrogen peroxide with EDTA are also applicable for the same purpose [126].

Extraction of Cellulose from Food Waste

In reference to the above-mentioned extraction techniques, the recovery rate of cellulose from various food waste sources is specified below. The source emphasized here is majorly the day-to-day primary food waste that includes, the peels, leaves, pomaces, bagasses of different fruits as well as vegetables and are enlisted in Table **3**:

Table 3. Extraction of cellulose from food waste.

Substrate	Cellulose recovery	References
Potato peel	41-42%	[127]
Tea leaf waste fibers	87.9%	[128]
Banana peel bran	5.1%	[125]
Orange peel	45.2%	[129]
Oil palm empty fruit bunches	64%	[130]
Distillers' dried grains with solubles	$80.98 \pm 0.04\%$	[131]
Mengkuang leaves	$81.6 \pm 0.6\%$	[132]
Rice husk	74%	[133]
Cassava bagasse	30%	[134]
Apple pomace	27.96 ± 0.78	[135]
Tomato plant residue	37%	[136]
Sugarcane bagasse	33%	[137]
Garlic stalk	49.8%	[138]
Onion stalk	45.5%	[138]
Onion skin	41.1%	[138]
Garlic skin	41.7%	[138]
Peanut shells	12.01 mg/ml	[139]

Application of Cellulose

Cellulose being highly abundant on this planet has emerged with extensive applications in various domains Fig. (**4**). The primary use of cellulose film has been for wrapping purposes, and further, its derivatives were employed as laminates, optical films, pharmaceuticals, and in food as well as textile industries [140]. Cellulose nanocrystals are produced from the fine pulp of cellulose through acid hydrolysis and possess enormous properties such as high surface area, tensile strength, stiffness, biodegradability, and biocompatibility [141, 142].

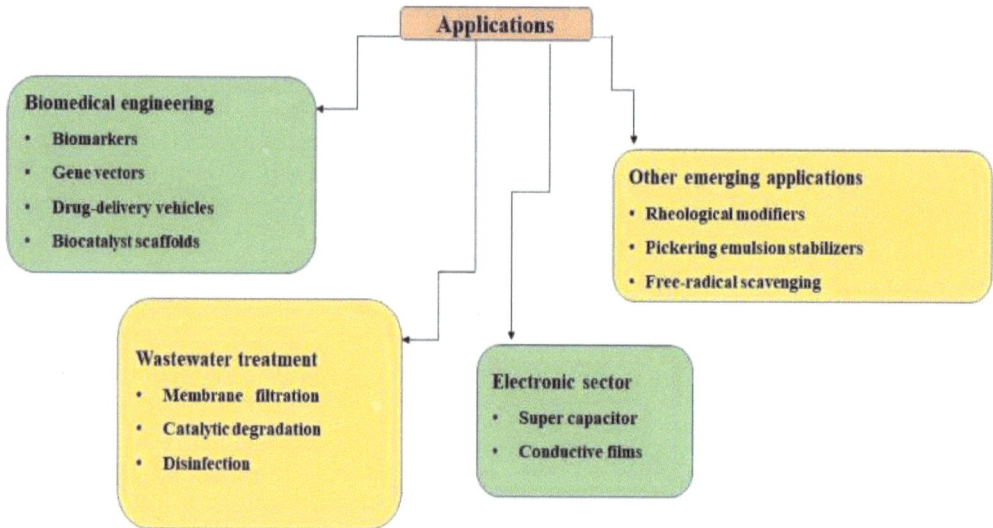

Fig. (4). Application of cellulose.

Cellulose and its derivatives are routinely used for various applications in biomedical engineering, *viz.* tissue engineering, antibacterial and antiviral agents, scaffolds, gene vectors, biomarkers or sensors, biocatalyst scaffolds, and drug delivery vehicles [143]. Cellulose nanocrystals (CNCs) have acquired attention on the aspect of water treatment systems that involve adsorption, absorption, flocculation, membrane filtration, catalytic degradation, and disinfection [144]. The use of promising nanomaterials like CNCs has diminished the dependence on activated carbon as these CNCs have tremendous adsorption capacity [145, 146]. Moreover, CNCs have an enormous role in electronics due to their large optical properties, surface area, and reinforcing capability, which enable them to be used as supercapacitors, conductive films, and so on. Similarly, cellulose nanofibrils (CNFs) are employed for the manufacture of conductive papers or energy storage systems due to their high aspect ratio and flexibility [147, 148]. The applications of CNCs is not limited to these sectors and have rather been implemented for other applications in various other domains, including rheological modifiers, pickering emulsion stabilizers, free radical scavenging, as well as a reinforcement in food packaging films. A number of substrates and methods are available to make cellulose composites [170 - 180]. Also, the research on green synthesis is enhancing, resulting in the diminution of organic solvents along with their toxic by-product. Further these CNCs have undergone modification for a sustainable approach towards the development of nanomaterials for future prospects [149].

Hydroxyapatite (HAp)

Hydroxyapatite (HAp) is a mineral and a well-known potent biomaterial. It is a naturally occurring biopolymer of calcium phosphate in the apatite family with the formula $Ca_{10}(PO_4)_6(OH)_2$. HAp is the major inorganic constituent of bone, teeth and hard tissues. HAp can be derived from a natural biological source and is quite economical. The conversion of trace amounts of beneficial ions present in the biological resources such as animal bones, sea-food shells, eggshells which are food wastes into a potentially clinically valuable biomaterial such as HAp is a remarkable strategy that will add more value to waste valorisation [150].

Properties of HAp

The crystal structure of HAp is a compact conglomeration of tetrahedral arrangement of phosphate (PO_4^{3-}) held together by Ca^{2+} ions in two possible sites, *i.e.*, Ca(I) and Ca(II) that are coordinated by oxygen atoms. At Ca(II) site, the oxygen atoms belong to both phosphate (PO_4^{3-}) and hydroxide ion (OH^-). HAp belongs to a hexagonal crystal system. The Ca/P molar ratio is 1.67 for HAp. The availability of trace elements in the biowastes serving as Ca-P sources and used for the production of HAp controls the Ca/P ratio. The surfaces of HAp show anisotropic adsorption profiles toward biomolecules. Numerous possibilities of cationic and anionic substitutions in HAp are possible. One such is the carbonate (CO_3^{2-}) substitution at the anionic sites to produce A, B or AB-type carbonated HAp. Such ion substitution alters lattice parameters, charge balance, crystallinity, solubility, particle size, morphology, stoichiometry and bioactivity of HAp. and thus, affects the activity of HAp in biomedical applications [151].

Production of Hap From Food Wastes

Food wastes from aquaculture industries, poultry and cattle farming serve as major sources for the HAp production owing to their rich content of calcium, phosphate and carbonate that can be easily and economically converted to HAp molecules. The excellent sources of HAp production are fish bones and scales, mussel shells from the aquaculture industry, eggshells, mammalian bones such as bovine bones from the poultry and cattle farming industry. Moreover, accumulation of such wastes becomes a major problem when it comes to waste disposal and management. Synthesis of HAp from these wastes contributes to the reduction in accumulation of wastes and can prove to be helpful in cost reduction of treatment in the bone repair and replacement contributing to medical therapy. A list of sources and methods for the production of HAp are exemplified in Table **4**.

Table 4. Extraction of HAp from food waste.

Source for HAp production	Method	Heat treatment (T°C)	Particle size	Ca/P ratio	References
Salmon bone	Alkaline hydrolysis	-	6–37 nm	-	[152]
Tilapia nilotica scales	Alkaline Hydrolysis	-	17 nm	1.67	[153]
Cirrhinus mrigala scales	Alkaline Hydrolysis	-	30–50 nm	1.67	[154]
Labeo rohita scales	Alkaline Hydrolysis	-	848 nm	1.69	[154]
Fish bone	Alkaline Hydrolysis	300 -750°C	-	-	[152]
Tilapia (*Oreochromis* sp.) scales	Enzymatic Hydrolysis	Sintered at 800°C for 4 h	719.8 nm	1.78	[155]
Mussel shells	Thermal calcination	800°C	-	-	[156]
Eggshells	Thermal calcination	900°C	100 nm	-	[157]
	Combustion method	250°C Sintered at 900°C for 2 h	44nm	1.6	[158]
Bovine bone	Alkaline Hydrolysis	250°C	-	1.86	[159]
	Thermal calcination	600-1100°C	133 nm	1.46 - 2.01	[160]

Extraction of HAp

Several synthesis methods to prepare HAp have been proposed. A comprehensive elucidation of the production of HAp biomaterial using food wastes as starting material is discussed.

Alkaline Hydrolysis Method: The fish bone and scale wastes were washed and dried to remove salts and other contaminants. Deproteinization of the wastes is done using an acid (usually HCl) followed by alkaline (NaOH) and heat treatment for several hours. Apart from NaOH, other alkaline solutions such as KOH or K_2CO_3 can be used [161]. This process results in a white precipitate subjected to another alkaline heat treatment. The resulting HAp was washed thoroughly with distilled water to remove the alkaline nature, followed by drying at a high temperature [162]. This method differs in alkali concentration, treatment temperature and time depending on the nature of the substrate [159].

Enzymatic Hydrolysis Method: For several hours, the pretreated crushed fish scales were hydrolyzed using protease. The hydrolysates, after heating are centrifuged and dried. Moreover, the HAp particles produced can be sintered at high temperatures for several hours [155].

Thermal Calcination Method: The food wastes such as fish bones were subjected to heat treatment at high temperatures for several hours using a muffle furnace. The products are crushed to form a fine powder of HAp [152]. While using food wastes such as mussel shells, the removal of organic contaminants is the first and foremost step, followed by drying and crushing them into powder. Either washing helps in the removal of organic contaminants with distilled water or during the initial hours of calcination [157]. $CaCO_3$ in the shell is converted into lime (CaO) through calcination in a tube furnace under flowing nitrogen at elevated temperatures. The calcined shells with distilled water form calcium hydroxide. The addition of phosphate solution assists in obtaining a 1.67 Ca/P ratio. The resultant solid is separated from the solution by centrifugation and is dried [156]. Agglomeration can be prevented by milling the crushed calcined shells in phosphoric acid. Food wastes such as bovine bones can be simply heated in the furnace at high temperatures for several hours [159, 160] to obtain white granular HAp . Calcining of bone wastes followed by heating of the calcined bone powders in the furnace at high temperatures produces HAp [163].

Wet Chemical Precipitation Method: In this method, the crushed deproteinized eggshell powder suspended in distilled water was treated with different concentrations of ammonium dihydrogen phosphate. HCl can also be used for the separation and removal of membranes. Prior to the precipitation process, the crushed dried eggshells can be treated with dilute nitric acid leading to froth formation [164]. At the end of the precipitation process, the pH needs to be adjusted to a moderate or high alkaline pH range followed by several hours of stirring. The precipitate obtained was subjected to centrifugation and drying to obtain powdered HAp. The dried mass of HAp can be ball milled for days followed by sintering but this is not an obligatory step. Thermal calcination method prior to wet chemical precipitation method to synthesize HAp with 1.67 molar ratio can be performed followed by sintering or further calcination.

Combustion Method: Food wastes such as eggshells are made free of organic contaminants and are dried and crushed. Addition of conc. acid such as HNO_3 leads to froth formation which is filtered and citric acid as a fuel for combustion is added. Adjustment of pH is a necessary step. The solution is stirred to form a gel at high temperature and subjected to combustion followed by sintering of the black coloured precursor at a very high temperature resulting in HAp [158, 165].

Microwave Irradiation Method: Microwave synthesis method is the best approach considering its advantages such as high homogeneity and lower crystallization temperature. The food wastes such as eggshells are calcined, crushed and mixed with distilled water and diammonium hydrogen phosphate solution similar to the wet chemical precipitation method. The mixture is subjected to microwave irradiation followed by thorough washing and drying to obtain HAp powder [166]. Sodium hypochlorite was also used for the removal of organic contaminants from the eggshell. The pre-treated eggshell powder is treated with EDTA and Na_2HPO_4. The reaction mixture needs pH adjustment and subjection to microwave irradiation which results in white precipitate. It is dried for several hours and HAp was obtained.

Hydrothermal Method: For the removal of organic components from eggshells, hexane can be used during the hydrothermal method of HAp production. Calcination and subsequent reaction with diammonium hydrogen phosphate hydrothermally lead to HAp production which needs drying at a certain temperature for a period of time. Other studies employed dicalcium phosphate dihydrate (DCPD) with pretreated eggshell powder in distilled water. This mixture when wet milled produces a resultant slurry that is dried and heated at high temperature for a day to obtain HAp. HAp can be made from food waste such as eggshells with the incorporation of fruit waste extract [167]. Diluted acids such as HCl are used for suspending the pre-treated eggshell and the fruit waste extract followed by the addition of phosphoric acid and the pH adjustment to alkaline range. The reaction mixture is subjected to hydrothermal transformation at an elevated temperature range. The resultant product is washed and dried to produce HAp powder. Acetone can also be used to remove fats, connective tissue, and other contaminants [168] from food wastes such as bovine bones.

Applications of HAp

In animals, the percentage of HAp in the bone differs. Synthesized HAp, which is chemically similar to the mineral component of bones [164] can be used as a graft material to replace damaged bone or can be used as a coating on implants to assist in the bone in-growth when used in orthopaedic, dental and maxillofacial applications. Thus, HAp finds its use in the fields of therapeutics such as biomedical and dental applications owing to its good biocompatibility [166], bioactivity, high osteo-conductive and osteo-inductive, non-toxicity, non-inflammatory behaviour and non-immunogenicity properties [152]. HAp is known to induce cell proliferation and differentiation and is non cytotoxic to cell lines [153]. Different biowastes-derived HA were also used in drug delivery to treat bone-related diseases (e.g. bone tumour and osteomyelitis) due to their capability to chemically bond with the host bone tissues while simultaneously improving drug efficiency by sustained release of drugs at the specific site [151]. Due to the high adsorption capacity, ion-exchange capability and low water solubility of HAp, it can act as an adsorbent material for the removal of hazardous pollutants (such as heavy metals, radionuclides, inorganic and organic pollutants) in water and wastewater treatment applications thus aiding in remediation of environmental pollution. HAp has been used for the removal of selenite from aqueous solution and hence can be applied as adsorbent for wastewater treatment [153].

HAp synthesized from the food wastes has been employed to prepare low-density polyethylene (LDPE) blends and composites by incorporating HAp into it. Such LDPE blends and composites exhibit good mechanical and surface properties. The incorporation of HAp in synthetic polymers maximizes the use of waste materials thus reducing environmental impact and the use of synthetic polymers [169]. However, the applications of HAp is limited to non-load bearing applications due its poor mechanical property [163]. Recent advances in the development of nano structured HAp having better bioactivity is focused on overcoming the shortcomings such as large surface area to volume ratio of their conventional form and thus resolving the issue. Nano HAp was found to promote osteoblast adhesion and proliferation, osseointegration and thus proves to be ideal for biomedical applications.

Existing Technologies and Future Research Perspective

Although the research about biopolymer production in tune with waste valorization is well discussed in the literature, most of them fail in achieving the commercial scale up majorly due to the techno economic constraints associated with the production process [2]. Variability in the quality and quantity of the waste generated per year, the problems associated with their transport and storage, difficulties in optimizing the production process when scaling up to an industrial scale further decline the commercial significance of these ecofriendly strategies [3]. The lack of proper policies and guidelines, improper awareness among the stakeholders and general public and lack of interest in investment by industrial partners also hinder the commercialization of a waste based bio refinery approach. Further the data regarding the life cycle assessment (LCA) of the biopolymers derived from waste is scarce in the literature and requires critical attention and intensive research to prove their ecofriendly tag to be reliable over the generations [4].

CONCLUSION AND PERSPECTIVES

Generation of enormous amounts of food waste is an inevitable part of day to day human life. The alarming concern over the mitigation of environmental pollution associated with food waste disposal into landfills has paved the way for development of sustainable solutions for food waste valorization. Production of commercially important bioplastics in tune with food waste valorization ensures zero waste to landfill concept. Moreover, this strategy can clearly reduce the cost of production of bioplastics which is the major obstacle in their commercialization. Development of biorefinery concept using food waste can be a key strategy in increasing the payback from the entire bioprocess chain. Moreover, the excellent biodegradability and biocompatibility of these biopolymers can truly make them serve as sustainable alternatives to their synthetic petrochemical based counterparts.

REFERENCES

[1]   Tacon AGJ. Trends in Global Aquaculture and Aquafeed Production: 2000–2017. Rev Fish Sci Aquacult 2020; 28(1): 43-56.
       [http://dx.doi.org/10.1080/23308249.2019.1649634]

[2]   Dietrich K, Dumont MJ, Del Rio LF, Orsat V. Producing PHAs in the bioeconomy — Towards a sustainable bioplastic. Sustainable Production and Consumption 2017; 9: 58-70.
       [http://dx.doi.org/10.1016/j.spc.2016.09.001]

[3]   Mitchell P, James K. Economic growth potential of more circular economies Waste and Resources Action Programme. Banbury, UK: WRAP 2015.

[4]   Muñoz I, Rodríguez C, Gillet DM, Moerschbacher B. Life cycle assessment of chitosan production in India and Europe. Int J Life Cycle Assess 2018; 23: 1151-60.
       [http://dx.doi.org/10.1007/s11367-017-1290-2]

[5]   Arrouze F, Desbrieres J, Rhazi M, Essahli M, Tolaimate A. Valorization of chitins extracted from North Morocco shrimps: Comparison of chitin reactivity and characteristics. J Appl Polym Sci 2019; 136(30): 47804.
       [http://dx.doi.org/10.1002/app.47804]

[6]   O'Hare P, Meenan BJ, Burke GA, Byrne G, Dowling D, Hunt JA. Biological responses to hydroxyapatite surfaces deposited via a co-incident microblasting technique. Biomaterials 2010; 31(3): 515-22.
       [http://dx.doi.org/10.1016/j.biomaterials.2009.09.067] [PMID: 19864018]

[7]   Panda NN, Pramanik K, Sukla LB. Extraction and characterization of biocompatible hydroxyapatite from fresh water fish scales for tissue engineering scaffold. Bioprocess Biosyst Eng 2014; 37(3): 433-40.
       [http://dx.doi.org/10.1007/s00449-013-1009-0] [PMID: 23846299]

[8]   Kwiecien I, Adamus G, Jiang G, *et al.* Biodegradable PBAT/PLA Blend with Bioactive MCPA-PHBV Conjugate Suppresses Weed Growth. Biomacromolecules 2018; 19(2): 511-20.
       [http://dx.doi.org/10.1021/acs.biomac.7b01636] [PMID: 29261293]

[9]   Muzzarelli RAA, Muzzarelli C. Chitosan Chemistry: Relevance to the Biomedical Sciences. Polysaccharides I: Structure, Characterization and Use. Berlin, Heidelberg: Springer Berlin Heidelberg 2005; pp. 151-209.
       [http://dx.doi.org/10.1007/b136820]

[10]   Croisier F, Jérôme C. Chitosan-based biomaterials for tissue engineering. Eur Polym J 2013; 49(4): 780-92.
       [http://dx.doi.org/10.1016/j.eurpolymj.2012.12.009]

[11]   Gonil P, Sajomsang W. Applications of magnetic resonance spectroscopy to chitin from insect cuticles. Int J Biol Macromol 2012; 51(4): 514-22.
       [http://dx.doi.org/10.1016/j.ijbiomac.2012.06.025] [PMID: 22732132]

[12]   Mourya VK, Inamdar NN. Chitosan-modifications and applications: Opportunities galore. React Funct Polym 2008; 68(6): 1013-51.
       [http://dx.doi.org/10.1016/j.reactfunctpolym.2008.03.002]

[13]   Rinaudo M. Chitin and chitosan: Properties and applications. Prog Polym Sci 2006; 31(7): 603-32.

[http://dx.doi.org/10.1016/j.progpolymsci.2006.06.001]

[14]	Aranaz I, Mengibar M, Harris R, Panos I, Miralles B, Acosta N, *et al.* Functional Characterization of Chitin and Chitosan. Curr Chem Biol 2009; 3: 203-30.

[15]	Huang M, Khor E, Lim LY. Uptake and cytotoxicity of chitosan molecules and nanoparticles: effects of molecular weight and degree of deacetylation. Pharm Res 2004; 21(2): 344-53.
[http://dx.doi.org/10.1023/B:PHAM.0000016249.52831.a5] [PMID: 15032318]

[16]	Islam MS, Khan S, Tanaka M. Waste loading in shrimp and fish processing effluents: potential source of hazards to the coastal and nearshore environments. Mar Pollut Bull 2004; 49(1-2): 103-10.
[http://dx.doi.org/10.1016/j.marpolbul.2004.01.018] [PMID: 15234879]

[17]	Kumari S, Rath P, Sri Hari Kumar A, Tiwari TN. Extraction and characterization of chitin and chitosan from fishery waste by chemical method. Environmental Technology & Innovation 2015; 3: 77-85.
[http://dx.doi.org/10.1016/j.eti.2015.01.002]

[18]	Wu T, Zivanovic S, Draughon FA, Sams CE. Chitin and chitosan--value-added products from mushroom waste. J Agric Food Chem 2004; 52(26): 7905-10.
[http://dx.doi.org/10.1021/jf0492565] [PMID: 15612774]

[19]	Santos VP, Marques NSS, Maia PCSV, Lima MAB, Franco LO, Campos-Takaki GM. Seafood Waste as Attractive Source of Chitin and Chitosan Production and Their Applications. Int J Mol Sci 2020; 21(12): 4290.
[http://dx.doi.org/10.3390/ijms21124290] [PMID: 32560250]

[20]	Teli MD, Sheikh J. Extraction of chitosan from shrimp shells waste and application in antibacterial finishing of bamboo rayon. Int J Biol Macromol 2012; 50(5): 1195-200.
[http://dx.doi.org/10.1016/j.ijbiomac.2012.04.003] [PMID: 22522048]

[21]	Manni L, Ghorbel-Bellaaj O, Jellouli K, Younes I, Nasri M. Extraction and characterization of chitin, chitosan, and protein hydrolysates prepared from shrimp waste by treatment with crude protease from Bacillus cereus SV1. Appl Biochem Biotechnol 2010; 162(2): 345-57.
[http://dx.doi.org/10.1007/s12010-009-8846-y] [PMID: 19960271]

[22]	Thomas S, Pius A, Gopi S. Handbook of Chitin and Chitosan: Volume 3: Chitin- and Chitosan-based Polymer Materials for Various Applications. Elsevier; 2020.

[23]	El Knidri H, El Khalfaouy R, Laajeb A, Addaou A, Lahsini A. Eco-friendly extraction and characterization of chitin and chitosan from the shrimp shell waste via microwave irradiation. Process Saf Environ Prot 2016; 104: 395-405.
[http://dx.doi.org/10.1016/j.psep.2016.09.020]

[24]	Cai J, Yang J, Du Y, *et al.* Enzymatic preparation of chitosan from the waste Aspergillus niger mycelium of citric acid production plant. Carbohydr Polym 2006; 64(2): 151-7.
[http://dx.doi.org/10.1016/j.carbpol.2005.11.004]

[25]	Iqbal J, Wattoo FH, Wattoo MHS, *et al.* Adsorption of acid yellow dye on flakes of chitosan prepared from fishery wastes. Arab J Chem 2011; 4(4): 389-95.
[http://dx.doi.org/10.1016/j.arabjc.2010.07.007]

[26]	Muley AB, Chaudhari SA, Mulchandani KH, Singhal RS. Extraction and characterization of chitosan from prawn shell waste and its conjugation with cutinase for enhanced thermo-stability. Int J Biol Macromol 2018; 111: 1047-58.
[http://dx.doi.org/10.1016/j.ijbiomac.2018.01.115] [PMID: 29366886]

[27]	Nouri M, Khodaiyan F, Razavi SH, Mousavi M. Improvement of chitosan production from Persian Gulf shrimp waste by response surface methodology. Food Hydrocoll 2016; 59: 50-8.
[http://dx.doi.org/10.1016/j.foodhyd.2015.08.027]

[28]	Sedaghat F, Yousefzadi M, Toiserkani H, Najafipour S. Bioconversion of shrimp waste Penaeus merguiensis using lactic acid fermentation: An alternative procedure for chemical extraction of chitin

and chitosan. Int J Biol Macromol 2017; 104(Pt A): 883-8.
[http://dx.doi.org/10.1016/j.ijbiomac.2017.06.099] [PMID: 28663153]

[29]   Hafsa J, Smach MA, Charfeddine B, Limem K, Majdoub H, Rouatbi S. Antioxidant and antimicrobial proprieties of chitin and chitosan extracted from Parapenaeus Longirostris shrimp shell waste. Ann Pharm Fr 2016; 74(1): 27-33.
[http://dx.doi.org/10.1016/j.pharma.2015.07.005] [PMID: 26687000]

[30]   Hajji S, Younes I, Ghorbel-Bellaaj O, Hajji R, Rinaudo M, Nasri M. Structural differences between chitin and chitosan extracted from three different marine sources. *Int J Biol Macromol* 2014;65:298–306. "http://paperpile.com/b/ypowKr/QNcPj

[31]   Baron RD, Pérez LL, Salcedo JM, Córdoba LP, Sobral PJA. Production and characterization of films based on blends of chitosan from blue crab (Callinectes sapidus) waste and pectin from Orange (Citrus sinensis Osbeck) peel. Int J Biol Macromol 2017; 98: 676-83.
[http://dx.doi.org/10.1016/j.ijbiomac.2017.02.004] [PMID: 28189792]

[32]   Shavandi A, Hu Z, Teh S, Zhao J, Carne A, Bekhit A. Antioxidant and functional properties of protein hydrolysates obtained from squid pen chitosan extraction effluent. Food Chem 2017;227:194–201. http://paperpile.com/b/ypowKr/E4tn0

[33]   Kucukgulmez A, Celik M, Yanar Y, Sen D, Polat H, Kadak AE. Physicochemical characterization of chitosan extracted from Metapenaeus stebbingi shells. Food Chem 2011; 126(3): 1144-8.
[http://dx.doi.org/10.1016/j.foodchem.2010.11.148]

[34]   Poverenov E, Arnon-Rips H, Zaitsev Y, Bar V, Danay O, Horev B. Potential of chitosan from mushroom waste to enhance quality and storability of fresh-cut melons. Food Chem 2018;268:233–41. "http://paperpile.com/b/ypowKr/QCCYX

[35]   Metin C, Alparslan Y, Baygar T, Baygar T. Physicochemical, Microstructural and Thermal Characterization of Chitosan from Blue Crab Shell Waste and Its Bioactivity Characteristics. J Polym Environ 2019; 27(11): 2552-61.
[http://dx.doi.org/10.1007/s10924-019-01539-3]

[36]   El-Naggar MM, Abou-Elmagd WSI, Suloma A, El-Shabaka HA, Khalil MT, Abd El-Rahman FA. Optimization and Physicochemical Characterization of Chitosan and Chitosan Nanoparticles Extracted from the Crayfish Procambarus clarkii Wastes. J Shellfish Res 2019; 38(2): 385-95.
[http://dx.doi.org/10.2983/035.038.0220]

[37]   Takarina ND, Nasrul AA, Nurmarina A. Degree of Deacetylation of Chitosan Extracted from White Snapper (Lates sp.) Scales Waste. International Journal of Pharma Medicine and Biological Sciences 2017; 6(1): 16-9.
[http://dx.doi.org/10.18178/ijpmbs.6.1.16-19]

[38]   Marzieh MN, Zahra F, Tahereh E, Sara KN. Comparison of the physicochemical and structural characteristics of enzymatic produced chitin and commercial chitin. Int J Biol Macromol 2019; 139: 270-6.
[http://dx.doi.org/10.1016/j.ijbiomac.2019.07.217] [PMID: 31376451]

[39]   Santos VP, Maia P, Alencar N de S, Farias L, Andrade RFS, Souza D, *et al.* Recovery of chitin and chitosan from shrimp waste with microwave technique and versatile application. Arq Inst Biol 2019.

[40]   Zeng X, Li P, Chen X, Kang Y, Xie Y, Li X. Effects of deproteinization methods on primary structure and antioxidant activity of Ganoderma lucidum polysaccharides. *Int J Biol Macromol* 2019;126:867–76. http://paperpile.com/b/ypowKr/oPqMv

[41]   Seenuvasan M, Sarojini G, Dineshkumar M. Chapter 6 - Recovery of chitosan from natural biotic waste. In: Varjani S, Pandey A, Gnansounou E, Khanal SK, Raveendran S, editors. Current Developments in Biotechnology and Bioengineering, Elsevier; 2020, p. 115–33. http://paperpile. com/b/ypowKr/pRc6O

[42]   Percot A, Viton C, Domard A. Characterization of shrimp shell deproteinization. Biomacromolecules

2003; 4(5): 1380-5.
[http://dx.doi.org/10.1021/bm034115h] [PMID: 12959609]

[43]   Hamed I, Özogul F, Regenstein JM. Industrial applications of crustacean by-products (chitin, chitosan, and chitooligosaccharides): A review. Trends Food Sci Technol 2016; 48: 40-50.
[http://dx.doi.org/10.1016/j.tifs.2015.11.007]

[44]   Sugiyanti D, Darmadji P, Anggrahini S, Anwar C, Santoso U. Preparation and characterization of chitosan from Indonesian Tambak Lorok shrimp shell waste and crab shell waste. Pak J Nutr 2018; 17(9): 446-53.
[http://dx.doi.org/10.3923/pjn.2018.446.453]

[45]   Samrot AV, Burman U, Philip SA, N S, Chandrasekaran K. Synthesis of curcumin loaded polymeric nanoparticles from crab shell derived chitosan for drug delivery. Informatics in Medicine Unlocked 2018; 10: 159-82.
[http://dx.doi.org/10.1016/j.imu.2017.12.010]

[46]   Ali M, Shakeel M, Mehmood K. Extraction and characterization of high purity chitosan by rapid and simple techniques from mud crabs taken from Abbottabad. Pak J Pharm Sci 2019; 32(1): 171-5.
[PMID: 30772806]

[47]   Palpandi C, Shanmugam V, Shanmugam A. Extraction of chitin and chitosan from shell and operculum of mangrove gastropod Nerita (Dostia) crepidularia Lamarck. McGill J Med 2009; 1: 198-205.

[48]   Tolaimate A, Desbrieres J, Rhazi M, Alagui A. Contribution to the preparation of chitins and chitosans with controlled physico-chemical properties. Polymer (Guildf) 2003; 44(26): 7939-52.
[http://dx.doi.org/10.1016/j.polymer.2003.10.025]

[49]   Abdulkarim A, Isa MT, Abdulsalam S, Muhammad AJ, Ameh AO. Extraction and characterisation of chitin and chitosan from mussel shell. Extraction 2013; 3: 108-14.

[50]   "http://paperpile.com/b/ypowKr/SI7zt"Namboodiri MMT, Pakshirajan K. Chapter 10 - Valorization of waste biomass for chitin and chitosan production. In: Bhaskar T, Pandey A, Rene ER, Tsang DCW, editors. Waste Biorefinery, Elsevier; 2020, p. 241–66.

[51]   Zamani A, Taherzadeh M. Production of low molecular weight chitosan by hot dilute sulfuric acid. BioResources 2010.

[52]   Thirunavukkarasu N, Dhinamala K, Inbaraj RM. Production of chitin from two marine stomatopods Oratosquilla spp.(Crustacea). J Chem Pharm Res 2011; 3: 353-9.

[53]   Tsigos I, Martinou A, Kafetzopoulos D, Bouriotis V. Chitin deacetylases: new, versatile tools in biotechnology. Trends Biotechnol 2000; 18(7): 305-12.
[http://dx.doi.org/10.1016/S0167-7799(00)01462-1] [PMID: 10856926]

[54]   Baranwal A, Kumar A, Priyadharshini A, et al. Chitosan: An undisputed bio-fabrication material for tissue engineering and bio-sensing applications. Int J Biol Macromol 2018; 110: 110-23.
[http://dx.doi.org/10.1016/j.ijbiomac.2018.01.006] [PMID: 29339286]

[55]   Lopes PP, Tanabe EH, Bertuol DA. hitosan as biomaterial in drug delivery and tissue engineering.Handbook of Chitin and Chitosan.Chapter 13Gopi S., Thomas S., Pius A.Elsevier2 2020; pp. 407-31.
[http://dx.doi.org/10.1016/B978-0-12-817966-6.00013-3]

[56]   Ameeduzzafar , Imam SS, Abbas Bukhari SN, Ahmad J, Ali A. Formulation and optimization of levofloxacin loaded chitosan nanoparticle for ocular delivery: In-vitro characterization, ocular tolerance and antibacterial activity. Int J Biol Macromol 2018; 108: 650-9.
[http://dx.doi.org/10.1016/j.ijbiomac.2017.11.170] [PMID: 29199125]

[57]   Shariatinia Z. Pharmaceutical applications of chitosan. Adv Colloid Interface Sci 2019; 263: 131-94.
[http://dx.doi.org/10.1016/j.cis.2018.11.008] [PMID: 30530176]

[58]   Yang Y, Wu J, Wang X, Liu J, Ding F, Gu X. Fabrication and Evaluation of Chitin-Based Nerve Guidance Conduits Used to Promote Peripheral Nerve Regeneration. Adv Eng Mater 2009; 11(11): B209-18.
[http://dx.doi.org/10.1002/adem.200900120]

[59]   Moradi Dehaghi S, Rahmanifar B, Moradi AM, Azar PA. Removal of permethrin pesticide from water by chitosan–zinc oxide nanoparticles composite as an adsorbent. J Saudi Chem Soc 2014; 18(4): 348-55.
[http://dx.doi.org/10.1016/j.jscs.2014.01.004]

[60]   Kwok KCM, Koong LF, Chen G, McKay G. Mechanism of arsenic removal using chitosan and nanochitosan. J Colloid Interface Sci 2014; 416: 1-10.
[http://dx.doi.org/10.1016/j.jcis.2013.10.031] [PMID: 24370394]

[61]   Sethulekshmi C. Chitin and its benefits. International Journal of Advanced Research in Biological Sciences 2014; 1: 171-5.

[62]   Okamoto Y, Kawakami K, Miyatake K, Morimoto M, Shigemasa Y, Minami S. Analgesic effects of chitin and chitosan. Carbohydr Polym 2002; 49(3): 249-52.
[http://dx.doi.org/10.1016/S0144-8617(01)00316-2]

[63]   Muzzarelli RAA, Orlandini F, Pacetti D, Boselli E, Frega NG, Tosi G. Chitosan taurocholate capacity to bind lipids and to undergo enzymatic hydrolysis: An in vitro model. Carbohydr Polym 2006;66:363–71. http://paperpile.com/b/ypowKr/9TkNz

[64]   Sedlacek P, Slaninova E, Koller M, *et al.* PHA granules help bacterial cells to preserve cell integrity when exposed to sudden osmotic imbalances. N Biotechnol 2019; 49: 129-36.
[http://dx.doi.org/10.1016/j.nbt.2018.10.005] [PMID: 30389520]

[65]   Lee SY. Bacterial polyhydroxyalkanoates. Biotechnol Bioeng 1996; 49(1): 1-14.
[http://dx.doi.org/10.1002/(SICI)1097-0290(19960105)49:1<1::AID-BIT1>3.0.CO;2-P]   [PMID: 18623547]

[66]   Hiraishi T, Taguchi S. Protein Engineering of Enzymes Involved in Bioplastic Metabolism.Protein Engineering - Technology and Application. InTech 2013.
[http://dx.doi.org/10.5772/55552]

[67]   Ong SY, Chee JY, Sudesh K. Degradation of Polyhydroxyalkanoate (PHA): a Review. J Sib Fed Univ. Biol 2017; 10: 21-225.

[68]   HM, Blue MKD, Cobbs BD, Ricotilli TA. Characterization of an Extracellular Polyhydroxyalkanoate Depolymerase from Streptomyces sp. SFB5A. J bioremediat biodegrad 2018; 9(5):1-11.

[69]   Lemes AP, Montanheiro TLA, Passador FR, Durán N. Nanocomposites of Polyhydroxyalkanoates Reinforced with Carbon Nanotubes: Chemical and Biological Properties.Eco-friendly Polymer Nanocomposites: Processing and Properties. New Delhi: Springer India 2015; pp. 79-108.
[http://dx.doi.org/10.1007/978-81-322-2470-9_3]

[70]   Khosravi-Darani K, Bucci DZ. Application of poly (hydroxyalkanoate) in food packaging: Improvements by nanotechnology. Chem Biochem Eng Q 2015; 29(2): 275-85.
[http://dx.doi.org/10.15255/CABEQ.2014.2260]

[71]   Koller M. Poly (hydroxyalkanoates) for food packaging: Application and attempts towards implementation 2014.

[72]   Chee JY, Lakshmanan M, Jeepery IF, Hairudin NHM, Sudesh K. The potential application of Cupriavidus necator as polyhydroxyalkanoates producer and single cell protein: A review on scientific, cultural and religious perspectives. Applied Food Biotechnology 2019; 6: 19-34.

[73]   Umesh M, Priyanka K, Thazeem B, Preethi K. Biogenic PHA nanoparticle synthesis and characterization from *Bacillus subtilis* NCDC0671 using orange peel medium. Int J Polym Mater 2018; 67(17): 996-1004.

[http://dx.doi.org/10.1080/00914037.2017.1417284]

[74] Umesh M, Thazeem B. Biodegradation Studies of Polyhydroxyalkanoates extracted from Bacillus Subtilis NCDC 0671. Res J Chem Environ 2019; 23: 107-14.

[75] Butt FI, Muhammad N, Hamid A, Moniruzzaman M, Sharif F. Recent progress in the utilization of biosynthesized polyhydroxyalkanoates for biomedical applications – Review. Int J Biol Macromol 2018; 120(Pt A): 1294-305.
[http://dx.doi.org/10.1016/j.ijbiomac.2018.09.002] [PMID: 30189278]

[76] Jacquel N, Lo CW, Wei YH, Wu HS, Wang SS. Isolation and purification of bacterial poly(3-hydroxyalkanoates). Biochem Eng J 2008; 39(1): 15-27.
[http://dx.doi.org/10.1016/j.bej.2007.11.029]

[77] Castilho LR, Mitchell DA, Freire DMG. Production of polyhydroxyalkanoates (PHAs) from waste materials and by-products by submerged and solid-state fermentation. Bioresour Technol 2009; 100(23): 5996-6009.
[http://dx.doi.org/10.1016/j.biortech.2009.03.088] [PMID: 19581084]

[78] Sindhu R. Solid-state Fermentation for the Production of Poly(hydroxyalkanoates). Chem Biochem Eng Q 2015; 29(2): 173-81.
[http://dx.doi.org/10.15255/CABEQ.2014.2256]

[79] Hafuka A, Sakaida K, Satoh H, Takahashi M, Watanabe Y, Okabe S. Effect of feeding regimens on polyhydroxybutyrate production from food wastes by Cupriavidus necator. Bioresour Technol 2011; 102(3): 3551-3.
[http://dx.doi.org/10.1016/j.biortech.2010.09.018] [PMID: 20870404]

[80] Tripathi AD, Raj Joshi T, Kumar Srivastava S, Darani KK, Khade S, Srivastava J. Effect of nutritional supplements on bio-plastics (PHB) production utilizing sugar refinery waste with potential application in food packaging. Prep Biochem Biotechnol 2019; 49(6): 567-77.
[http://dx.doi.org/10.1080/10826068.2019.1591982] [PMID: 30929621]

[81] Farah NO, Norrsquo Aini AR, Halimatun SH, Tabassum M, Phang LY, Mohd AH. Utilization of kitchen waste for the production of green thermoplastic polyhydroxybutyrate (PHB) by Cupriavidus necator CCGUG 52238. Afr J Microbiol Res 2011; 5(19): 2873-9.
[http://dx.doi.org/10.5897/AJMR11.156]

[82] Eshtaya MK, Nor , Rahman AA, Hassan MA. Bioconversion of restaurant waste into Polyhydroxybutyrate (PHB) by recombinant E. coli through anaerobic digestion. Int J Environ Waste Manag 2013; 11(1): 27.
[http://dx.doi.org/10.1504/IJEWM.2013.050521]

[83] Verlinden RAJ, Hill DJ, Kenward MA, Williams CD, Piotrowska-Seget Z, Radecka IK. Production of polyhydroxyalkanoates from waste frying oil by Cupriavidus necator. AMB Express 2011; 1(1): 11.
[http://dx.doi.org/10.1186/2191-0855-1-11] [PMID: 21906352]

[84] Getachew A, Woldesenbet F. Production of biodegradable plastic by polyhydroxybutyrate (PHB) accumulating bacteria using low cost agricultural waste material. BMC Res Notes 2016; 9(1): 509.
[http://dx.doi.org/10.1186/s13104-016-2321-y] [PMID: 27955705]

[85] Mohapatra S, Sarkar B, Samantaray DP, Daware A, Maity S, Pattnaik S. Bioconversion of fish solid waste into PHB using Bacillus subtilis based submerged fermentation process. Environ Technol 2017;38:3201–8. http://paperpile.com/b/ypowKr/q5dJI

[86] Umesh M, Priyanka K, Thazeem B, Preethi K. Production of Single Cell Protein and Polyhydroxyalkanoate from Carica papaya Waste. Arab J Sci Eng 2017; 42(6): 2361-9.
[http://dx.doi.org/10.1007/s13369-017-2519-x]

[87] Hamieh A, Olama Z, Holail H. Microbial production of polyhydroxybutyrate, a biodegradable plastic using agro-industrial waste products. Glo Adv Res J Microbiol 2013; 2: 54-64.

[88] Sathiyanarayanan G, Kiran GS, Selvin J, Saibaba G. Optimization of polyhydroxybutyrate production

by marine Bacillus megaterium MSBN04 under solid state culture. Int J Biol Macromol 2013; 60: 253-61.
[http://dx.doi.org/10.1016/j.ijbiomac.2013.05.031] [PMID: 23748002]

[89]   Shivakumar S. Polyhydroxybutyrate (PHB) production using agro-industrial residue as substrate by Bacillus thuringiensis IAM 12077. Int J Chemtech Res 2012; 4: 1158-62.

[90]   Umesh M, Preethi K. Fabrication of antibacterial bioplastic sheet using orange peel medium and its antagonistic effect against common clinical pathogens. Res J Biotechnol 2017; 12: 67-74.

[91]   Oliveira FC, Freire DMG, Castilho LR. Production of poly(3-hydroxybutyrate) by solid-state fermentation with Ralstonia eutropha. Biotechnol Lett 2004; 26(24): 1851-5.
[http://dx.doi.org/10.1007/s10529-004-5315-0] [PMID: 15672227]

[92]   Gowda V, Shivakumar S. Agrowaste-based Polyhydroxyalkanoate (PHA) production using hydrolytic potential of Bacillus thuringiensis IAM 12077. Braz Arch Biol Technol 2014; 57(1): 55-61.
[http://dx.doi.org/10.1590/S1516-89132014000100009]

[93]   Thammasittirong A, Saechow S, Thammasittirong SNR. Efficient polyhydroxybutyrate production from Bacillus thuringiensis using sugarcane juice substrate. Turk J Biol 2017; 41: 992-1002.
[http://dx.doi.org/10.3906/biy-1704-13] [PMID: 30814863]

[94]   v D. Enhancement of Polyhydroxybutyrate (PHB) Production using Organic Waste as Substrate. Int J Res Appl Sci Eng Technol 2020; 8(5): 1012-6.
[http://dx.doi.org/10.22214/ijraset.2020.5160]

[95]   Wang B, Sharma-Shivappa RR, Olson JW, Khan SA. Production of polyhydroxybutyrate (PHB) by Alcaligenes latus using sugarbeet juice. Ind Crops Prod 2013; 43: 802-11.
[http://dx.doi.org/10.1016/j.indcrop.2012.08.011]

[96]   Sukruansuwan V, Napathorn SC. Use of agro-industrial residue from the canned pineapple industry for polyhydroxybutyrate production by *Cupriavidus necator* strain A-04. Biotechnol Biofuels 2018; 11(1): 202.
[http://dx.doi.org/10.1186/s13068-018-1207-8] [PMID: 30061924]

[97]   Vijay R, Tarika K. Microbial Production of Polyhydroxy alkanoates (PHAs) using Kitchen Waste as an Inexpensive Carbon Source. Biosci Biotechnol Res Asia 2019; 16(1): 155-66.
[http://dx.doi.org/10.13005/bbra/2733]

[98]   Kumalaningsih S, Hidayat N. Optimization of polyhydroxyalkanoates (PHA) production from liquid bean curd waste by Alcaligenes latus bacteria. J Agric Food Technol 2011.

[99]   Priyanka K, Umesh M, Thazeem B, Preethi K. Polyhydroxyalkanoate biosynthesis and characterization from optimized medium utilizing distillery effluent using Bacillus endophyticus MTCC 9021: a statistical approach. Biocatal Biotransformation 2020; pp. 1-13.

[100] Kunasundari B, Sudesh K. Isolation and recovery of microbial polyhydroxyalkanoates. Express Polym Lett 2011; 5(7): 620-34.
[http://dx.doi.org/10.3144/expresspolymlett.2011.60]

[101] Madkour MH, Heinrich D, Alghamdi MA, Shabbaj II, Steinbüchel A. PHA recovery from biomass. Biomacromolecules 2013; 14(9): 2963-72.
[http://dx.doi.org/10.1021/bm4010244] [PMID: 23875914]

[102] Ramsay JA, Berger E, Voyer R, Chavarie C, Ramsay BA. Extraction of poly-3-hydroxybutyrate using chlorinated solvents. Biotechnol Tech 1994; 8(8): 589-94.
[http://dx.doi.org/10.1007/BF00152152]

[103] Lafferty RM, Heinzle E. Use of cyclic carbonic acid esters as solvents for poly-(.beta.-hydroxybutyric acid). CA:1091600:A, 1980.

[104] Hocking PJ, Marchessault RH. Chemistry and Technology of Biodegradable Polymers ed GJL Griffin 1994.

[105] Koller M. A Review on Established and Emerging Fermentation Schemes for Microbial Production of Polyhydroxyalkanoate (PHA) Biopolyesters. Fermentation (Basel) 2018; 4(2): 30.
[http://dx.doi.org/10.3390/fermentation4020030]

[106] Oliveira FC, Dias ML, Castilho LR, Freire DMG. Characterization of poly(3-hydroxybutyrate) produced by Cupriavidus necator in solid-state fermentation. Bioresour Technol 2007; 98(3): 633-8.
[http://dx.doi.org/10.1016/j.biortech.2006.02.022] [PMID: 16580194]

[107] Arun A, Arthi R, Shanmugabalaji V, Eyini M. Microbial production of poly-β-hydroxybutyrate by marine microbes isolated from various marine environments. Bioresour Technol 2009; 100(7): 2320-3.
[http://dx.doi.org/10.1016/j.biortech.2008.08.037] [PMID: 19101142]

[108] Sun Z, Ramsay J, Guay M, Ramsay B. Increasing the yield of MCL-PHA from nonanoic acid by co-feeding glucose during the PHA accumulation stage in two-stage fed-batch fermentations of Pseudomonas putida KT2440. J Biotechnol 2007; 132(3): 280-2.
[http://dx.doi.org/10.1016/j.jbiotec.2007.02.023] [PMID: 17442441]

[109] Bonartsev AP, Bonartseva GA, Reshetov IV, Shaitan KV, Kirpichnikov MP. Application of Polyhydroxyalkanoates in Medicine and the Biological Activity of Natural Poly(3-Hydroxybutyrate). Acta Nat (Engl Ed) 2019; 11(2): 4-16.
[http://dx.doi.org/10.32607/20758251-2019-11-2-4-16] [PMID: 31413875]

[110] Elmowafy E, Abdal-Hay A, Skouras A, Tiboni M, Casettari L, Guarino V. Polyhydroxyalkanoate (PHA): applications in drug delivery and tissue engineering. Expert Rev Med Devices 2019; 16(6): 467-82.
[http://dx.doi.org/10.1080/17434440.2019.1615439] [PMID: 31058550]

[111] Bucci DZ, Tavares LBB, Sell I. PHB packaging for the storage of food products. Polym Test 2005; 24(5): 564-71.
[http://dx.doi.org/10.1016/j.polymertesting.2005.02.008]

[112] Philip S, Keshavarz T, Roy I. Polyhydroxyalkanoates: biodegradable polymers with a range of applications. J Chem Technol Biotechnol 2007; 82(3): 233-47.
[http://dx.doi.org/10.1002/jctb.1667]

[113] Grigore ME, Grigorescu RM, Iancu L, Ion RM, Zaharia C, Andrei ER. Methods of synthesis, properties and biomedical applications of polyhydroxyalkanoates: a review. J Biomater Sci Polym Ed 2019; 30(9): 695-712.
[http://dx.doi.org/10.1080/09205063.2019.1605866] [PMID: 31012805]

[114] Kulasinski K, Keten S, Churakov SV, Derome D, Carmeliet J. A comparative molecular dynamics study of crystalline, paracrystalline and amorphous states of cellulose. Cellulose 2014; 21(3): 1103-16.
[http://dx.doi.org/10.1007/s10570-014-0213-7]

[115] Sun SN, Cao XF, Xu F, Sun RC, Jones GL, Baird M. Structure and thermal property of alkaline hemicelluloses from steam exploded Phyllostachys pubescens. Carbohydr Polym 2014; 1(01): 1191-7.
[http://dx.doi.org/10.1016/j.carbpol.2013.09.109] [PMID: 24299891]

[116] Ravindran R, Jaiswal AK. A comprehensive review on pre-treatment strategy for lignocellulosic food industry waste: Challenges and opportunities. Bioresour Technol 2016; 199: 92-102.
[http://dx.doi.org/10.1016/j.biortech.2015.07.106] [PMID: 26277268]

[117] Duval A, Lawoko M. A review on lignin-based polymeric, micro- and nano-structured materials. React Funct Polym 2014; 85: 78-96.
[http://dx.doi.org/10.1016/j.reactfunctpolym.2014.09.017]

[118] Stephen AM, Phillips GO. Food Polysaccharides and Their Applications. CRC Press, Taylor & Francis 2016.
[http://dx.doi.org/10.1201/9781420015164]

[119] Heinze T. Cellulose: Structure and Properties. Adv Polym Sci 2015; 271: 1-52.
[http://dx.doi.org/10.1007/12_2015_319]

[120] Floros JD, Wetzstein H, Chinnan MS. Chemical (NaOH) Peeling as Viewed by Scanning Electron Microscopy: Pimiento Peppers as a Case Study. J Food Sci 1987; 52(5): 1312-6.
[http://dx.doi.org/10.1111/j.1365-2621.1987.tb14071.x]

[121] García JC, Díaz MJ, Garcia MT, Feria MJ, Gómez DM, López F. Search for optimum conditions of wheat straw hemicelluloses cold alkaline extraction process. Biochem Eng J 2013; 71: 127-33.
[http://dx.doi.org/10.1016/j.bej.2012.12.008]

[122] Tian SQ, Zhao RY, Chen ZC. Review of the pretreatment and bioconversion of lignocellulosic biomass from wheat straw materials. Renew Sustain Energy Rev 2018; 91: 483-9.
[http://dx.doi.org/10.1016/j.rser.2018.03.113]

[123] Li X, Kim TH, Nghiem NP. Bioethanol production from corn stover using aqueous ammonia pretreatment and two-phase simultaneous saccharification and fermentation (TPSSF). Bioresour Technol 2010; 101(15): 5910-6.
[http://dx.doi.org/10.1016/j.biortech.2010.03.015] [PMID: 20338749]

[124] Bhattacharya D, Germinario LT, Winter WT. Isolation, preparation and characterization of cellulose microfibers obtained from bagasse. Carbohydr Polym 2008; 73(3): 371-7.
[http://dx.doi.org/10.1016/j.carbpol.2007.12.005]

[125] Pelissari FM, Sobral PJA, Menegalli FC. Isolation and characterization of cellulose nanofibers from banana peels. Cellulose 2014; 21(1): 417-32.
[http://dx.doi.org/10.1007/s10570-013-0138-6]

[126] Rehman N, Alam S, Amin NU, Mian I, Ullah H. Ecofriendly Isolation of Cellulose from *Eucalyptus lenceolata* : A Novel Approach. Int J Polym Sci 2018; 2018: 1-7.
[http://dx.doi.org/10.1155/2018/8381501]

[127] Chen D, Lawton D, Thompson MR, Liu Q. Biocomposites reinforced with cellulose nanocrystals derived from potato peel waste. Carbohydr Polym 2012; 90(1): 709-16.
[http://dx.doi.org/10.1016/j.carbpol.2012.06.002] [PMID: 24751097]

[128] Abdul Rahman NH, Chieng BW, Ibrahim NA, Abdul Rahman N. Extraction and Characterization of Cellulose Nanocrystals from Tea Leaf Waste Fibers. Polymers (Basel) 2017; 9(11): 588.
[http://dx.doi.org/10.3390/polym9110588] [PMID: 30965890]

[129] Bicu I, Mustata F. Cellulose extraction from orange peel using sulfite digestion reagents. Bioresour Technol 2011; 102(21): 10013-9.
[http://dx.doi.org/10.1016/j.biortech.2011.08.041] [PMID: 21893413]

[130] Nazir MS, Wahjoedi BA, Yussof AW, Abdullah MA. Eco-Friendly Extraction and Characterization of Cellulose from Oil Palm Empty Fruit Bunches. BioResources 2013; 8(2): 2161-72.
[http://dx.doi.org/10.15376/biores.8.2.2161-2172]

[131] Xu W, Reddy N, Yang Y. Extraction, characterization and potential applications of cellulose in corn kernels and Distillers' dried grains with solubles (DDGS). Carbohydr Polym 2009; 76(4): 521-7.
[http://dx.doi.org/10.1016/j.carbpol.2008.11.017]

[132] Sheltami RM, Abdullah I, Ahmad I, Dufresne A, Kargarzadeh H. Extraction of cellulose nanocrystals from mengkuang leaves (Pandanus tectorius). Carbohydr Polym 2012; 88(2): 772-9.
[http://dx.doi.org/10.1016/j.carbpol.2012.01.062]

[133] Rosa SML, Rehman N, de Miranda MIG, Nachtigall SMB, Bica CID. Chlorine-free extraction of cellulose from rice husk and whisker isolation. Carbohydr Polym 2012; 87(2): 1131-8.
[http://dx.doi.org/10.1016/j.carbpol.2011.08.084]

[134] Pasquini D, Teixeira EM, Curvelo AAS, Belgacem MN, Dufresne A. Extraction of cellulose whiskers from cassava bagasse and their applications as reinforcing agent in natural rubber. Ind Crops Prod 2010; 32(3): 486-90.
[http://dx.doi.org/10.1016/j.indcrop.2010.06.022]

[135] Melikoğlu AY, Bilek SE, Cesur S. Optimum alkaline treatment parameters for the extraction of cellulose and production of cellulose nanocrystals from apple pomace. Carbohydr Polym 2019; 215: 330-7.
[http://dx.doi.org/10.1016/j.carbpol.2019.03.103] [PMID: 30981362]

[136] Kassab Z, Kassem I, Hannache H, Bouhfid R, Qaiss AEK, El Achaby M. Tomato plant residue as new renewable source for cellulose production: extraction of cellulose nanocrystals with different surface functionalities. Cellulose 2020; 27(8): 4287-303.
[http://dx.doi.org/10.1007/s10570-020-03097-7]

[137] Kassab Z, Aziz F, Hannache H, Ben Youcef H, El Achaby M. Improved mechanical properties of k-carrageenan-based nanocomposite films reinforced with cellulose nanocrystals. Int J Biol Macromol 2019; 123: 1248-56.
[http://dx.doi.org/10.1016/j.ijbiomac.2018.12.030] [PMID: 30529205]

[138] Reddy JP, Rhim JW. Extraction and Characterization of Cellulose Microfibers from Agricultural Wastes of Onion and Garlic. J Nat Fibers 2018; 15(4): 465-73.
[http://dx.doi.org/10.1080/15440478.2014.945227]

[139] Ganguly P, Sengupta S, Das P, Bhowal A. Valorization of food waste: Extraction of cellulose, lignin and their application in energy use and water treatment. Fuel 2020; 280118581.
[http://dx.doi.org/10.1016/j.fuel.2020.118581]

[140] Sindhu KA, Prasanth R, Thakur VK. Medical Applications of Cellulose and its Derivatives: Present and Future. In: Thakur V K Eds. Nanocellulose Polymer Nanocomposites 2014: 437–77.
[http://dx.doi.org/10.1002/9781118872246.ch16]

[141] Elazzouzi-Hafraoui S, Nishiyama Y, Putaux JL, Heux L, Dubreuil F, Rochas C. The shape and size distribution of crystalline nanoparticles prepared by acid hydrolysis of native cellulose. Biomacromolecules 2008; 9(1): 57-65.
[http://dx.doi.org/10.1021/bm700769p] [PMID: 18052127]

[142] Lam E, Male KB, Chong JH, Leung ACW, Luong JHT. Applications of functionalized and nanoparticle-modified nanocrystalline cellulose. Trends Biotechnol 2012; 30(5): 283-90.
[http://dx.doi.org/10.1016/j.tibtech.2012.02.001] [PMID: 22405283]

[143] Mahmoud KA, Mena JA, Male KB, Hrapovic S, Kamen A, Luong JHT. Effect of surface charge on the cellular uptake and cytotoxicity of fluorescent labeled cellulose nanocrystals. ACS Appl Mater Interfaces 2010; 2(10): 2924-32.
[http://dx.doi.org/10.1021/am1006222] [PMID: 20919683]

[144] Peng BL, Dhar N, Liu HL, Tam KC. Chemistry and applications of nanocrystalline cellulose and its derivatives: A nanotechnology perspective. Can J Chem Eng 2011; 89(5): 1191-206.
[http://dx.doi.org/10.1002/cjce.20554]

[145] Batmaz R, Mohammed N, Zaman M, Minhas G, Berry RM, Tam KC. Cellulose nanocrystals as promising adsorbents for the removal of cationic dyes. Cellulose 2014; 21(3): 1655-65.
[http://dx.doi.org/10.1007/s10570-014-0168-8]

[146] Mohammed N, Grishkewich N, Berry RM, Tam KC. Cellulose nanocrystal–alginate hydrogel beads as novel adsorbents for organic dyes in aqueous solutions. Cellulose 2015; 22(6): 3725-38.
[http://dx.doi.org/10.1007/s10570-015-0747-3]

[147] Zhu H, Fang Z, Preston C, Li Y, Hu L. Transparent paper: fabrications, properties, and device applications. Energy Environ Sci 2014; 7(1): 269-87.
[http://dx.doi.org/10.1039/C3EE43024C]

[148] Pérez-Madrigal MM, Edo MG, Alemán C. Powering the future: application of cellulose-based materials for supercapacitors. Green Chem 2016; 18(22): 5930-56.
[http://dx.doi.org/10.1039/C6GC02086K]

[149] Grishkewich N, Mohammed N, Tang J, Tam KC. Recent advances in the application of cellulose

nanocrystals. Curr Opin Colloid Interface Sci 2017; 29: 32-45.
[http://dx.doi.org/10.1016/j.cocis.2017.01.005]

[150] Akram M, Ahmed R, Shakir I, Ibrahim WAW, Hussain R. Extracting hydroxyapatite and its precursors from natural resources. J Mater Sci 2014; 49(4): 1461-75.
[http://dx.doi.org/10.1007/s10853-013-7864-x]

[151] Bee SL, Hamid ZAA. Hydroxyapatite derived from food industry bio-wastes: Syntheses, properties and its potential multifunctional applications. Ceram Int 2020; 46(11): 17149-75.
[http://dx.doi.org/10.1016/j.ceramint.2020.04.103]

[152] Senthil R. Hydroxyapatite and Demineralized Bone Matrix from Marine Food Waste – A Possible Bone Implant. American Journal of Materials Synthesis and Processing 2018; 3(1): 1.
[http://dx.doi.org/10.11648/j.ajmsp.20180301.11]

[153] Venkatesan J, Lowe B, Manivasagan P, *et al.* Isolation and Characterization of Nano-Hydroxyapatite from Salmon Fish Bone. Materials (Basel) 2015; 8(8): 5426-39.
[http://dx.doi.org/10.3390/ma8085253] [PMID: 28793514]

[154] Sathiskumar S, Vanaraj S, Sabarinathan D, *et al.* Green synthesis of biocompatible nanostructured hydroxyapatite from Cirrhinus mrigala fish scale – A biowaste to biomaterial. Ceram Int 2019; 45(6): 7804-10.
[http://dx.doi.org/10.1016/j.ceramint.2019.01.086]

[155] Huang YC, Hsiao PC, Chai HJ. Hydroxyapatite extracted from fish scale: Effects on MG63 osteoblast-like cells. Ceram Int 2011; 37(6): 1825-31.
[http://dx.doi.org/10.1016/j.ceramint.2011.01.018]

[156] Jones MI, Barakat H, Patterson DA. Production of hydroxyapatite from waste mussel shells. IOP Conf Series. Mater Sci Eng 2002; 18: 19.

[157] Gergely G, Wéber F, Lukács I, *et al.* Preparation and characterization of hydroxyapatite from eggshell. Ceram Int 2010; 36(2): 803-6.
[http://dx.doi.org/10.1016/j.ceramint.2009.09.020]

[158] Sasikumar S, Vijayaraghavan R. Low Temperature Synthesis of Nanocrystalline Hydroxyapatite from Egg Shells by Combustion Method 2006.

[159] Barakat NAM, Khil MS, Omran AM, Sheikh FA, Kim HY. Extraction of pure natural hydroxyapatite from the bovine bones bio waste by three different methods. J Mater Process Technol 2009; 209(7): 3408-15.
[http://dx.doi.org/10.1016/j.jmatprotec.2008.07.040]

[160] Bahrololoom ME, Javidi M, Javadpour S. Characterisation of natural hydroxyapatite extracted from bovine cortical bone ash. Plann Perspect 2009; 12: 9-138.

[161] Iriarte-Velasco U, Sierra I, Zudaire L, Ayastuy JL. Conversion of waste animal bones into porous hydroxyapatite by alkaline treatment: effect of the impregnation ratio and investigation of the activation mechanism. J Mater Sci 2015; 50(23): 7568-82.
[http://dx.doi.org/10.1007/s10853-015-9312-6]

[162] Sathiskumar S, Vanaraj S, Sabarinathan D, *et al.* Green synthesis of biocompatible nanostructured hydroxyapatite from Cirrhinus mrigala fish scale – A biowaste to biomaterial. Ceram Int 2019; 45(6): 7804-10.
[http://dx.doi.org/10.1016/j.ceramint.2019.01.086]

[163] Toque JA, Herliansyah MK, Hamdi M, Ide-Ektessabi A, Wildan MW. The effect of sample preparation and calcination temperature on the production of hydroxyapatite from bovine bone powders. Plann Perspect 2007; 152: 155.

[164] Bardhan R, Mahata S, Mondal B. Processing of natural resourced hydroxyapatite from eggshell waste by wet precipitation method. Adv Appl Ceramics 2011; 110(2): 80-6.
[http://dx.doi.org/10.1179/1743676110Y.0000000003]

[165] Krupa-uczek K, Kowalski Z, Wzorek Z. Manufacturing of phosphoric acid from hydroxyapatite, contained in the ashes of the incinerated meat-bone wastes. Pol J Chem Technol 2008; 10.
[http://dx.doi.org/10.2478/v10026-008-0030-6Pol]

[166] Siva Rama Krishna D, Siddharthan A, Seshadri SK, Sampath Kumar TS. A novel route for synthesis of nanocrystalline hydroxyapatite from eggshell waste. J Mater Sci Mater Med 2007; 18(9): 1735-43.
[http://dx.doi.org/10.1007/s10856-007-3069-7] [PMID: 17483877]

[167] Wu SC, Tsou HK, Hsu HC, Hsu SK, Liou SP, Ho WF. A hydrothermal synthesis of eggshell and fruit waste extract to produce nanosized hydroxyapatite. Ceram Int 2013; 39(7): 8183-8.
[http://dx.doi.org/10.1016/j.ceramint.2013.03.094]

[168] Amna T. Valorization of Bone Waste of Saudi Arabia by Synthesizing Hydroxyapatite. Appl Biochem Biotechnol 2018; 186(3): 779-88.
[http://dx.doi.org/10.1007/s12010-018-2768-5] [PMID: 29740796]

[169] Boaventura T, Peres A, Gil V, Gil C, Oréfice R, Luz R. Reuse of collagen and hydroxyapatite from the waste processing of fish to produce polyethylene composites. Quim Nova 2020; 43: 168-74.
[http://dx.doi.org/10.21577/0100-4042.20170475]

[170] M. K. Awasthi et al., Current state of the art biotechnological strategies for conversion of watermelon wastes residues to biopolymers production: A review, Chemosphere, p. 133310, 2021.

[171] Liu H, Kumar V, Jia L, Sarsaiya S, Kumar D, Juneja A, Zhang Z, Sindhu R, Binod P, Bhatia SK, Awasthi MK. Biopolymer poly-hydroxyalkanoates (PHA) production from apple industrial waste residues: A review. Chemosphere. 2021 Dec 1;284:131427.

[172] Awasthi SK, Kumar M, Kumar V, Sarsaiya S, Anerao P, Ghosh P, Singh L, Liu H, Zhang Z, Awasthi MK. A comprehensive review on recent advancements in biodegradation and sustainable management of biopolymers. Environmental Pollution. 2022 Aug 15;307:119600.

[173] Duan, Y., Tarafdar, A., Kumar, V., Ganeshan, P., Rajendran, K., Giri, B.S., Gomez-Garcia, R., Li, H., Zhang, Z., Sindhu, R. and Binod, P., 2022. Sustainable biorefinery approaches towards circular economy for conversion of biowaste to value added materials and future perspectives. Fuel, 325, p.124846.

[174] Kumar V, Sharma N, Umesh M, Selvaraj M, Al-Shehri BM, Chakraborty P, Duhan L, Sharma S, Pasrija R, Awasthi MK, Lakkaboyana SR. Emerging challenges for the agro-industrial food waste utilization: A review on food waste biorefinery. Bioresource Technology. 2022 Aug 13:127790.

[175] Awasthi SK, Sarsaiya S, Kumar V, Chaturvedi P, Sindhu R, Binod P, Zhang Z, Pandey A, Awasthi MK. Processing of municipal solid waste resources for a circular economy in China: An overview. Fuel. 2022 Jun 1;317:123478.

[176] Kumar V, Sharma N, Maitra SS. In vitro and in vivo toxicity assessment of nanoparticles. Int Nano Lett 2017; 7(4): 243-56.
[http://dx.doi.org/10.1007/s40089-017-0221-3]

[177] Vinay K, Neha S, Maitra S. Protein and Peptide Nanoparticles: Preparation and Surface Modification, in Functionalized Nanomaterials I. CRC Press 2020; pp. 191-204.

[178] S. Vallinayagam.,Recent developments in magnetic nanoparticles and nano-composites for wastewater treatment, Journal of Environmental Chemical Engineering, vol. 9, no. 6, p. 106553, 2021

[179] Kumar V, Sharma N, Lakkaboyana SK, Maitra SS. Silver nanoparticles in poultry health: Applications and toxicokinetic effects, in Silver Nanomaterials for Agri-Food Applications.. Elsevier 2021; pp. 685-704.
[http://dx.doi.org/10.1016/B978-0-12-823528-7.00005-6]

[180] Egbosiuba T C. Biochar and bio-oil fuel properties from nickel nanoparticles assisted pyrolysis of cassava peel 2022.
[http://dx.doi.org/10.1016/j.heliyon.2022.e10114]

CHAPTER 11

Waste Valorization Technologies for Egg and Broiler Industries

Jithin Thomas[1,*] and **Sruthi Sunil**[1]

¹ Department of Biotechnology, Mar Athanasius College, Kerala, India

Abstract: The poultry industry is one of the fastest-growing markets at the global level. As the industry expands, the solid waste generated from the poultry sector increases. However, a large amount of waste are generated in poultry farms which needs proper management and disposal to avoid many serious issues like environmental pollution, the spread of diseases due to pathogens residing in the waste as well as breeding of flies and rodents near the waste. Several methods are implemented for the proper utilization and disposal of residues produced in the farms. The methodology used for management varies widely based on many factors like the type of waste generated, nutritional value, and potential hazards to humans and the environment. The techniques adapted for utilization or disposal of the waste generated have evolved from simple conventional methods to highly advanced and more reliable methods (Pyrolysis, anaerobic digestion and catalytic pyrolysis), which are practiced increasingly nowadays, especially in large-scale poultry farms. Many projects and research are being held to improvise waste management techniques in the coming years. The appropriate processing, utilization and disposal of waste and its by-products are important to prevent unwanted side effects and increase the pecuniary output.

Keywords: Anaerobic digestion, Bio-diesel, Bio-char, Bio-filters, Catalytic pyrolysis, Composting, Incineration, Litter, Manure, Poultry waste, Pyrolysis, Rendering, Zeolites.

INTRODUCTION

Poultry Farming is one of the rapidly emerging industries at the global level, involving raising birds domestically or commercially for products such as meat and egg. According to the food and agricultural organization of the United States, poultry products form a significant part of animal-based food eaten by people following different religions, castes, cultures, traditions and beliefs [1]. In the present scenario, the critical role played by small-scale poultry production in reducing and eliminating major problems like poverty and unemployment in rural

* **Corresponding author Jithin Thomas:** Department of Biotechnology, Mar Athanasius College, Kerala, India; Email: jithinthomas@macollege.in

Vinay Kumar, Sivarama Krishna Lakkaboyana & Neha Sharma (Eds.)

areas is gaining recognition. The fact that it has excellent employment opportunities, especially in rural areas for people belonging to diverse categories such as youngsters, middle-aged men and women, small as well as marginal farmers, *etc.*, is gaining momentary recognition. The emergence of the poultry industry has also helped upgrade nutrition levels by ensuring food security to citizens in rural areas. Vocational training is an essential tool that helps farmers by providing them with knowledge, which can ensure the success of rural poultry, including information regarding poultry management, use of locally available feed resources, disease control through vaccination and hygienic management practices [2, 3].

Poultry farming is a much more favorable source of income for rural farmers as compared to urban farmers because the latter face various problems that include highly compact living conditions with limited surrounding space and closely located houses in most of the residential areas which is not an issue faced by most people living in the rural environment. Some municipalities and cities have prohibitions on backyard poultry farming while others have strict rules that must be followed to begin a farm. The owners have to take into consideration the discomfort which may occur to the neighbors in an urban setting due to the noise, odour, flies and insects which need to be controlled by taking proper measures. These problems are less to be faced by rural farmers due to the availability of more open space with fewer houses and people residing nearby [4].

With the increasing population, the food requirements also undergo a steep rise. Though crop production is the major food source, animal husbandry contributes significantly to fulfilling the increasing demand. Poultry farming has several advantages over crop production and other animal-rearing practices for farmers. The primary benefit is the low capital requirement to start small-scale poultry farming compared to an agricultural field for crop production or breeding of other animals like cattle. Another major benefit is the absence of seasonal breeding in poultry which ensures continuous income.

Crop production can put farmers at risk of seasonal unemployment caused by several factors like crop selection, nature of the soil, methods of farming, the possibility of multiple cropping, *etc.* Farming practices that involve poultry and crop production simultaneously have been of great benefit to the farmers. This practice is profitable because some of the crops like wheatgrass, corn, barley, peas, oats, *etc.*, can be used as poultry feed. This also helps in the management of waste produced in poultry farming as it can be processed to make organic fertilizers that can be utilised for crops which are discussed in detail in the following sections [3, 5].

The rapid growth of the poultry industry is mainly driven by the countries that are the largest poultry meat producers, exporters, and importers, an overview of which suggests an annual growth rate of 3.0% in the market value by 2027 [6]. Advancements in technology, improved breeding methods, modern ways of farming, increasing population, and urbanization act as a driving force in the intensification of poultry farming in developing countries. Another major reason that favors poultry growth is the product's affordability and high nutritional value [7].

According to the statistics of the international poultry council, countries like the USA, Canada, Russia, Israel, Saudi Arabia, Iraq, Brazil, China, Japan, India, and the European Union currently form the hub of the poultry industry. The United States of America is the largest meat producer in the world followed by China, Brazil, and Russia. Poultry is considered one of India's most organized sectors, worth about 14,500 million €. In India, the need for processed meat has increased by about 15-20% per annum [2, 6].

China is the largest egg producer followed by USA and India. World poultry meat and egg production escalated from 9 to 122 million tonnes and 15 to 87 million tonnes, respectively, between 1961 and 2017. Apart from soaring as a large-scale industry, traditional small-scale poultry plays a crucial role in encouraging income in rural parts of developing countries [1].

A typical poultry industry produces a huge amount of solid waste materials. These waste materials cause severe environmental problems, producing extremely offensive odours and promoting rodent breeding and flies. Also, derisory methods and careless disposal of waste products will eventually lead to increased disease ailments among the birds. Thus, these wastes have to be managed properly, in order to protect society from unwanted side effects.

Kinds of Waste Generated and its Nutritional Value

Poultry waste is known by several names such as chicken litter, poultry litter, layer litter, dry broiler litter, poultry compost, poultry excreta and broiler excreta. Poultry wastes mainly include, bedding material or litter, a mixture of urinary or faecal excreta, broken eggs, dead birds, wasted feeds, and feathers. Basically, poultry excreta can be classified into poultry litter and confined layers. Confined layer wastes are from the concerned animal and poultry excreta consist of waste from sheet material and excreta.

One of the major reasons for a steep rise in poultry production worldwide is the high nutritional value at an affordable expense, making it available for people from a broader range of socio-economic backgrounds. The nutrient content of poultry products depends upon various factors including the genotype of the breed, age, sex, feed, health, breeding and processing methods [7, 8].

Poultry waste matter has also been proven to have sufficient nutritional value which is responsible for its use in various purposes like feeding livestock, composting or making fertilisers using the residual materials (poultry litter, feathers, manure, *etc.*) that can improve crop production and processing the waste to be used as materials for soil amendment. The major waste materials produced in the poultry industry that have high nutritional value and can be processed for the above-mentioned applications before decomposition are feathers, poultry offal and manure or litter.

Manure and Litter

The three main wastes of concern are manure (resulting from poultry production), dead birds and finally bedding materials that are used for poultry housing. The term 'Manure' refers to the urine and faeces produced by animals. It is organic in nature and nutritious, which makes it a perfect fertilizer.

On the other hand, the poultry industry's waste by-product consisting of bedding materials, feathers and wasted feed, is referred to as 'Litter'. It mainly contains pesticide residues, plant nutrients (N, P and K), trace elements (Zn, Cu, and As), microorganisms and pharmaceuticals like endocrine disruptors and coccidiostats. Similar to other organic wastes, pH, moisture content, elemental composition and soluble salt level of litter and manure vary. Variability in nutrient concentration is attributed to factors such as the type of bedding material, feed use efficiency, litter management practices, *etc.*

Trimethylamine, Dimethylamine (DMA) and ammonia are the most common odorous compounds in poultry manure. Recent studies conducted in this area, state that these compounds may lead to the cause of cell death by apoptosis and necrosis. Also, these chemical compounds in poultry manure cause environmental-related problems such as land, air or water pollution [9].

Poultry Offal

Offal consists of all the organic-solid by-products and waste produced in the poultry industry, including feet, head, bones, blood and other organs. Offal is rich in proteins and lipids, which constitute approximately 32% and 54% of its total nutritional content, respectively. It also possesses a small quantity of kjeldahl nitrogen (5.3%) and has the potential for methane production ranging between 0.6% and 0.9%.

Feathers

Feathers (mostly chicken) have a very rich quantity of protein (approximately 91%) along with a small quantity of lipids (1%) and water (8%). The protein that makes up the feather has the maximum quantity of keratin fibres. Its structure constitutes α- helical and some β- sheet conformations. It is rich in many amino acids, out of which, the most abundant ones are serine, proline, cystine and glutamine which constitute about 16%, 12%, 8.85%, and 7.63% of the total protein sequence, respectively. Along with these, the fibre sequence contains amino acids including arginine, aspartic acid, threonine, tyrosine, leucine, isoleucine, valine, alanine, phenylalanine, methionine and asparagine in small quantities. Along with feathers, keratin is also present in other waste parts like beak and claws.

The presence of keratin makes these parts mechanically stable and resistant to proteolytic degradation due to strong bonds (disulphide and hydrogen bonds) and cross-linkages. Due to its rich protein content, it is processed to make feather meal, organic fertilizers and feed supplements. It is also decomposed using chemical hydrolysis or utilised through various methods like, bioconversion or bio-diesel production.

Poultry excreta, bedding materials, spilled feed and feathers together make up the poultry litter which is the major waste matter in the poultry industry as it is produced in very large quantities. It is a source of several minerals, proteins, carbohydrates, and lipids. The utilisation of poultry litter is carried out by making fertilizers, composting and biogas production through anaerobic digestion.

The minerals present predominantly in the litter include carbon, phosphorous, nitrogen, chlorine, calcium, magnesium, iron, copper, sodium, zinc and arsenic. Proteins constitute a small portion of the total nutrient content ranging between 15.0% - 41.5% (crude protein) and 1.4% -13.2% (bound protein). Other nutrients like lipids and carbohydrates are present in very small quantities as compared to minerals and protein [10, 11].

Potential Hazards and Issues Related to Poultry Wastes

Improper treatment of waste generated from the poultry industry gives rise to possible human health and environmental concerns. Some of the main documented concerns, are mentioned below,

Water and Soil Pollutants Released From Poultry Farm

In most developing countries, poultry litter and manure are applied on land near the production farm. This kind of land management of the poultry by-products invites the risk of groundwater and surface contamination from the pollutants present in the litter and manure.

Factors such as geological conditions of the land in which the farm is built, the agronomical potential of the crop to take in the waste nutrients, soil type, riparian buffers or vegetative areas near the surface water, distance to the water bodies and the climate, play a huge role in the contamination caused by the poultry waste. It gets even worse if poultry waste by-products affect drinking water supplies. Nutrient build-up or loading, within a particular geological region will create an impact on the productivity and diversity of living organisms in that region [12].

Excessive loading of nutrients (potassium, phosphorous and nitrogen) in the nearby surface or groundwater, creates excessive water pollution. Poultry and livestock waste enters the nearby water bodies mainly through farmland surface runoff, underground seepage, volatilization and farmland drainage.

Excreta from poultry and livestock farming consist of a large quantity of undigested phosphorous and nitrogen compounds, drug residues and heavy metals. Excreta discharge soil and surface elements (copper and calcium complexes) to form insoluble complexes, which decreases water permeability, soil consolidation and air permeability.

Soil pollution due to an increase in nitrate content and alteration of the physical and chemical properties of the soil may happen, if poultry and livestock manure are applied directly to the soil without maturation [13].

Air Quality Impacts of Poultry and Livestock Pollutants

Ammonia released into the atmosphere is the most concerned aerial pollutant related to poultry and livestock production. High concentrations of ammonia may trigger environmental issues related to impacts on human health and local systems. Also, poultry production results in the emission of other gases such as hydrogen sulphide, dust particulates, and volatile organic compounds (VOCs), which causes air pollution. Also, greenhouse gas emissions and health problems connected with nuisance odorants are considered relevant issues due to global climatic change and increased human populations in close proximity to poultry industries, respectively [14].

Dust or particulate matter is produced in typical poultry operations, where a number of birds are constricted. Dust mainly contains faecal matters and includes endotoxin, mites, bacteria and moulds.

The impact of aerial poultry pollutants significantly depends on the climatic conditions. For instance, an extremely dry climatic condition (especially in litter) causes an increased respiratory condition that affects the bird's productivity [15]. In case of wet litter, ammonia concentration increases, this is also considered detrimental to productivity [16].

Environmental management of poultry and livestock waste has been questionable in China, because of the several environmental pollution cases reported in the past 20 years. There had been suggestions, regarding the shifting of waste management policy from a single government controlled way to a multi-cooperative way.

One of the main drawbacks of poultry and livestock waste management is that most of the cooperative management of manure remains at the theoretical level and very little has been put forward in the practical applications. Also, it has been spotted that very less research has been proposed on the mechanism of biogas fertilizer recycling for various farming subjects.

Utilisation of Poultry Wastes

Waste valorization is a process of transforming waste materials into valuable products. The large quantity of waste produced by the poultry industry needs to undergo proper disposal or management to avoid any potential harm that it can cause to the environment. Efficient utilisation of waste products helps to avoid the risk of causing environmental pollution (offensive odours, increasing flies, mosquito, rodent breeding, *etc.*), harm to humans or animals and also enables the use of its nutrient content for various purposes [17].

The utilisation of waste can be broadly divided into two types: energetic and non-energetic methods. Energetic methods involve the use of waste for the production of materials with a high amount of stored energy, while non-energetic methods make use of the material and convert them into products that are less harmful, environmentally friendly and thus, suitable for use by humans, animals or plants [11].

Non-Energetic Waste Utilisation

There are numerous non-energetic methods of waste utilisation, which tend to outnumber the energetic methods. They include rendering, fertilizer production, producing livestock feed, soil amendment, and processing of fly larvae on waste to produce a substitute for chicken and fish feed.

Rendering

Rendering is one of the most beneficial methods of recycling poultry waste that involves converting waste material into environmentally acceptable, protein-rich by-products, suitable for fertilizing and feeding purposes by applying high temperature and pressure. It can breakdown waste material like feathers, feet, viscera, blood, head and other parts which are difficult to decompose in normal conditions. The waste matter is exposed to a temperature as high as 133°C for a minimum of 20 min at a pressure of 3 bars to convert all the organic residues into various by-products.

The by-products formed can be used as poultry and other animal feed, fertilizers, or further processed using methods like anaerobic digestion. The wastes produced in broiler slaughtering, like blood, skin, viscera, bones, *etc.*, are organic solids that are converted into protein-rich chicken meal and fat using rendering. Solid waste has high nutritional value with 32% proteins and 54% lipids. Rendering of good quality inedible visceral waste is used for fish culture in some regions without extensive processing, provided adequate biosecurity precautions are taken.

Feathers are a rich source of crude protein (75-90%) but are not preferred to produce animal feed for various reasons like the extensively complex structure of keratin and destruction of many essential amino acids under high thermal treatment during rendering, which make it difficult to digest. However, it is infrequently used to produce feather meals due to its rich protein and nitrogen content. It is mostly used for bedding, clothing, and preparing other materials for humans. Rendering is one of the most preferred techniques in the present scenario, which has been discussed further in the following sections [18, 19].

Fertilizer

Poultry waste is rich in nutrients that are highly beneficial for crop production and improvement like nitrogen, potassium, and phosphorus which are present with an approximated quantity of 65.5%, 83.5% and 68.5%, respectively. It also contains other elements deficient in other commercial fertilizers like calcium, magnesium, boron, cobalt, iron, copper, manganese, sulphur, zinc, and molybdenum. Thus, poultry litter can be used to produce good quality fertilizers at a low cost.

The fertilizer value of poultry litter depends on the type of bird, moisture content and how old the litter is. The soil scheduled to receive fertilizer made of poultry litter and the litter sample are tested to determine the existing fertility of the soil and the amount of fertiliser used. Poultry litter is not applied to soil beyond the limits of the crop's nutrient requirements to ensure the efficient use of nutrient content and restrict nutrient leaching into the soil surface or groundwater [11, 20].

Livestock Feed

Poultry waste is used extensively as livestock feed in many countries like Israel and some states in the United States of America. One of the oldest and most used techniques for processing poultry waste to be used as livestock feed is drying. It is stated that dried poultry waste is a source of proteins (28%), ash (30%) and minerals like calcium, zinc, potassium, iron and phosphorus. Cage layer waste (collection of excreta, spilled feed, and feathers) contains nitrogen, 40-60% of which is from a non-protein source and the remaining is from the amino acids (approximately 37-40%).

Uric acid forms the major non-protein nitrogen (NPN) source in poultry which is degraded to produce ammonia when fed to ruminants by the action of rumen microbes. The addition of broiler litter at a level of 20% or greater into beef cattle rations has proved to be beneficial in satisfying the requirement of crude protein and minerals like calcium and phosphorus. Studies also state that including poultry waste in sheep feed at levels higher than 35% covers the requirement of total protein in the animals.

Poultry waste when fed at a substantially higher level can cause health hazards in cattle due to pathogens, residues of pesticides in the waste, and nutrient loss, mainly due to ammonia volatilization. Thus, using poultry waste as livestock feed depends on the pattern and standard of cattle management [11, 21].

Soil Amendment

Poultry manure is also used as a soil amendment material because it improves water holding capacity and lateral water movement, which boosts irrigation possibilities, reducing soil's dryness and enhancing crop production. It also plays a role in increasing the diversity of essential microorganisms that reside in the soil. There is no definite prescription for the limit of the manure that needs to be applied because of the varying demands for every type of soil. However, it is preferred to spread thin layers of manure rather than dumping it in a heap. It has been observed that continuous application of poultry manure (composted or non-composted) results in the up-gradation of the physical properties of degraded soil and subsequently increases the yield [17, 22].

Fly Larvae Processing

Housefly larvae on poultry waste were found to be of sufficient quality that can be used as a substitute for soybean meal in chicken feed. It has a high protein and fat content of 63.1% and 15.5%, respectively. In dried maggots and pupae, the crude protein and fat content range between 56.9% – 60.7% and 19.2% - 20.9%, respectively. Innovations and emerging technologies have a high potential to find ways to convert complete organic wastes into feed materials for poultry, livestock or fish farming [19, 23].

Energetic Waste Utilisation Methods

This involves techniques to convert poultry residues into energetic sources. Biogas is the most common method of production of biodiesel and methane (biogas). Several technologies are being implemented to convert poultry biomass into electrical energy considering its high energetic potential.

Bio-diesel Production

Bio-diesel can be produced from animal fat obtained from slaughterhouse wastes like feathers, blood, and other such residues. Producing diesel from animal fat serves as a good alternative for the non-renewable diesel produced by fractional distillation of crude oil [24]. Feather meal contains 12% of fat which is extracted using boiling water. It is then processed by trans-esterification using potassium, nitrogen and methane to produce bio-diesel. ASTM (analytical chemistry standard methods) analysis has confirmed that biodiesel produced using feather meal is of better quality when compared to the biodiesel made from other feedstock [10, 25].

Methane Production

The rate and yield of methane production vary based on different poultry residues. The presence of rich protein and lipid content in poultry blood, bone and offal makes it suitable for methane production. Anaerobic digestion is a widely used method that degrades the organic waste matter produced from poultry converting it into methane. It has got several advantages over other conventional methods which have been discussed in further detail later in the chapter [10].

Conventional Methods of Poultry Waste Management

There are many approaches for effectively disposing of poultry waste which is necessary for controlling the hazardous effect it can have on living beings and the environment, collectively. As mentioned earlier, inefficacious and careless measures in disposal methods can also cause many infectious diseases. It can also cause serious ailments in poultry animals which can have a deleterious effect on farm productivity due to mortality or reduced quality of products like meat and eggs. Therefore, early and proper disposal of waste is done using conventional techniques, including rendering, composting, drying, burning, landfills and burial.

Composting

Composting is an aerobic biological process that breaks down complex organic compounds and decomposes them. The organic material is converted into carbon dioxide, water, minerals, and other stable compounds. It is a common method used to decompose poultry slaughterhouse waste, carcasses, litter, feathers, and manure. It is carried out in the presence of numerous microorganisms which feed upon the organic waste when provided with appropriate environmental conditions of temperature, moisture content, and oxygen. The optimum temperature for composting depends on the temperature released by the specific microorganism during the process, which is let out through aeration or surface cooling. Microbes work efficiently when provided with an efficient temperature mostly ranging between 40-60°C, 40-60% moisture, >5% oxygen and a carbon to nitrogen ratio of 21:1 to 35:1.

Composting is an easy method that can be done at any time of the year with equipment available. It is applicable to any farm and is considered economical by the farmers. It is an effective method as it helps in reducing the disease-causing pathogens due to the high amount of heat produced during the process and produces by-products that can be effectively used as organic fertilisers to improve soil conditions for the production of crops with a better yield and quality.

Composting immobilizes nitrogen and phosphorous present in the waste, reducing the risk of these nutrients leaching into the groundwater and thus restricting the potential hazard it can have on living beings. The loss of nitrogen is also a disadvantage of composting as plants require nitrogen for growth, which when immobilised would not reach adequately to the roots for utilisation by plants. Other disadvantages include the requirement of land, odour problems and emission of greenhouse gases such as methane and nitrous oxide, which absorb radiation from the sun, increasing the earth's temperature and resulting in major environmental problems like global warming and acid rain [11, 18].

Drying

Drying of poultry dropping is one of the most feasible, applicable methods, requiring minimum equipment and cost-effectiveness. It can be done using natural conditions *i.e.*, drying under the sun to remove the waste's moisture content or with the help of solar heaters. Poultry waste has a high-water content that can trigger the growth of many pathogens. Drying of the waste under the sunlight helps to remove the excessive moisture and eliminate harmful pathogens. Drying methods that make use of more advanced technologies, are the provision of warm air flow in the poultry houses, manure drying on steel plates, forced tilting method, *etc.*

The drying results with conventional methods vary based on factors like the thickness of the manure layer, temperature range and exposure time. Thin layer of droppings approximately 1-3 cm thickness undergo appropriate drying within the temperature range provided by solar heaters *i.e.*, 40-60°C. Exposure to higher temperatures leads to a greater loss of nitrogen with a reduction in N: P: K from 4.58:1.29:1 to values ranging between, 2.07:1.30:1 – 2.57:1.28:1. This N: P: K value is suitable for the growth of plants, thus it can be used as a fertiliser for crop production. It helps in reducing the malodourous odour by 69.3% and the elimination of microorganisms-like bacteria by 65.6-99.8% (99.97% of *E.coli* is eliminated), and yeast and mold by 74.1-99.6% [11, 26].

Burning

This is one of the most common conventional methods of waste disposal extensively used by small-scale farmers. This involves the complete burning of the waste at a relatively high temperature using fuels. The major drawback associated with burning is the atmospheric pollution it causes especially when burning large volumes of waste. Burning also leads to the emission of carcinogens like dioxins and furans due to incomplete combustion. These can affect reproduction in humans, their immune system and can cause poor development. The burning site is chosen because it is located away from public places like roads and residential areas, to be careful to stay away from gas lines, electrical lines and wires [11].

Burial

Burial has been used for a long time for the disposal of poultry waste mostly flesh and bones, which serves as a feasible method for waste management. It is one of the most effective and inexpensive methods for disposal at times of mass mortality losses caused by diseases like avian influenza, swine flu, *etc.*

The earlier system of burial made use of an open hole or trench that was dug in the farm which was filled with waste and packed with the soil to cover up. This technique has many hazardous effects if the pit is left open or covered poorly, the chances of which are high as most of the work is done manually or with simple equipment. It may lead to soil and groundwater contamination unless the pits made are well packed, to prevent any diffusion of contents from the waste into the surrounding soil. These harmful effects can be avoided by following strict safety measures like proper site selection (avoid areas with sandy soils and high-water tables) and following the burial pit construction guidelines set by the concerned authorities.

There have been different calculations made through previous works done by various organisations or companies like 'Payne and Anon'. These organizations provide the guidelines, including the architectural details and minimum distance required to construct a pit from water sources and other public places to restrict any harmful effect the waste contents can have on the living beings and their nearby environment. Payne indicated that the burial pits constructed need to maintain a minimum distance of 99.44m from water sources like wells, nearby residences, public areas and property lines. It was also suggested that the bottom of the pit needs to be built 60.96 cm above the SHWT (seasonal high-water tables) and 30.48 cm above floodplain level. If there is bedrock in the area, the pit must be 60.96 cm above it. Payne recommended that the pit must be closed with at least 76.2 cm of topsoil to cover the waste.

On the other hand, Anon recommended the pit to be constructed at a distance greater than 30.48m from any existing water source and 4.47m horizontally away from the edge of dams or embankments. They also suggested that the waste should be buried at least 0.91m below the ground level, not exceeding 2.44m.

The pit can be constructed with materials like concrete blocks, treated lumber, *etc., and* a concrete cover fitted with PVC (polyvinyl chloride) pipe drop chute at the centre. Another low-cost method is the implementation of precast, open-bottom tanks in the farm which can reduce labour and money spend on constructing a pit. Waste matter undergoes decomposition by an array of anaerobic processes, producing foul smells around the pits. On the contrary, aerobic fermentations do not produce objectionable odour, and thus are more desirable than anaerobic decomposition.

There are several places that have passed legislation, that mandate the prohibition of using burial pits for poultry carcass disposal taking into account the environmental hazard they possess. Arkansas is the first state in the U.S.A to implement the above legislation on July 1st 1994, followed by Alabama in the year 2000 [11].

Landfills

Municipal landfills are predominant options for poultry waste disposal which are used extensively for handling catastrophic poultry mortalities. Decomposition in landfills takes place slowly at a relatively low temperature ranging from 54 to 65°C. It was one of the most used waste disposal methods during the avian influenza outbreak in the last few decades. During the outbreaks, one of the major concerns raised about the potential biosecurity risks, that persists in the disposal of the infected flocks off-site (*i.e.* away from the farm where outbreak was reported) was that the chance of increase in the spreading of the virus to other farms. Thus, the off-site disposal was made more effective by using sealed, leak proof transportation facilities.

The containers used for transport are double-lined, made of waterproof material, and absorbent to retain body fluids of organs. The fee for disposing of waste in landfills varies with an approximate amount of $77 per ton. The cost primarily depends on transportation and the volume of waste to be disposed of. Landfills, similar to burial methods, threaten the safety of the environment and living beings if proper measures are not taken. Landfills need to be properly managed with supervised transportation, instant covering of the waste and bagging of the infected bodies during transportation to restrict the risk associated with the spread of pathogens [18].

Modern Techniques in Poultry Waste Management

Incineration

Incineration has been considered as effective, environmentally aware and cost efficient poultry waste disposal method. It is considered the safest method of waste disposal, as it eliminates all the threats of diseases. Incineration has been used to control excess litter, including feathers, bedding, manure and spilled feed. It is stated that, the first poultry litter incinerator was developed at United Kingdom in 1993, in order to find a solution to an excess dumping of poultry litter from Industrial poultry operations. For application of poultry waste as a fertilizer, incineration is an ideal solution. Incinerated chicken waste consists of 3% of bottom ash, which is extremely rich in vitamins and nutrients and is suitable for spreading [27].

The type of incinerator to be used, depends on stock size and the mode of operation. A typical incinerator has a burn rate of 50kg/hour, and is considered a perfect choice for free range – farm holdings, hatcheries and broilers. Large-scale poultry industries require models with higher capacity, but to operate these, it requires additional sanctions from government authorities.

But, several disadvantages have been mentioned in the application of this technique. Some of the major concerns in the process are the emitted air, maintaining the process condition and controlled disposal of the solid and liquid residues. 'Payne' has stated about the high operational costs involved in the process and also, if not properly maintained, it may contribute to air pollution.

Another drawback regarding the incineration process is that, loading decomposed carcass requires 0.3 tonnes of ash per tonne of carcass. This requires, appropriate supervision of the entire process and selection of perfect fuel, otherwise, it may produce unpleasant odour and smoke.

Anaerobic Digestion

Biological process of anaerobic digestion degrades organic matter into methane, which can be used as a source of bio-energy. The main advantage of anaerobic digestion is that it reduces odour and pathogens. Also, anaerobic digestion requires comparatively very little space for the treatment and can easily degrade pasty and wet poultry wastes. Anaerobic digestion has controlled releases to land, water and air. Another advantage is that, nutrients remain in the treated material and later recovered for application in the agricultural fields or as feed use.

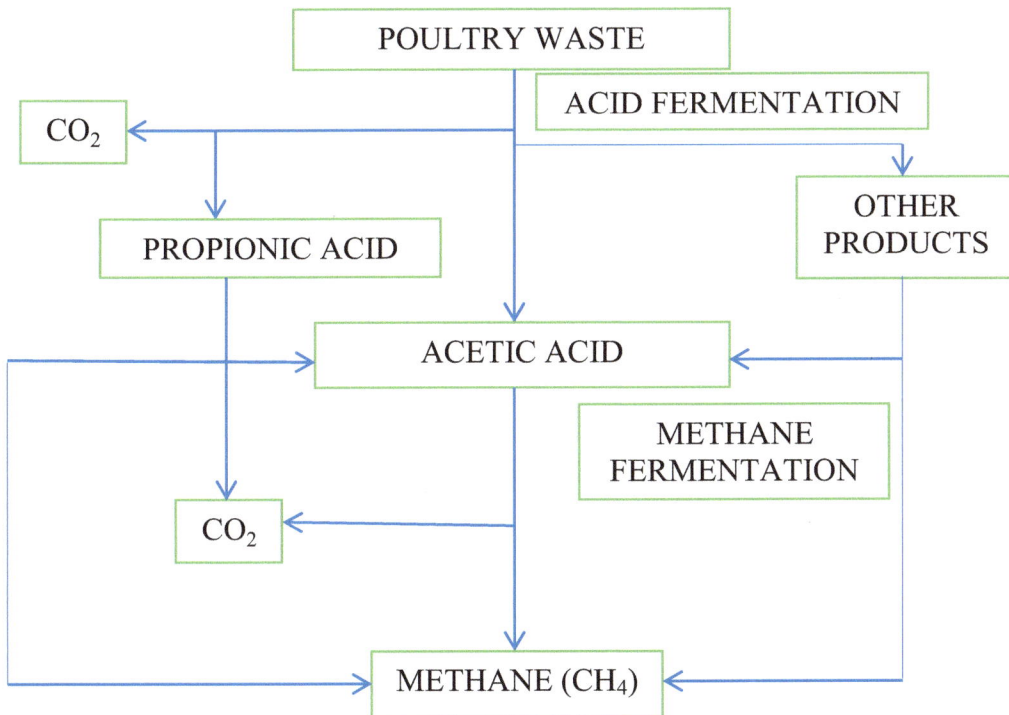

The treatment mainly involves the degradation and stabilization of an organic material in the absence of oxygen with the help of micro-organisms to form methane and some inorganic compounds such as carbon dioxide.

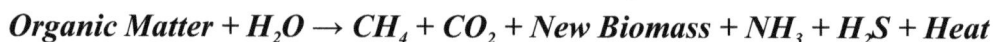

Organic Matter + $H_2O \rightarrow CH_4 + CO_2$ + New Biomass + $NH_3 + H_2S$ + Heat

The anaerobic treatment of the poultry litter mainly consist of two distinct phases, in the initial stage, complex components such as proteins, polysaccharides and fats are hydrolysed and split into sub units. The initial stage is facilitated by anaerobic and facultative bacteria subsequently by hydrolysis to fermentation and other metabolic processes that convert it into simple organic compounds. This initial stage is called 'acid fermentation', in which organic compounds are transformed into organic acids, bacterial cells and alcohols. In the second stage, hydrolysis products are converted into methane and carbon dioxide with the help of strictly anaerobic microbes, which is known as methane fermentation. Two stages of anaerobic fermentation are presented in the flow chart below [10]:

The methane production rate and yield differ according to the various slaughtering residues. High methane yields are shown by blood, poultry offal, and bone meal that are rich in lipids and proteins [28, 29].

Pyrolysis

Pyrolysis is an alternative recent technology developed for producing biofuels from poultry litter. The process transforms biomass and organic matter into bio-oil, bio-char, biogas, or syngas. In other words, it is a thermal decomposition process that transforms biomass into high-energy-density products without an oxidising agent [30]. The pyrolysis process mainly depends on various parameters such as residence time, reactor temperature, feedstock heating rate, the particle size of the feedstock, the pressure within the reactor, and reactor configuration [31].

The process of pyrolysis is categorised into – fast, medium, and slow pyrolysis. Fast pyrolysis results in the production of aerosols, vapour, bio-char, and gases. Fast pyrolysis is mainly performed using a fluidised bed reactor due to the stability, ease of operation, the highest bio-oil yield and scaling-up potential [32].

Poultry litter is considered to have extremely high ash content and during elevated temperatures, this gives rise to ash fusion and sintering, along with phosphorous, potassium and other several alkali present in the ash leading to bed agglomeration or de-fluidization problems. As the pyrolysis temperature is 400 – 600°C, problems such as bed agglomeration or defluidization normally do not occur [33].

Research works on pyrolysis technology in feedstock that originates from livestock are comparatively less. In an investigation, dairy manure, paved feedlot manure, turkey litter, swine-separated solids, and poultry litter were pyrolysed at a temperature range of 300 – 700°C. Out of these, poultry litter and char showed maximum electronegativity, which is a parameter that measures nutrients present in the substrate [34].

Batch-type tubular pyrolysis reactor is used to produce bio-diesel (fatty acid methyl ester) from bio-char from chicken manure. Pseudo-catalytic trans-esterification reaction of bio-char resulted in 95.6% fatty acid methyl ester yield at 350° C, when compared to the commercial porous material. The thermal cracking of Calcium in bio-char can be completely avoided, if it is produced at a temperature below 350° C [35]. In addition, various recent methods of biomass valorization are also available [38 - 48].

Product gas generated during gasification and pyrolysis of chicken manure at various temperatures showed that energy yield during the process of gasification was higher than pyrolysis. Slow pyrolysis of chicken litter can be performed in a fixed bed reactor at different temperatures [36]. Adequate pyrolytic gases were generated for the entire process to be self-sufficient. Analysis of energy transfer indicated that about one-third of the heating value of poultry litter has been transferred to the organic liquid condensate and a similar ratio has been retained in the bio-char. At a temperature of 550°C, the maximum fraction of liquid condensate could be exalted to biofuels [37, 49].

Catalytic Pyrolysis

Products formed during the process of 'pyrolysis' can be enhanced by catalytic pyrolysis. It is a comparatively easy method to transform lignocellulosic biomass into toluene, xylene, and benzene (aromatic compounds) with the help of selected catalysts. Pyrolysis oil produced as a by-product of pyrolysis; consists of various classes of oxygenated organic compounds, which can be removed by catalytic pyrolysis. Shape-selective catalysts remove oxides in the form of carbon and water. These shape-selective catalysts are known as 'Zeolites' [50].

Existing Limitations and Future Aspects of Waste Disposal Methods

Livestock breeding has been developing at a very fast pace due to its high-quality food products. This development has also increased the demand for improved and more reliable techniques to implement better waste utilisation and management practices. Researchers are finding ways to upgrade the existing waste valorization methods accompanied by the effort to invent new techniques.

Poultry farming makes a major contribution to the anthropogenic emissions of ammonia accompanied by foul odour that affects the surrounding environment and is a major issue for people living nearby. Thus, the farmers need mitigation techniques that are cost-effective, have low maintenance risks, and give efficient results. In order to keep the earlier mentioned factors in mind and find a suitable method, the efficiency of which needs to be verified by comparison with already existing ones, a multinational collaboration was established between Germany, Netherlands, and Denmark called the verification of environmental technologies for agricultural production (VERA) test protocol.

One of the methods that have been in the interests of researchers is the use of bio-filters for waste management. Different materials were used to make the filters and experiments were conducted to find the performance by comparing parameters like ammonia and odour emission in raw or untreated state of air from the farm as well as through the different variants of filters in identical conditions. An experiment conducted at a broiler fattening facility in Lower Saxony, Germany, made use of 2 variants in which half of the bio-filter was filled with root wood and the other half with honeycombed paper pads. The results showed a significant reduction in ammonia as well as odour emission. In the root wood filter, Ammonia concentration and odour emission were reduced by 71% and 97 OUE m^{-3} (OUE - European odour units) respectively. Honeycombed paper pad filters were found to be more effective when compared to the other variant with ammonia and odour reduction by 68% and 131 OUE m^{-3}, respectively [14, 16].

Some of the methods that are used largely for waste management in poultry like composting, pyrolysis, anaerobic digestion and drying are also under constant experimentation to eliminate the limitations and improvise the existing strategies to make the waste less harmful to the environment and thus the living beings. Many different strategies are employed to obtain these products from varied sources.

Composting poultry litter and manure is an extensively used method which is accompanied by challenges such as the existence of a large amount of nitrogen as ammonia and its fate, higher temperature requirement for wetter material, pH level and moisture maintenance which increases the demand for labour, time and money.

Anaerobic digestion is another general method with limitations such as inhibition in the process or acclimation of microorganisms due to high ammonia content, scum formation, difficulty in transportation, collection, and storage of manure and the unawareness of the fate of micropollutants like insecticides during the digestion process.

Drying is a cheap and effective method, which has drawbacks that include the requirement of long time periods and high dependence on temperature for carrying out natural drying. Techniques like using a dryer solve the former problems to a certain extent but introduce additional energy costs [51].

Some of the recent projects that were initiated to improve waste management methods include research on innovative technology for the conversion of poultry manure and whey through fermentation that mainly aims at reducing the nitrogen content by uric acid precipitation to produce biogas which was carried out at Poznan University of Life Science, Poland. Other research projects include the production of soil improver with animal manure (primarily poultry) that can enhance soil structure and activate mineral components, the production of energy in biogas plants making use of poultry manure by converting plant substrates into algae as well as the development of more reliable techniques to recover nutrients and energy from poultry manure through composting, pyrolysis, and anaerobic digestion [52].

Leading Poultry Waste Management Organizations

Several NGOs play a crucial role in collecting poultry and other animal wastes for the purpose of reduction in environmental pollution. Organizations such as 'Global waste cleaning network' and 'Round glass foundations', aim to collect various animal wastes for reducing pollution in lands, oceans, and the atmosphere. These organizations mainly reduce pollution by the implementation and advocacy of farmers regarding the applications of various machinery. These types of machinery are supplied by various groups for setting up a small-scale or large-scale recycling unit in the farm.

Some of the most prominent poultry waste management machinery are supplied by Haat poultry waste incinerator, Hebei Chengzhu Group and Dingli group, with headquarters in China. For a customizable pyrolysis plant, it is better to rely on 'BIOFABRIK', who provides low tech and large-scale pyrolysis plants for small-scale and large-scale poultry farmers, respectively. Another prominent poultry machinery and electronic product supplying firm is 'Dorset', which was founded by Mr. Henk Haaring in 1984. Today, the Dorset group is globally active with Netherlands as the headquarters. Dorset group has two main divisions, one concentrated on the production of 'Green Machines' and another on 'Electronic devices'.

REFERENCES

[1]   Products and processing | Gateway to poultry production and products | Food and Agriculture Organization of the United Nations", Fao.org, 2020. [Online]. Available: http://www.fao.org/ poultry-production-products/products-processing/en/ [Accessed: 27- Nov- 2020].

[2]   Singh NK, Singh C, Sharma D, Tiwari S. Role of Poultry Production in Rural Development the poultry 2020.
https://thepoultrypunch.com/2019/08/role-of-poultry-production-in-rural-development/ [Accessed: 27-Nov- 2020]

[3]   Sonkar N, Singh N, Santra A, Prakash Verma L, Soni A. Backyard poultry farming: A source of livelihood and food security in rural India The Pharma Innovation Journal 2020; 9(4): 28-32. [Accessed 27 November 2020].

[4]   Hady AA, Kean R. Poultry in urban areas 2011.https ://learningstore .extension .wisc.edu/ collections/poultry/products/poultry-in-urban-areas-p1463

[5]   Marangoni F, Corsello G, Cricelli C, *et al.* Role of poultry meat in a balanced diet aimed at maintaining health and wellbeing: an Italian consensus document. Food Nutr Res 2015; 59(1): 27606.
[http://dx.doi.org/10.3402/fnr.v59.27606] [PMID: 26065493]

[6]   Top Poultry Meat Producing Countries 2020 | Largest Poultry Meat Exporters and Importers | Global Poultry Meat Industry Factsheet", Bizvibe Blog, 2020. [Online]. Available: https://www.bizvibe.com/ blog/top-poultry-meat-producing-countries [Accessed: 27- Nov- 2020].

[7]   Gündüz S, Aslanova F, Abdullah KSH. Poultry Waste Management Techniques in Urban Agriculture and its Implications: A Case Study of Tripoli, Libya. Ekoloji 2019; 28(107): 4077-84.

[8]   Kralik G, Kralik Z. Poultry products enriched with nutricines have beneficial effects on human health. Med Glas 2017; 14(1): 1-7.
[PMID: 27917847]

[9]   Lu L, Liao X, Luo X. Nutritional strategies for reducing nitrogen, phosphorus and trace mineral excretions of livestock and poultry. J Integr Agric 2017; 16(12): 2815-33.
[http://dx.doi.org/10.1016/S2095-3119(17)61701-5]

[10]   Thyagarajan D, Barathi M, Sakthivadivu R. Scope of Poultry Waste Utilization. IOSR J Agric Vet Sci 2013; 6(5): 29-35.
[http://dx.doi.org/10.9790/2380-0652935]

[11]   Singh P, Mondal T, Sharma R, Mahalakshmi N, Gupta M. Poultry Waste Management. Int J Curr Microbiol Appl Sci 2018; 7(8): 701-12.
[http://dx.doi.org/10.20546/ijcmas.2018.708.077]

[12]   Rodic V, Peric L, Djukic-Stojcic M, Vukelic N. The environmental impact of poultry production. Biotechnol Anim Husb 2011; 27(4): 1673-9.
[http://dx.doi.org/10.2298/BAH1104673R]

[13]   Pollution and Cooperative Treatment of Livestock and Poultry Waste: A Review of the Literature", in IOP conference series: material science and engineering. 2019.

[14]   Strohmaier C, Krommweh MS, Büscher W. Suitability of Different Filling Materials for a Biofilter at a Broiler Fattening Facility in Terms of Ammonia and Odour Reduction. Atmosphere (Basel) 2019; 11(1): 13.
[http://dx.doi.org/10.3390/atmos11010013]

[15]   Ismein R, Klimov D, Mikhalev A, Milovanov O. Torrefaction – the New Method for Decontamination of Poultry Litter Food and nutrition science 2017; vol. 2 [Accessed 27 November 2020]

[16]   Janni KA, Nicola RE, Hof SJ, Stenglein RM. Air Quality Education in Animal Agriculture: Biofilters for Odor and Air Pollution Mitigation in Animal Agriculture. Lincoln, NE: Agricultural and Biosystems Engineering Extension and Outreach Publications 2011; pp. 1-8.

[17]   Tiwari R, Dev G. Scope of Poultry Waste Management and Utilization - Poultry Punch 2020.
https://thepoultrypunch.com/2020/04/scope-of-poultry-waste-management-and-utilization/ [Accessed: 27- Nov- 2020]

[18]   Blake JP. Poultry Carcass Disposal Options for Routine and Catastrophic Mortality. CAST 2008.

[19]   O. F. Sari, S. Ozdemir and A. Celebi, "Utilization and Management of Poultry Slaughterhouse Wastes with New Methods", in Eurasia 2016 Waste Management Symposium, istanbul, 2016.

[20]   Zhang H, Hamilton DW, Payne J. Using Poultry Litter as Fertilizer - Oklahoma State University 2020. https://extension.okstate.edu/fact-sheets/using-poultry-litter-as-fertilizer.html [Accessed: 27- Nov-2020]

[21]   C. Michael Williams, poultry waste management in developing countries, the role of poultry in human nutrition, 2013. [Accessed 27 November 2020].

[22]   Sellami F, Jarboui R, Hachicha S, Medhioub K, Ammar E. Co-composting of oil exhausted olive-cake, poultry manure and industrial residues of agro-food activity for soil amendment. Bioresour Technol 2008; 99(5): 1177-88.
[http://dx.doi.org/10.1016/j.biortech.2007.02.018] [PMID: 17433668]

[23]   Lynch D, Henihan AM, Kwapinski W, Zhang L, Leahy JJ. Ash Agglomeration and Deposition during Combustion of Poultry Litter in a Bubbling Fluidized-Bed Combustor. Energy Fuels 2013; 27(8): 4684-94.
[http://dx.doi.org/10.1021/ef400744u]

[24]   Jung JM, Lee SR, Lee J, Lee T, Tsang DCW, Kwon EE. Biodiesel synthesis using chicken manure biochar and waste cooking oil. Bioresour Technol 2017; 244(Pt 1): 810-5.
[http://dx.doi.org/10.1016/j.biortech.2017.08.044] [PMID: 28841785]

[25]   Barclay E. Fat's Chance as a Renewable Diesel Fuel 2010. https://www.nationalgeographic.com/news/energy/2010/12/101222/animal-fat-tyson-renewable-fuel/#:~:text=But%20diesel%20fuel%20made%20from,to%20form%20the%20biodiesel%20molecule [Accessed: 27- Nov- 2020]

[26]   Drying Poultry Manure | STRONGA, STRONGA, 2020. [Online]. Available: https://stronga.com/drying-material/drying-poultry-manure/. [Accessed: 27- Nov- 2020].

[27]   Stingone JA, Wing S. Poultry litter incineration as a source of energy: reviewing the potential for impacts on environmental health and justice. New Solut 2011; 21(1): 27-42.
[http://dx.doi.org/10.2190/NS.21.1.g] [PMID: 21411424]

[28]   Singh K, Risse LM, Das KC, Worley J, Thompson S. Effect of fractionation and pyrolysis on fuel properties of poultry litter. J Air Waste Manag Assoc 2010; 60(7): 875-83.
[http://dx.doi.org/10.3155/1047-3289.60.7.875] [PMID: 20681435]

[29]   Sharara M, Sadaka S. Opportunities and Barriers to Bioenergy Conversion Techniques and Their Potential Implementation on Swine Manure. Energies 2018; 11(4): 957.
[http://dx.doi.org/10.3390/en11040957]

[30]   Mante OD, Agblevor FA. Influence of pine wood shavings on the pyrolysis of poultry litter. Waste Manag 2010; 30(12): 2537-47.
[http://dx.doi.org/10.1016/j.wasman.2010.07.007] [PMID: 20688503]

[31]   Burra KG, Hussein MS, Amano RS, Gupta AK. Syngas evolutionary behavior during chicken manure pyrolysis and air gasification. Appl Energy 2016; 181: 408-15.
[http://dx.doi.org/10.1016/j.apenergy.2016.08.095]

[32]   Akdeniz N. A systematic review of biochar use in animal waste composting. Waste Manag 2019; 88: 291-300.
[http://dx.doi.org/10.1016/j.wasman.2019.03.054] [PMID: 31079642]

[33]   Turan NG. The effects of natural zeolite on salinity level of poultry litter compost. Bioresour Technol 2008; 99(7): 2097-101.
[http://dx.doi.org/10.1016/j.biortech.2007.11.061] [PMID: 18248810]

[34]   Hadroug S, Jellali S, Leahy JJ. Marzena K, Mejdi J, Helmi H and Witold K Pyrolysis Process as a Sustainable Management Option of Poultry Manure: Characterization of the Derived Biochars and Assessment of their Nutrient Release Capacities. Water 2019; Vol. 11.

[35] Cantrell KB, Hunt PG, Uchimiya M, Novak JM, Ro KS. Impact of pyrolysis temperature and manure source on physicochemical characteristics of biochar. Bioresour Technol 2012; 107: 419-28.
[http://dx.doi.org/10.1016/j.biortech.2011.11.084] [PMID: 22237173]

[36] Hussein MS, Burra KG, Amano RS, Gupta AK. Temperature and gasifying media effects on chicken manure pyrolysis and gasification. Fuel 2017; 202: 36-45.
[http://dx.doi.org/10.1016/j.fuel.2017.04.017]

[37] Bridgwater AV. Review of fast pyrolysis of biomass and product upgrading. Biomass Bioenergy 2012; 38: 68-94.
[http://dx.doi.org/10.1016/j.biombioe.2011.01.048]

[38] Awasthi MK, Kumar V, Yadav V, *et al.* Current state of the art biotechnological strategies for conversion of watermelon wastes residues to biopolymers production: A review. Chemosphere 2022; 290133310.
[http://dx.doi.org/10.1016/j.chemosphere.2021.133310] [PMID: 34919909]

[39] Liu H, Kumar V, Jia L, *et al.* Biopolymer poly-hydroxyalkanoates (PHA) production from apple industrial waste residues: A review. Chemosphere 2021; 284131427.
[http://dx.doi.org/10.1016/j.chemosphere.2021.131427] [PMID: 34323796]

[40] Awasthi SK, Kumar M, Kumar V, *et al.* A comprehensive review on recent advancements in biodegradation and sustainable management of biopolymers. Environ Pollut 2022; 307119600.
[http://dx.doi.org/10.1016/j.envpol.2022.119600] [PMID: 35691442]

[41] Duan Y, Tarafdar A, Kumar V, *et al.* Sustainable biorefinery approaches towards circular economy for conversion of biowaste to value added materials and future perspectives. Fuel 2022; 325124846.
[http://dx.doi.org/10.1016/j.fuel.2022.124846]

[42] Kumar V, Sharma N, Umesh M, *et al.* Emerging challenges for the agro-industrial food waste utilization: A review on food waste biorefinery. Bioresour Technol 2022; 362127790.
[http://dx.doi.org/10.1016/j.biortech.2022.127790] [PMID: 35973569]

[43] Awasthi SK, Sarsaiya S, Kumar V, *et al.* Processing of municipal solid waste resources for a circular economy in China: An overview. Fuel 2022; 317123478.
[http://dx.doi.org/10.1016/j.fuel.2022.123478]

[44] Kumar V, Sharma N, Maitra SS. *In vitro* and *in vivo* toxicity assessment of nanoparticles. Int Nano Lett 2017; 7(4): 243-56.
[http://dx.doi.org/10.1007/s40089-017-0221-3]

[45] Vinay K, Neha S, Maitra S. Protein and Peptide Nanoparticles: Preparation and Surface Modification, in Functionalized Nanomaterials I. CRC Press 2020; pp. 191-204.

[46] Vallinayagam S, *et al.* Recent developments in magnetic nanoparticles and nano-composites for wastewater treatment. J Environ Chem Eng 2021; 9(6)106553.
[http://dx.doi.org/10.1016/j.jece.2021.106553]

[47] Kumar V, Sharma N, Lakkaboyana SK, Maitra SS. Silver nanoparticles in poultry health: Applications and toxicokinetic effects, in Silver Nanomaterials for Agri-Food Applications. Elsevier 2021; pp. 685-704.
[http://dx.doi.org/10.1016/B978-0-12-823528-7.00005-6]

[48] Egbosiuba T C. Biochar and bio-oil fuel properties from nickel nanoparticles assisted pyrolysis of cassava peel Heliyon, 2022; 8(8)
[http://dx.doi.org/10.1016/j.heliyon.2022.e10114]

[49] Shankar Pandey D, Katsaros G, Lindfors C, Leahy J J, Tassou S A. Fast Pyrolysis of Poultry Litter in a Bubbling Fluidised Bed Reactor: Energy and Nutrient Recovery sustainability 2019; 11 [Accessed 27 November 2020].

[50] Vries T, Scott E, van Haasterecht T. Pyrolysis and Catalytic Upgrading of Poultry Litter to Produce

Chemicals 2017.

[51]   Santos Dalólio F, da Silva JN, Carneiro de Oliveira AC, *et al.* Poultry litter as biomass energy: A review and future perspectives. Renew Sustain Energy Rev 2017; 76: 941-9.
[http://dx.doi.org/10.1016/j.rser.2017.03.104]

[52]   Dróżdż D, Wystalska K, Malińska K, Grosser A, Grobelak A, Kacprzak M. Management of poultry manure in Poland – Current state and future perspectives. J Environ Manage 2020; 264110327.
[http://dx.doi.org/10.1016/j.jenvman.2020.110327] [PMID: 32217329]

CHAPTER 12

Valorization of Sugar Industry Waste for Value-Added Products

Neha Kumari[1] and **Saurabh Bansal**[1,*]

[1] *Department of Biotechnology and Bioinformatics, Jaypee University of Information Technology, Waknaghat, Distt. Solan, Himachal Pradesh, India*

Abstract: India is the second-largest cultivator of sugarcane worldwide, the primary source of refined sugar. Increased demand for sugar has driven this industry as a mainstream pollutant-generating industry. Every year, a tremendous amount of liquid (molasses) and solid wastes (sugarcane bagasse, filter cake) are generated, posing a major bottleneck for waste management. Although there exist traditional approaches like incineration, landfills are being employed for handling sugarcane waste which leads to the emission of greenhouse gases, and foul odour and adds more cost to running a sustainable industry. Moreover, no value-added product is formed from such traditional approaches resulting in an immense loss of bioenergy. Researchers have emphasized transforming waste into a sustainable economic generation of higher\-value products over the past few decades. Sugarcane industrial waste is a rich source of lignocellulosic organic biomass, which is used as a raw material for the production of biofuel (bioethanol, biogas), single cells proteins, enzymes, organic acids, food additives and nutraceuticals. Day by day, with advanced technology, novel applications are evolving, adding more thrust to this area. In this chapter, the potential of valorization of sugarcane waste to value-added products is discussed comprehensively.

Keywords: Biochemical, Biofuel, Lignocellulosic, Sugarcane waste, Value-added products.

INTRODUCTION

Agro-industrial residues are generated in vast quantities and pose major issues in handling and disposing of waste into the environment. These residues are either burnt openly or dumped directly into the environment owing to their biodegradable nature. Nowadays, these organic, renewable, energy-rich agro-industrial residues are bio-transformed into a wide array of valuable products. Sugarcane is a tropical crop with a planting and harvesting cycle of 12 months. It

* **Corresponding author Saurabh Bansal:** Department of Biotechnology and Bioinformatics, Jaypee University of Information Technology, Waknaghat, Distt. Solan, Himachal Pradesh, India; Email: saurab.bansal02@gmail.com

Vinay Kumar, Sivarama Krishna Lakkaboyana & Neha Sharma (Eds.)

has high sucrose content and yields a large amount of sugarcane organic biomass ideal for bioconversion into many important industrial products [1, 2]. Asia is the biggest producer of sugarcane and contributes to 44% of global production [3]. Brazil and India are the largest producers of sugarcane and thus, generating the sugarcane industry waste [4]. The sugarcane industry is one of the mainstream industries which generate a large amount of solid, liquid and gaseous waste. Waste is generated at every step of sugarcane processing, from its harvesting to the final stage of packaging. Solid (Sugarcane Bagasse, Bagasse fly ash, Press mud) and liquid waste (Molasses) generated during processing need proper management for the sustainable sugar industry. Wastewater has high BOD (Biological oxygen demand) and COD (Chemical oxygen demand), thus need to be treated before disposing of into water bodies. This industry is generating a large quantity of sugarcane trash (leaves, dried stalk and roots) which has tremendous potential as fuel and feedstock; water effluent is generated, which is worrisome to handle and disposed off. Sugarcane trash is either burnt in open fields or dumped as it is, so causing problems of pollution and health risks. Sugarcane bagasse is the biggest agro-industrial fibrous waste left after the crushing of sugar stalks [5]. Molasses are dark colour nutrient-rich waste generated during the final stage of sugar syrup processing [6]. Vinasse waste is generated from the sugar-alcohol industry and has great potential for its bioconversion into valuable high-demand products [7, 8]. Nowadays, much emphasis is on the conversion of waste generated through the sugarcane industry to value-added products for maintaining the socioeconomic sector and sustainability of the industry. Valorization of sugarcane industry waste could solve the problem of pollution generated through this industry to a large extent. Biorefinery emergence in sugarcane resulted in combinatorial approaches for the sustainable sugar industry. Sugarcane bagasse has several applications in the bioenergy sector, paper industry, feed industry, enzymes, antibiotics, organic acid, alkaloids and other biochemical productions [2, 5, 9, 10]. In this chapter, we are going to study different waste generated so far by the sugarcane industry and their utilization as a raw material for the production of various high-value products.

Sugarcane Processing

Solid and liquid waste is generated during the processing of sugarcanes Fig. (**1**). The processing of sugarcane starts right from its harvesting as some stalk and dry leaves left behind, are either burnt or dumped in the field to be used as biofertilizers. The canes are washed, shredded and crushed to extract the juice. The juice is separated from solid organic waste termed sugarcane bagasse (SB). Further raw sugarcane is concentrated and precipitated to form clear sugarcane juice at the top and slurry left at the bottom called sugarcane filter cake (pressed). The clear sugarcane juice is heated to form a thick syrup catalyzed by sugar granules. Finally, after crystallization, the mixture was spun to separate the remaining syrup, termed sugarcane molasses (SCM) [1, 5]. The wastewater composition generated during each step is variable and contributes to a vast wide array of products Fig. (**1**). Composition shown in Table. **1**.

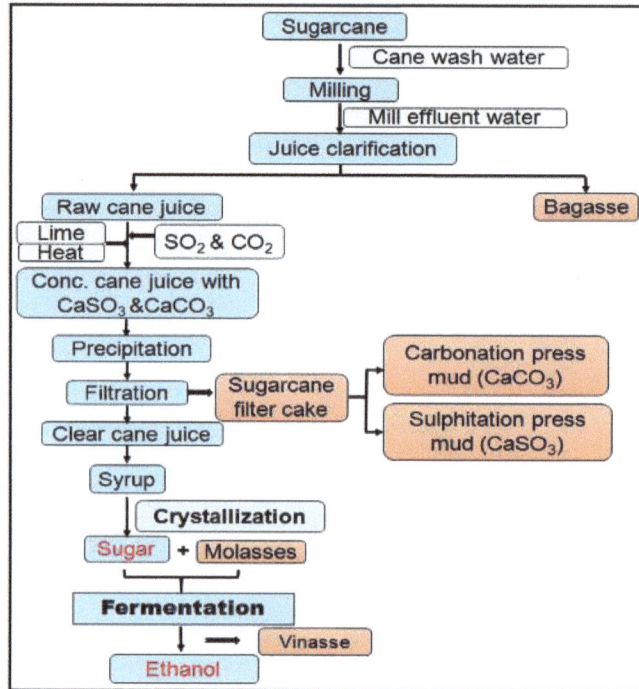

Fig. (1). Schematic presentation of sugarcane processing and waste generated during various steps of processing.

Table 1. Composition of sugarcane waste (Sugarcane bagasse, molasses, press-mud and Vinasse) where cellulose: C; Hemicellulose: HC; Lignin: L; Saccharose: S; Other Polysaccharides: OP; Ash: A.

Waste% dry weight											References
Sugarcane bagasse	**pH**	**Fibers**	**Water**	**Soluble solids**	**Lignocellulosic content**						
					C	HC	L	S	OP	A	
	4.5-5.5	48	50	2	42	28	20	4.6	3	2.4	[7, 9, 11, 12]
Molasses	5-5.5	-	-	46	-	-	-	49.9	-	10.25	
Vinasse	4.8	-	93	5.3	-	-	-	-	-	21.4	
Sugarcane filter cake (Press-mud)	4.95	15-30	75	-	11.4	27.1	9.3	1-15	-	9-20	

Sugarcane Bagasse Valorization

This is the solid fibrous waste left over after squeezing all the juice of sugarcane. This waste generated is rich in nutrition and contributes to 30% weight of sugarcane. Bagasse is a major lignocellulosic waste generated based on a dry weight basis rich in cellulose (42%), hemicellulose (28%), lignin (20%), other polysaccharides (4.6%), saccharose (3%) and ash (2.4%) [5, 9]. One of its major applications is in the production of second-generation bioethanol and biofuel. The energy balance potential of sugarcane bagasse produced bioethanol compared to corn-b sed bioethanol. Moreover, the operational cost is also less for sugarcane-produced biofuel as no pretreatment is required for processing as in case the starchy waste prerequisite is to pre-treat the waste before further processing [13, 14]. Incineration of sugarcane bagasse is a common practice as it is used as fuel for jaggery preparation and generation of thermal and electrical energy. Nevertheless, this approach resulted in tremendous energy loss, generated pollution and was economically inappropriate. Due to its surplus generation, it has a wide array of other applications in the paper and pulp industry, biofuel industry, and production of biochemicals like single-cell proteins, enzymes and food additives [9, 11, 15] Fig. (2). It is dumped in agricultural fields as it conditions the soil, improves its texture and ameliorates its nutritional status Listed in Table. 2.

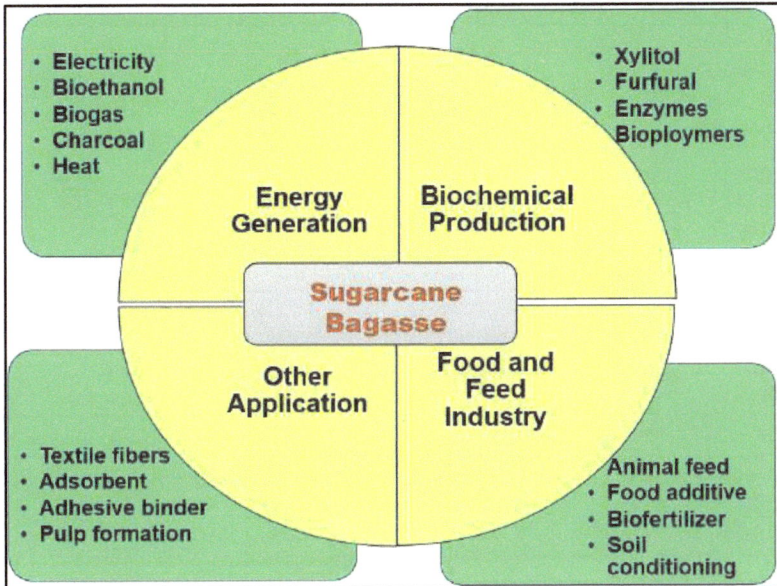

Fig. (2). Schematic presentation showing potential applications of sugarcane bagasse in a wide array of sectors.

Table 2. Various applications of sugarcane bagasse for bioconversion to value-added products

Biochemicals/ Enzymes produced	Process	Microorganisms/Meditator	References
Inulinase	SSF	*Kluyveromyces marxianus*Y-7571	[16]
Ergot Alkaloid	SSF	*Claviceps purpurea* 1029c	[17]
Xylanase	SSF	*Trichoderma harzianum* Rifai	[18]
Lipase	SSF	*Rhizomucor pusillus, Rhizopus rhizopodiformis, Rhizopus oryzae*	[19, 20]
Cellulase, xylanases	SSF	*Penicillium echinulatum* 9A02S1 *Botryosphaeria* sp. AM01 and *Saccharicola* sp. EJC 04	[21, 22]
Cellulase	SSF & SMF	*Aspergillus niger* A12	[23]
Composite	Organosolv process	-	[24, 25]
Xylitol	SMF	*Candida guilliermondii* FTI 20037,*Candida tropicalis*	[26 - 28]
Poly β-hydroxybutyrate (PHB)	SSF	*Halomonas campisalis* MCM B-1027 and*Bacillus* spp.	[29, 30]
Laccases	SSF	*Pleurotus ostreatus*	[31]
Laccase. endo-β-1-4-xylanase and β-xylosidase	SMF	*Aspergillus japonicas*	[32]
Cellulase, hemicellulase, pectinase, esterase, amylases	SSF	*Penicillium echinulatum*	[33]
Endoglucanase (CMCase), total cellulase (FPase) and xylanase	SMF	*Aspergillus flavus* KUB2	[34]
Laccase	SSF	*Pleurotus ostreatus,* Lentinus crinitus	[31, 35]
Citric acid	SSF	*Aspergillus niger* DS 1	[36]
Lactic acid	SSF	*Lactobacillus delbrueckii*	[37]
Amylase	SSF	*Aspergillus niger* strain UO-01	[38]
Surfactin	SSF	*Bacillus pumilus* UFPEDA 448	[39]
Polyphenol oxidase (PPO) and manganese peroxidase (MnP)	SSF	*Phanerochaete chrysosporium* PC2, *Lentinula edode* LE16 and *Pleurotus ostreatus* PO45	[40]
Invertase	SSF	*Aspergillus niger* GH1	[41]
Pectinase	SSF	*Aspergillus oryzae*	[42]
Gluconic acid, Xylooligosaccharides	SSF	*Gluconobacter oxydans* ATCC 621H	[43, 44]
Cellulose nanocrystals	SSF	*Aspergillus fumigatus* CCT 7873	[45]

(Table 2) cont.....

Biochemicals/ Enzymes produced	Process	Microorganisms/Meditator	References
Ethanol	SSF	*Pycnoporus sanguineus MCA 16, Pycnoporus sanguineus, Candida shehatae*	[46 - 48]
Formaldehyde resin	Alkaline organosolv process	laccase mediated	[49]
Aroma compounds	SSF	*Kluyveromyces marxianus*	[50, 51]
Sugarcane bagasse nanocellulose-based hydrogel	TEMPO-mediated oxidation	Bromothymol blue/methyl red	[52]
Levulinic acid	NA	1-ethyl-3-methylimidazolium hydrogen sulfate	[53]
Immobilization matrix	SSF	*Streptomyces* sp. SR13−2	[54]

Valorization of Sugarcane Bagasse (SB) to Form Pulp

SB has a high fiber content of 48% making it compatible to be used as an alternative sustainable source of raw material for pulp formation, relieving the stress of deforestation. Firstly the bagasse fibers were isolated from SB using spray water and dewatering units, followed by treating the bagasse in a steam boiler for 10-15 minutes. After heat treatment, the pulp left behind was screened, and cleaned with water to eliminate its blackish coloration. Finally, the pulp was thickened by water removal and underwent whitening treatment. Pulp generated from SB undergoes various processes to be used as paper, paperboard, newsprint and cardboard [55, 56].

Valorization of SB to Biochar (Charcoal)

Biochar is another attractive value-added product formed from SB processed by pyrolysis for complete carbonization of raw materials followed by its activation. SB provides an economically sustainable, environmentally friendly raw material for biochar generation [10]. The biochar resulting from SB has many applications as fuel, as an excellent absorbent of toxic metal ions, dyes and for electrochemical desalination [57 - 60].

SB as Inert Material and Carbon/Energy Source for Industrial Enzymes /Biochemical Production

The chemical constituents of SB make it an excellent source of carbon in SSF and SMF for the growth of microbes which produce important industrial enzymes. SB proved to be the most promising cheap agro-industrial waste for producing hydrolytic enzymes (cellulase, xylanase, laccase, ligninase, amylase, invertase and many others) [2, 9, 32]. In contrast, some studies suggested that SB is an inert support material for microbial growth and as an immobilization matrix (Listed in Table **1**).

Valorization of SB to Biofuel (Bioethanol, Biodiesel, 2,3-Butanediol and H_2)

Increased population has also impacted the increased demand for fuel which resulted in overexploitation of fossil reservoirs of energy. SB is the most attractive raw material for bioethanol production due to its lignocellulosic nature, commonly termed as second generation (2G) of ethanol, biodiesel, 2,3-butanediol and H_2 production. To achieve the maximum conversion of SB to biofuel, lignin needs to be hydrolyzed as it is most resistant to degradation. Pretreatment of SB is done through various physical, chemical and biological processes [40, 47] Fig. (**3**). The cellulosic hemicellulose content of SB undergoes fermentation of a commonly used industrial strain of yeast *Saccharomyces cerevisiae* [61, 62], *Candida tropicalis* and bacteria *Zymomonas mobilis* [63] to form second-generation ethanol. There are very few studies conducted for biodiesel production using SB as a carbon source. Studies carried out by Rattanapoltee and Kaewkannetra 2014, demonstrate the biodiesel production potential by microalgae *Scenedesmus acutus* using SB and pineapple peel as a sole carbon source [64, 65]. SB hydrolysate is a potential feedstock for fermentation by *Clostridium acetobutylicum* to produce biobutanol [66]. 2,3-butanediol, a liquid biofuel, is also produced from pretreated SB by *Klebsiella pneumonia* [67]. Xylose and XOS obtained from SB pretreatment are hydrolyzed to simpler sugars by genetic engineered microbes and undergo dark fermentation to produce biohydrogen gas; dark fermentation is employed by mixed hydrolytic and acidogenic communities in aerobic conditions [68]. Pretreated SB undergoes fermentation by a microbial consortium of *Clostridium* and *Tepidimicrobium* to produce clean fuel H_2 along with organic acids (butyric acid, acetic acid and propionic acid) [69].

SB as Enriched Animal Feed

Owing to its nutritional properties, it is being utilized as feed in the feedstock relieving the burden of fodder cost. SB acts as the source of roughage and is cost-effective as compared to other forage sources. This cheap source of forage helps to overcome forage shortage, and enhances the digestibility and fermentation potential of ruminates [70]. SB is being utilized as a substrate for fermentation (SSF) by microbes to produce enriched animal feed [71, 72].

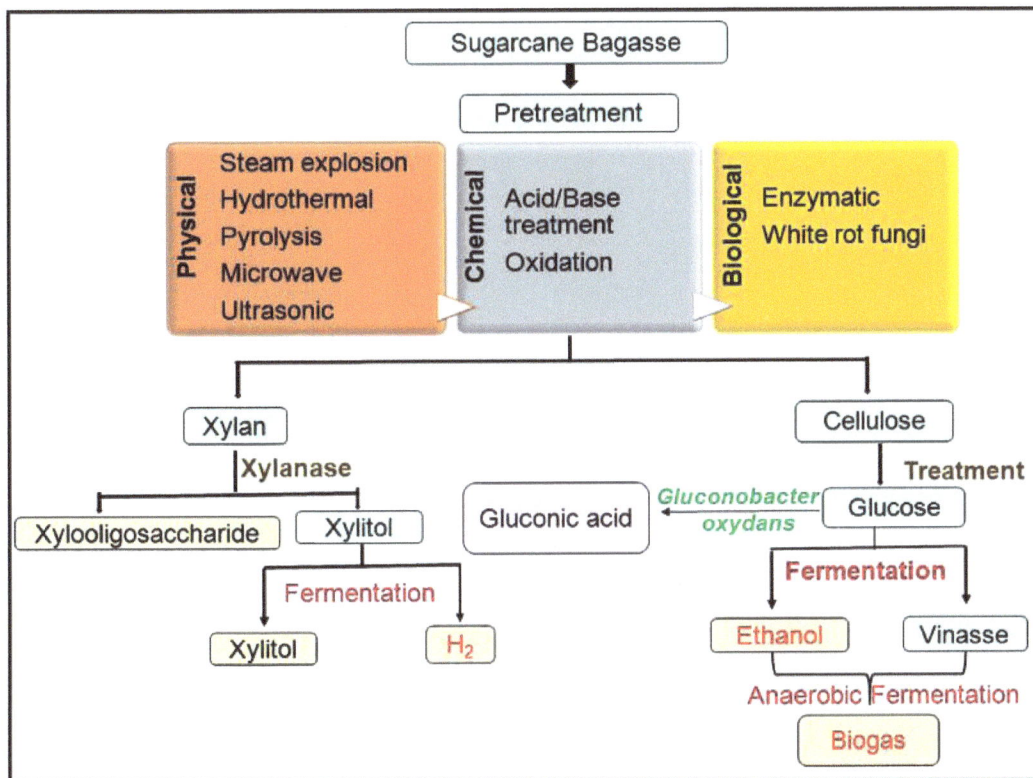

Fig. (3). Schematic representation of fractionation of sugarcane bagasse for biofuel production.

SB Bioconversion to Xylooligosaccharides (XOS)

XOS is a major prebiotic that helps in the growth of bifidobacteria. SB is rich in xylan content, which makes it a potential candidate for XOS production. SB pretreatment is a prerequisite for its conversion to XOS [73]. The physical method of steam explosion is the most conventional method adopted due to its ease in the purification step. Other chemical methodologies are also opted for SB pretreatment to form XOS which include hydrogen peroxide treatment, acetic acid-mediated [43], gluconic acid-mediated and alkaline/oxidation treatment [74, 75]. But none of them is fully satisfactory as they not only require special equipment and high temperature, but also result in the release of cellulose-derived species. The enzymatic approach is much desirable as it is much controlled and leads to the maximum leaching of xylan from SB without any contaminants [75]. New technologies and approaches emerged day by day, like microwaves irritated with more controllable factors and xylan extraction [76].

Valorization of SB to Biopolymer

Chemical polymers pose major health and ecological problems due to their non-biodegradable nature. Bioplastic is a biodegradable alternative that overcomes the problem of plastic waste management. Bioplastic has one limitation as it requires carbon-rich nutrients, which could be tackled by using agro-industrial waste. Sugar industry waste is an attractive raw material for bioplastic production as it is rich in carbon contents. Effluent sugar waste is an inexpensive medium for biodegradable poly β-hydroxybutyrate (PHB) production by *Bacillus subtilis* NG220 isolated from sugarcane fields. The yield of PHB is enhanced by supplementing media with carbon (maltose) and nitrogen sources (ammonium sulphate) [77]. Bacillus spp., *Bacillus safensis* EBT1 and *Halomonas campisalis* MCM B-1027 also proved to be excellent PHB producers by utilizing SB as a raw source of carbon [29, 30, 78]. There are few reports where sugarcane juice was utilized as a cheap carbon source by *Alcaligenes latus* for PHB production [79]. Some research groups also concluded the production of other biopolymers like β-glucan by using sugarcane straw as a raw material for its synthesis. Sugarcane straw undergoes acid and alkaline pretreatment followed by enzymatic (cellulase) treatment. Finally, pretreated sugarcane straw undergoes fermentation by *Lasiodiplodia theobromae* CCT3966 to form β-glucan [80].

SB as Composite

The natural fiber is fascinating and attracts research groups for more extensive studies as it is light-weighted, renewable, biodegradable, less costly and environmentally friendly. Pretreatment of SB was carried out as it ameliorated the adhesive properties of fiber [81]. Mainly lignin content of SB extracted undergo processing to form resins and fibers [25].

SB as Carbon and Inert Source for Organic Acid Synthesis

SB proved to be an excellent carbon source for the production of organic acids like citric, lactic, butyric, propionic acid and others by microbial fermentation. Researchers have concluded that replacing expensive chemical sugar sources with inexpensive SB resulted in itaconic acid production at a low cost with high productivity by *Aspergillus spp.* mediated fermentation [82 - 84]. Succinic acid is the high-value food additive and SB pretreated hydroxylate serves as a carbon and nitrogen source for its synthesis by *Actinobacillus succinogenes*, *Yarrowiali polytica* [85 - 87]. Citric acid, an important organic acid, has several applications in the food industry as a flavouring agent and preservative. Microbial cultures, especially *Aspergillus spp.* utilize SB as an inert material for its growth. Not only it could produce citric acid at sustainable rates but biomass left after fermentation can be effectively utilized as an animal feed [36, 88]. In recent times, lactic acid has gained lots of sight from researchers as it is being used for poly-lactic synthesis. SB provides a way for the green synthesis of lactic acid without hampering environmental conditions [89]. Butyric acid is an important nutraceutical produced from pretreated SB by *Clostridium tyrobutyricum* [90]. Some studies concluded pretreated SB as a carbon source by *Propionibacterium acidipropionici* CGMCC 1.2230 strain [91] and as an immobilized matrix by *Propionibacterium freudenreichii* CCTCC M207015 for propionic acid production [92].

Bagasse Fly Ash (BFA) and its Valorization to Value Added Products

Sugarcane solid waste (bagasse) is used for energy generation resulting in porous, small-sized (0.5-300 microns) particles called bagasse fly ash [93]. Chemically BFA is composed of an elemental carbon and contains oxides of metals (silicon, aluminium, iron and others). BFA could lead to respiratory problems due to its inhalation and also pose a major threat to agricultural land as it hardens the soil due to its metal oxides [5, 94]. BFA is extensively employed as an adsorbent for effluent treatment of wastewater of the pulp and paper industry [95]. Prior studies also suggested that BFA is an excellent adsorbent for toxic metals and dyes like malachite green [96]. Rheological studies of BFA suggested that it enhances the mechanical strength and durability of construction material (cement and bricks) [97] Fig. (**4**).

Sugarcane Filter Cake and its Applications

It is the solid colloidal byproduct formed after the clarification of precipitated cane juice. This is also termed as pressmud, and it is a rich source of organic carbon, which could be composted by microbes into a biofertilizer [12]. Most conventionally, pressmud cake is used as a fertilizer as it ameliorates the nutritional status of the soil. But earlier studies suggested that pretreated pressmud has enhanced nutrients availability and composting capability [98]. It is also utilized as a carrier of biofertilizers [99]. Pressmud has a very good adsorbent property which leads to its applications for the removal of dyes, and toxic metal ions [100]. *Aspergillus niger* also utilizes pressmud for citric acid synthesis [101]. Pressmud is being utilized as a cheap source of animal feed as it enhances the meat quality of lamb [102]. Several studies conducted concluded pressmud as a potential substrate for biofuel production. A recently conducted study concluded methane production by utilizing vinasse and pressmud as substrates [103]. Pretreated pressmud undergoes fermentation by *Clostridium acetobutylicum* NRRL B-527 to form butanol [104]. Pressmud is also a potential source for ethanol production with its subsequent treatment to form clean fuel H_2 [105] Fig. (**4**).

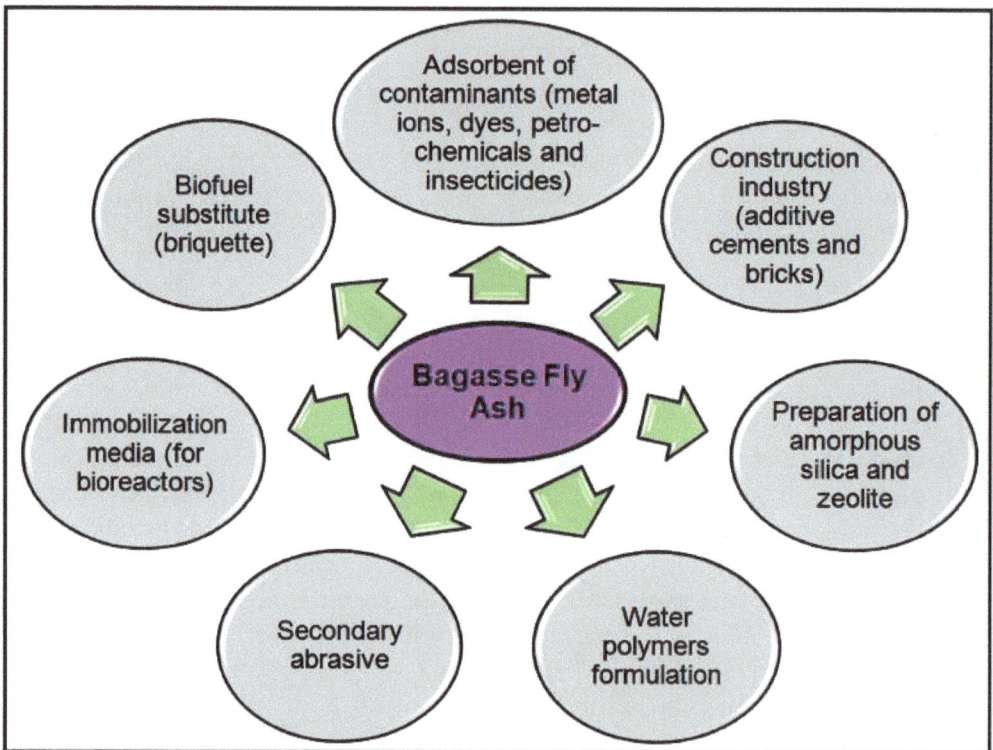

Fig. (4).   Various Applications of bagasse fly ash.

Molasses Valorization to High-End Product

Molasses is the dark sugar-rich liquid leftover after the crystallization of sucrose. It was nutritionally very rich in fermentable sugar and ash, which is nowadays well utilized by industries for value-added product formation. The chemical composition varied with the type of sugarcane but the main ingredients are sucrose along with other reducing sugars (glucose). It has major applications in the biofuel industry, feed industry and provides an excellent medium for microbial growth [2, 6]. Vinasses, molasses and pretreated molasses are being utilized as carbon and nitrogen source for PHB production up to 12g/l by *Cupriavidus necator* [106, 107]. *Xanthomonas campestris* strain IBRC-M 10644 also utilized cane molasses as production media for Xanthan biopolymer synthesis [108]. Several new applications like the synthesis of carbon dots activated carbons and nanocomposites are emerging day by day by technological advancement, making the sugar industry attractive and feasible for the long economic run [148 - 158]. Some applications of molasses are listed in Table **3**.

Table 3. Value-added products produced by bioconversion of molasses.

Product	Process	Microorganism	Application of Molasses	References
Carbon dots	Hydrothermal method	NA	Molasses as carbon source	[109]
Activated carbon	Carbonization at high temperature (400–800 °C) in the presence of KOH	NA	Substrate for bioconversion	[110]
Single-cell oil *(polyunsaturated fatty acids (PUFA) and biodiesel)*	Aerobic fermentation	*Mucor circinelloides*	Molasses as carbon source	[111]
Biosurfactants (Sophorolipid Glycolipid)	Fed-batch fermentation	*Starmerella bombicola NBRC 10243*	As a sole culture medium component for biosurfactant production	[112]
Hyaluronic acid	Batch fermentation	*Streptococcus zooepidemicus*	Carbon source	[113]
Welan gum	Batch fermentation	*Alcaligenes* sp. ATCC31555	Substrate	[114]
Succinoglycans	Aerobic submerged fermentation	*Agrobacterium radiobacter* NBRC 12665	Carbon source	[115]
Levan	SMF	*Zymomonas mobilis*	Low cost carbon source	[116]
Inulinase	SMF	*Kluyveromyces marxianus* NRRL Y-7571	Acts as carbon source along with corn steep liqour	[117, 118]
Protease	SMF	*Streptomyces sp.* 594	Carbon source	[119]
Bioethanol	SMF/Batch fermentation	*Saccharomyces cerevisiae,* Zymomonas mobilis	Provides sugar source for fermentation	[120, 121, 123, 124]
2,3-butanediol	SMF	*Bacillus subtilis*	Substrate	[125]
D- Lactic acid	Batch fermentation	*Lactobacillus delbrueckii*	Carbon source	[89]
Succinate	Anaerobic fermentation	*E. coli*	As a sucrose source	[126]
Gluconic acid	SSF	*Aspergillus niger* (ARNU-4)	Carbohydrate source	[127]

(Table 3) cont.....

Product	Process	Microorganism	Application of Molasses	References
Polymalic acid	SMF	*Aureobasidium pullulans*	Substrate	[128]
Fructooligo-saccharides (FOS)	SMF	*Aspergillus japonicus*-FCL 119T, *Aspergillus niger* ATCC 20611, *A. tubingensis* XG21	Act as a substrate to form FOS by β-Fructosyltransferase activity	[129, 130]
Microbial Oleochemicals	Fed-batch Fermentation	*Rhodosporidium toruloides* NRRL Y-27012and *R. kratochvilovae* Y-43	Carbon source	[131]
Laccase	SMF	*Agaricus blazei, Pycnoporus sanguineus*	Optimal substrate	[132, 133]
β-d-galactosidase	SMF	*Kluyveromyces marxianus*	Carbon source	[134, 135]
L-asparaginase	SMF	*Zymomonas mobilis*	Carbon source	[136]
Extracellular polysaccharide	SMF	*Zoogloea sp.*	Substrate	[137]
Mannitol	SMF	*Lactobacillus reuteri CRL 1101*	Carbon source	[138]
Invertase	SMF	*Aspergillus niger GH1*	Substrate	[41]
β-Carotene	SMF	*Rhodotorula glutinis*	As a sole nutrient source for production	[139]

Vinasse and Its Derived High-End Products

Vinasse is the byproduct generated from the processing of sugarcane waste to ethanol. It is the acidic (3.5-5), dark, unpleasant odour organic-rich waste of the sugar ethanol industry. It is worrisome as disposed of effluent as it results in a shift in the natural pH of the soil, salinization and water bodies. Its main constituent is water (93%) and solids (7%). 1litre of ethanol generated from sugar resulted in approximately 15litres of vinasse [2, 7]. The emerging concept of biorefinery helps in the sustainable production of sugar and its consequent valuable products from its waste. Its chemical richness contributes to its tremendous potential as growth media for a variety of microbes which in turn produce industrially important products. It has a wide array of applications as a biofertilizer, as a raw material for the growth of microbes (algae, yeast and fungi), feed production for livestock animals and as a bioenergy source (biogas) [8, 140, 141]. Listed in Table **4**.

Table 4. Vinasse and its applications for producing various high-end products.

Product	Microorganism	Application	References
Single-cell protein	*Saccharomycescerevisiae*(CCMA0187 and CCMA0188), *Candida glabrata* (CCMA0193) and *Candida parapsilosis*(CCMA0544)	Vinasse as raw material for the production of animal feed.	[142]
Hydrogen (H₂) and volatile fatty acids	*Clostridium, Bacillus,* and *Enterobacter*	As a substrate.	[143]
Methane	*Bacteroidia, Betaproteobacteria, Methanosaeta, Methanomassiliicoccaceae,* and *Methanobacterium.*	As a raw material for bioconversion.	[144]
Biostimulant	Soil microbiota	Vinasse enhances the biodegradation of oily sludge soil.	[145]
Fertilizer	*Desmodesmus subspicatus*	Stabilizer for high methoxyl pectin particleand microbes for slow release of micro-nutrients.	[141]
Bioenergy (Bioethanol	*Chlorella vulgaris, Chlamydomonas reinhardtii CC-1093,*	Anerobic digested vinasse acts as growth material for algae.	[146, 147]
Lignocellulolytic enzymes (Laccase, peroxidase, manganese peroxidase)	*Pleurotus and Trichoderma reesei*	As energy and inert media for the production of enzymes.	[159, 160]
Citric acid	*Aspergillus niger* and *Trichoderma*	Vinasse and SB as inert material for citric acid production.	[88, 161]
PHB	*Cupriavidus necator*	Vinasse and sugarcane molasses acts as the substrate	[106]

CONCLUSION

The sugarcane industry is the utmost industry as it has applicability in many processes from providing sweetening agent as the main product to generate a large quantity of raw materials for various biochemicals. New technologies and integrated processes need to be searched and evaluated in terms of their sustainability, self-sufficient operational capability, and the potential of producing many highly valuable products from the same pipeline. Biorefinery concept and the emergence of new technology and innovation resulted in concurrent utilization of sugarcane waste with a generation of several high-end valuable products at once. Recently this area gained attention and researchers have already pipelined multiple products together by biorefinery concept. These up-listed approaches not only tackle the bottleneck of sugar industry waste but also resulted in the complete utilization of the sugarcane industry with a generation of new revenues. Still, new technologies and biotechnological intervention are required for the complete utilization of sugarcane waste. New research and development approaches could be helpful to generate sustainable methodologies, recombinant microbial strain to withstand end-product inhibition, and toxic metabolites secreted during fermentation. Another problem that needs to be addressed is the poor biomass conversion to value-added products. New novel integrated technologies, metabolic engineering and strain development could address these issues.

REFERENCES

[1]   Nigam PS-N, Pandey A, Eds. Biotechnology for agro-industrial residues utilisation: utilisation of agro-residues. Springer Netherlands 2009.
        [http://dx.doi.org/10.1007/978-1-4020-9942-7]

[2]   Sindhu R, Gnansounou E, Binod P, Pandey A. Bioconversion of sugarcane crop residue for value added products – An overview. Renew Energy 2016; 98: 203-15.
        [http://dx.doi.org/10.1016/j.renene.2016.02.057]

[3]   de Matos M, Santos F, Eichler P. Chapter 1 - Sugarcane world scenario. In: Santos F, Rabelo SC, De Matos M, et al. (eds) Sugarcane biorefinery, technology and perspectives. Academic Press, pp. 1–19

[4]   Irmak S. Biomass as raw material for production of high-value products. biomass volume estimation and valorization for energy. Epub ahead of print 22 February 2017.
        [http://dx.doi.org/10.5772/65507]

[5]   Oliveira AD. Sugarcane: Technology and Research. BoD – Books on Demand 2018.
        [http://dx.doi.org/10.5772/intechopen.69564]

[6]   Sheth A, Borse P. Chapter 7 - Sugarcane vinasse, molasses, yeast cream: agricultural, environmental, and industrial aspects. In: Chandel AK, Luciano Silveira MH (eds) Advances in Sugarcane Biorefinery. Elsevier, pp. 153–161

[7]   Carrilho ENVM, Labuto G, Kamogawa MY. Destination of vinasse, a residue from alcohol industry: resource recovery and prevention of pollution.In: Prasad MNV, Shih K (eds) Environmental Materials and Waste. Academic Press,21-43.

[8]   Parsaee M, Kiani Deh Kiani M, Karimi K. A review of biogas production from sugarcane vinasse. Biomass Bioenergy 2019; 122: 117-25.
        [http://dx.doi.org/10.1016/j.biombioe.2019.01.034]

[9]   Pandey A, Soccol CR, Nigam P, Soccol VT. Biotechnological potential of agro-industrial residues. I: sugarcane bagasse. Bioresour Technol 2000; 74(1): 69-80.
        [http://dx.doi.org/10.1016/S0960-8524(99)00142-X]

[10]   Kwon G, Bhatnagar A, Wang H, Kwon EE, Song H. A review of recent advancements in utilization of biomass and industrial wastes into engineered biochar. J Hazard Mater 2020; 400123242..

[http://dx.doi.org/10.1016/j.jhazmat.2020.123242] [PMID: 32585525]

[11]   Santos F, Eichler P, Machado G, et al. Chapter 2 - By-products of the sugarcane industry. In: Santos F, Rabelo SC, De Matos M, et al. (eds) Sugarcane Biorefinery, Technology and Perspectives. Academic Press, pp. 21–48.

[12]   Katakojwala R, Naresh Kumar A, Chakraborty D, *et al.* 3 - Valorization of sugarcane waste: Prospects of a biorefinery.Industrial and Municipal Sludge. 47-60.
       [http://dx.doi.org/10.1016/B978-0-12-815907-1.00003-9]

[13]   Méjean A, Hope C. Modelling the costs of energy crops: A case study of US corn and Brazilian sugar cane
       https://ideas.repec.org/p/hal/journl/halshs-00736154.html2010.
       [http://dx.doi.org/10.1016/j.enpol.2009.10.006]

[14]   Waclawovsky AJ, Sato PM, Lembke CG, Moore PH, Souza GM. Sugarcane for bioenergy production: an assessment of yield and regulation of sucrose content. Plant Biotechnol J 2010; 8(3): 263-76.
       [http://dx.doi.org/10.1111/j.1467-7652.2009.00491.x] [PMID: 20388126]

[15]   Loh YR, Sujan D, Rahman ME, Das CA. Sugarcane bagasse—The future composite material: A literature review. Resour Conserv Recycling 2013; 75: 14-22.
       [http://dx.doi.org/10.1016/j.resconrec.2013.03.002]

[16]   Mazutti M, Bender JP, Treichel H, Luccio MD. Optimization of inulinase production by solid-state fermentation using sugarcane bagasse as substrate. Enzyme Microb Technol 2006; 39(1): 56-9.
       [http://dx.doi.org/10.1016/j.enzmictec.2005.09.008]

[17]   Trejo Hernández MR, Lonsane BK, Raimbault M, Roussos S. Spectra of ergot alkaloids produced by *Claviceps purpurea* 1029c in solid-state fermentation system: influence of the composition of liquid medium used for impregnating sugar-cane pith bagasse. Process Biochem 1993; 28(1): 23-7.
       [http://dx.doi.org/10.1016/0032-9592(94)80032-4]

[18]   Rezende MI, Barbosa AM, Vasconcelos AFD, Endo AS. Xylanase production by *Trichoderma harzianum rifai* by solid state fermentation on sugarcane bagasse. Braz J Microbiol 2002; 33(1): 67-72.
       [http://dx.doi.org/10.1590/S1517-83822002000100014]

[19]   Cordova J, Nemmaoui M, Ismaïli-Alaoui M, *et al.* Lipase production by solid state fermentation of olive cake and sugar cane bagasse. J Mol Catal, B Enzym 1998; 5(1-4): 75-8.
       [http://dx.doi.org/10.1016/S1381-1177(98)00067-8]

[20]   Vaseghi Z, Najafpour GD, Mohseni S, Mahjoub S. Production of active lipase by *Rhizopus oryzae* from sugarcane bagasse: solid state fermentation in a tray bioreactor. Int J Food Sci Technol 2013; 48(2): 283-9.
       [http://dx.doi.org/10.1111/j.1365-2621.2012.03185.x]

[21]   Marques NP, de Cassia Pereira J, Gomes E, *et al.* Cellulases and xylanases production by endophytic fungi by solid state fermentation using lignocellulosic substrates and enzymatic saccharification of pretreated sugarcane bagasse. Ind Crops Prod 2018; 122: 66-75.
       [http://dx.doi.org/10.1016/j.indcrop.2018.05.022]

[22]   Camassola M, Dillon AJP. Production of cellulases and hemicellulases by *Penicillium echinulatum* grown on pretreated sugar cane bagasse and wheat bran in solid-state fermentation. J Appl Microbiol 2007; 103(6): 2196-204.
       [http://dx.doi.org/10.1111/j.1365-2672.2007.03458.x] [PMID: 18045402]

[23]   Cunha FM, Esperança MN, Zangirolami TC, Badino AC, Farinas CS. Sequential solid-state and submerged cultivation of *Aspergillus niger* on sugarcane bagasse for the production of cellulase. Bioresour Technol 2012; 112: 270-4.
       [http://dx.doi.org/10.1016/j.biortech.2012.02.082] [PMID: 22409979]

[24]   Paiva JMF, Frollini E. Sugarcane bagasse reinforced phenolic and lignophenolic composites. J Appl

Polym Sci 2002; 83(4): 880-8.
[http://dx.doi.org/10.1002/app.10085]

[25] da Silva CG, Grelier S, Pichavant F, Frollini E, Castellan A. Adding value to lignins isolated from sugarcane bagasse and Miscanthus. Ind Crops Prod 2013; 42: 87-95.
[http://dx.doi.org/10.1016/j.indcrop.2012.04.040]

[26] Felipe MGA, Vitolo M, Mancilha IM, Silva SS. Environmental parameters affecting xylitol production from sugar cane bagasse hemicellulosic hydrolyzate by *Candida guilliermondii.* J Ind Microbiol Biotechnol 1997; 18(4): 251-4.
[http://dx.doi.org/10.1038/sj.jim.2900374]

[27] Rao RS, Jyothi CP, Prakasham RS, Sarma PN, Rao LV. Xylitol production from corn fiber and sugarcane bagasse hydrolysates by *Candida tropicalis.* Bioresour Technol 2006; 97(15): 1974-8.
[http://dx.doi.org/10.1016/j.biortech.2005.08.015] [PMID: 16242318]

[28] Antunes FAF, Thomé LC, Santos JC, *et al.* Multi-scale study of the integrated use of the carbohydrate fractions of sugarcane bagasse for ethanol and xylitol production. Renew Energy 2021; 163: 1343-55.
[http://dx.doi.org/10.1016/j.renene.2020.08.020]

[29] Kulkarni SO, Kanekar PP, Jog JP, Sarnaik SS, Nilegaonkar SS. Production of copolymer, poly (hydroxybutyrate-co-hydroxyvalerate) by *Halomonas campisalis* MCM B-1027 using agro-wastes. Int J Biol Macromol 2015; 72: 784-9.
[http://dx.doi.org/10.1016/j.ijbiomac.2014.09.028] [PMID: 25277119]

[30] Getachew A, Woldesenbet F. Production of biodegradable plastic by polyhydroxybutyrate (PHB) accumulating bacteria using low cost agricultural waste material. BMC Res Notes 2016; 9(1): 509.
[http://dx.doi.org/10.1186/s13104-016-2321-y] [PMID: 27955705]

[31] Karp SG, Faraco V, Amore A, *et al.* Characterization of laccase isoforms produced by *Pleurotus ostreatus* in solid state fermentation of sugarcane bagasse. Bioresour Technol 2012; 114: 735-9.
[http://dx.doi.org/10.1016/j.biortech.2012.03.058] [PMID: 22487128]

[32] Ferreira FL, Dall'Antonia CB, Shiga EA, Alvim LJ, Pessoni RAB. Sugarcane bagasse as a source of carbon for enzyme production by filamentous fungi1. Hoehnea 2018; 45(1): 134-42.
[http://dx.doi.org/10.1590/2236-8906-40/2017]

[33] Schneider WDH, Gonçalves TA, Uchima CA, *et al.* Comparison of the production of enzymes to cell wall hydrolysis using different carbon sources by *Penicillium echinulatum* strains and its hydrolysis potential for lignocelullosic biomass. Process Biochem 2018; 66: 162-70.
[http://dx.doi.org/10.1016/j.procbio.2017.11.004]

[34] Namnuch N, Thammasittirong A, Thammasittirong SNR. Lignocellulose hydrolytic enzymes production by *Aspergillus flavus* KUB2 using submerged fermentation of sugarcane bagasse waste. Mycology 2021; 12(2): 119-27.
[http://dx.doi.org/10.1080/21501203.2020.1806938] [PMID: 34026303]

[35] Tavares MF, Avelino KV, Araújo NL, *et al.* Decolorization of azo and anthraquinone dyes by crude laccase produced by *Lentinus crinitus* in solid state cultivation. Braz J Microbiol 2020; 51(1): 99-106.
[http://dx.doi.org/10.1007/s42770-019-00189-w] [PMID: 31776865]

[36] Kumar D, Jain VK, Shanker G, Srivastava A. Citric acid production by solid state fermentation using sugarcane bagasse. Process Biochem 2003; 38(12): 1731-8.
[http://dx.doi.org/10.1016/S0032-9592(02)00252-2]

[37] John RP, Nampoothiri KM, Pandey A. Solid-state fermentation for l-lactic acid production from agro wastes using *Lactobacillus delbrueckii.* Process Biochem 2006; 41(4): 759-63.
[http://dx.doi.org/10.1016/j.procbio.2005.09.013]

[38] Rosés RP, Guerra NP. Optimization of amylase production by *Aspergillus niger* in solid-state fermentation using sugarcane bagasse as solid support material. World J Microbiol Biotechnol 2009; 25(11): 1929-39.

[http://dx.doi.org/10.1007/s11274-009-0091-6]

[39]   Slivinski CT, Mallmann E, de Araújo JM, Mitchell DA, Krieger N. Production of surfactin by *Bacillus pumilus* UFPEDA 448 in solid-state fermentation using a medium based on okara with sugarcane bagasse as a bulking agent. Process Biochem 2012; 47(12): 1848-55.
       [http://dx.doi.org/10.1016/j.procbio.2012.06.014]

[40]   Dong XQ, Yang JS, Zhu N, Wang ET, Yuan HL. Sugarcane bagasse degradation and characterization of three white-rot fungi. Bioresour Technol 2013; 131: 443-51.
       [http://dx.doi.org/10.1016/j.biortech.2012.12.182] [PMID: 23376835]

[41]   Veana F, Martínez-Hernández JL, Aguilar CN, Rodríguez-Herrera R, Michelena G. Utilization of molasses and sugar cane bagasse for production of fungal invertase in solid state fermentation using *Aspergillus niger* GH1. Braz J Microbiol 2014; 45(2): 373-7.
       [http://dx.doi.org/10.1590/S1517-83822014000200002] [PMID: 25242918]

[42]   Biz A, Finkler ATJ, Pitol LO, Medina BS, Krieger N, Mitchell DA. Production of pectinases by solid-state fermentation of a mixture of citrus waste and sugarcane bagasse in a pilot-scale packed-bed bioreactor. Biochem Eng J 2016; 111: 54-62.
       [http://dx.doi.org/10.1016/j.bej.2016.03.007]

[43]   Zhou X, Xu Y. Integrative process for sugarcane bagasse biorefinery to co-produce xylooligosaccharides and gluconic acid. Bioresour Technol 2019; 282: 81-7.
       [http://dx.doi.org/10.1016/j.biortech.2019.02.129] [PMID: 30852335]

[44]   Li H, Chen X, Xiong L, *et al.* Stepwise enzymatic hydrolysis of alkaline oxidation treated sugarcane bagasse for the co-production of functional xylo-oligosaccharides and fermentable sugars. Bioresour Technol 2019; 275: 345-51.
       [http://dx.doi.org/10.1016/j.biortech.2018.12.063] [PMID: 30597396]

[45]   de Oliveira Júnior SD, Asevedo EA, de Araújo JS, *et al.* Enzymatic extract of Aspergillus fumigatus CCT 7873 for hydrolysis of sugarcane bagasse and generation of cellulose nanocrystals (CNC). Biomass Conver Bioref 2020.
       [http://dx.doi.org/10.1007/s13399-020-01020-5]

[46]   González-Bautista E, Alarcón-Gutierrez E, Dupuy N, Gaime-Perraud I, Ziarelli F, Farnet-da-Silva A-M. Influence of yeast extract enrichment and *Pycnoporus sanguineus* inoculum on the dephenolisation of sugar-cane bagasse for production of second-generation ethanol. Fuel 2020; 260116370.
       [http://dx.doi.org/10.1016/j.fuel.2019.116370]

[47]   Prajapati BP, Jana UK, Suryawanshi RK, Kango N. Sugarcane bagasse saccharification using *Aspergillus tubingensis* enzymatic cocktail for 2G bio-ethanol production. Renew Energy 2020; 152: 653-63.
       [http://dx.doi.org/10.1016/j.renene.2020.01.063]

[48]   Pereira Scarpa JC, Paganini Marques N, Alves Monteiro D, *et al.* Saccharification of pretreated sugarcane bagasse using enzymes solution from *Pycnoporus sanguineus* MCA 16 and cellulosic ethanol production. Ind Crops Prod 2019; 141111795.
       [http://dx.doi.org/10.1016/j.indcrop.2019.111795]

[49]   Raj A, Devendra LP, Sukumaran RK. Comparative evaluation of laccase mediated oxidized and unoxidized lignin of sugarcane bagasse for the synthesis of lignin-based formaldehyde resin. Ind Crops Prod 2020; 150112385.
       [http://dx.doi.org/10.1016/j.indcrop.2020.112385]

[50]   Martínez O, Sánchez A, Font X, Barrena R. Valorization of sugarcane bagasse and sugar beet molasses using *Kluyveromyces marxianus* for producing value-added aroma compounds via solid-state fermentation. J Clean Prod 2017; 158: 8-17.
       [http://dx.doi.org/10.1016/j.jclepro.2017.04.155]

[51]   Roy P, Kaur I, Singh S, *et al.* Beneficial microbes as alternative food flavour ingredients for achieving sustainability.Microbial Biotechnology: Basic Research and Applications. 79-90.

[52] Lu P, Yang Y, Liu R, *et al.* Preparation of sugarcane bagasse nanocellulose hydrogel as a colourimetric freshness indicator for intelligent food packaging. Carbohydr Polym 2020; 249116831.
[http://dx.doi.org/10.1016/j.carbpol.2020.116831] [PMID: 32933676]

[53] Mthembu LD, Lokhat D, Gupta R, *et al.* Optimization of Levulinic acid production from depithed sugarcane bagasse in 1- Ethyl-3-methylimidazolium hydrogen sulfate. Waste Biomass Valor 2020; pp. 1-13.

[54] Detraksa J. Evaluation of plant growth–promoting *Streptomyces* sp. SR13−2 immobilized with sugarcane bagasse and filter cake for promoting rice growth. Food Appl Biosci J 2020; 8: 1-13.

[55] Kumaraguru K, Rengasamy M, Kumar E, *et al.* Factors affecting printing quality of paper from bagasse pulp. Int J Chemtech Res 2014; 6(5): 2763-7.

[56] Mamaye M, Kiflie Z, Feleke S, *et al.* Valorization of Ethiopian Sugarcane Bagasse to Assess its Suitability for Pulp and Paper Production Sugar tech 2019. Accessed: Oct. 01, 2020

[57] Yi Y, Tu G, Zhao D, Tsang PE, Fang Z. Key role of FeO in the reduction of Cr(VI) by magnetic biochar synthesised using steel pickling waste liquor and sugarcane bagasse. J Clean Prod 2020; 245118886.
[http://dx.doi.org/10.1016/j.jclepro.2019.118886]

[58] Biswas S, Mohapatra SS, Kumari U, Meikap BC, Sen TK. Batch and continuous closed circuit semi-fluidized bed operation: Removal of MB dye using sugarcane bagasse biochar and alginate composite adsorbents. J Environ Chem Eng 2020; 8(1)103637.
[http://dx.doi.org/10.1016/j.jece.2019.103637]

[59] Khan AZ, Khan S, Ayaz T, *et al.* Popular wood and sugarcane bagasse biochars reduced uptake of chromium and lead by lettuce from mine-contaminated soil. Environ Pollut 2020; 263(Pt A)114446.
[http://dx.doi.org/10.1016/j.envpol.2020.114446] [PMID: 32283452]

[60] Tang YH, Liu SH, Tsang DCW. Microwave-assisted production of CO_2-activated biochar from sugarcane bagasse for electrochemical desalination. J Hazard Mater 2020; 383121192.
[http://dx.doi.org/10.1016/j.jhazmat.2019.121192] [PMID: 31539661]

[61] Martín C, Galbe M, Wahlbom CF, Hahn-Hägerdal B, Jönsson LJ. Ethanol production from enzymatic hydrolysates of sugarcane bagasse using recombinant xylose-utilising *Saccharomyces cerevisiae.* Enzyme Microb Technol 2002; 31(3): 274-82.
[http://dx.doi.org/10.1016/S0141-0229(02)00112-6]

[62] Singh A, Sharma P, Saran AK, Singh N, Bishnoi NR. Comparative study on ethanol production from pretreated sugarcane bagasse using immobilized *Saccharomyces cerevisiae* on various matrices. Renew Energy 2013; 50: 488-93.
[http://dx.doi.org/10.1016/j.renene.2012.07.003]

[63] da Silveira dos Santos D, Camelo AC, Rodrigues KCP, Carlos LC, Pereira N Jr. Ethanol production from sugarcane bagasse by *Zymomonas mobilis* using simultaneous saccharification and fermentation (SSF) process. Appl Biochem Biotechnol 2010; 161(1-8): 93-105.
[http://dx.doi.org/10.1007/s12010-009-8810-x] [PMID: 19876607]

[64] Rattanapoltee P, Kaewkannetra P. Utilization of agricultural residues of pineapple peels and sugarcane bagasse as cost-saving raw materials in *Scenedesmus acutus* for lipid accumulation and biodiesel production. Appl Biochem Biotechnol 2014; 173(6): 1495-510.
[http://dx.doi.org/10.1007/s12010-014-0949-4] [PMID: 24817554]

[65] Arora N, Patel A, Pruthi PA, Pruthi V. Boosting TAG Accumulation with improved biodiesel production from novel *Oleaginous* microalgae *Scenedesmus* sp. IITRIND2 utilizing waste sugarcane bagasse aqueous extract (SBAE). Appl Biochem Biotechnol 2016; 180(1): 109-21.
[http://dx.doi.org/10.1007/s12010-016-2086-8] [PMID: 27093970]

[66] Pang Z-W, Lu W, Zhang H, *et al.* Butanol production employing fed-batch fermentation by *Clostridium acetobutylicum* GX01 using alkali-pretreated sugarcane bagasse hydrolysed by enzymes

from *Thermoascus aurantiacus* QS 7-2-4. Bioresour Technol 2016; 212: 82-91.
[http://dx.doi.org/10.1016/j.biortech.2016.04.013] [PMID: 27089425]

[67]   Zhao X, Song Y, Liu D. Enzymatic hydrolysis and simultaneous saccharification and fermentation of alkali/peracetic acid-pretreated sugarcane bagasse for ethanol and 2,3-butanediol production. Enzyme Microb Technol 2011; 49(4): 413-9.
[http://dx.doi.org/10.1016/j.enzmictec.2011.07.003] [PMID: 22112569]

[68]   de Sá LRV, Faber MO, da Silva ASA, Cammarota MC, Ferreira-Leitão VS. Biohydrogen production using xylose or xylooligosaccharides derived from sugarcane bagasse obtained by hydrothermal and acid pretreatments. Renew Energy 2020; 146: 2408-15.
[http://dx.doi.org/10.1016/j.renene.2019.08.089]

[69]   Ratti RP, Delforno TP, Sakamoto IK, Varesche MBA. Thermophilic hydrogen production from sugarcane bagasse pretreated by steam explosion and alkaline delignification. Int J Hydrogen Energy 2015; 40(19): 6296-306.
[http://dx.doi.org/10.1016/j.ijhydene.2015.03.067]

[70]   Molavian M, Ghorbani GR, Rafiee H, Beauchemin KA. Substitution of wheat straw with sugarcane bagasse in low-forage diets fed to mid-lactation dairy cows: Milk production, digestibility, and chewing behavior. J Dairy Sci 2020; 103(9): 8034-47.
[http://dx.doi.org/10.3168/jds.2020-18499] [PMID: 32684450]

[71]   Zadrazil F, Puniya AK. Studies on the effect of particle size on solid-state fermentation of sugarcane bagasse into animal feed using white-rot fungi. Bioresour Technol 1995; 54(1): 85-7.
[http://dx.doi.org/10.1016/0960-8524(95)00119-0]

[72]   So S, Cherdthong A, Wanapat M, *et al.* Fermented sugarcane bagasse with *Lactobacillus* combined with cellulase and molasses promotes *in vitro* gas kinetics, degradability, and ruminal fermentation patterns compared to rice straw. Anim Biotechnol 2020; 0: 1-12.
[PMID: 32567474]

[73]   Poletto P, Pereira GN, Monteiro CRM, Pereira MAF, Bordignon SE, de Oliveira D. Xylooligosaccharides: Transforming the lignocellulosic biomasses into valuable 5-carbon sugar prebiotics. Process Biochem 2020; 91: 352-63.
[http://dx.doi.org/10.1016/j.procbio.2020.01.005]

[74]   Li H, Chen X, Xiong L, *et al.* Production, separation, and characterization of high-purity xylobiose from enzymatic hydrolysis of alkaline oxidation pretreated sugarcane bagasse. Bioresour Technol 2020; 299122625.
[http://dx.doi.org/10.1016/j.biortech.2019.122625] [PMID: 31881437]

[75]   Ávila PF, Martins M, Goldbeck R. Enzymatic Production of Xylooligosaccharides from alkali-solubilized arabinoxylan from sugarcane straw and coffee husk. Bioenerg Res 2020; 3 Epub ahead of print.
[http://dx.doi.org/10.1007/s12155-020-10188-7]

[76]   Bian J, Peng P, Peng F, Xiao X, Xu F, Sun RC. Microwave-assisted acid hydrolysis to produce xylooligosaccharides from sugarcane bagasse hemicelluloses. Food Chem 2014; 156: 7-13.
[http://dx.doi.org/10.1016/j.foodchem.2014.01.112] [PMID: 24629931]

[77]   Singh G, Kumari A, Mittal A, Yadav A, Aggarwal NK. Poly β-hydroxybutyrate production by *Bacillus subtilis* NG220 using sugar industry waste water. BioMed Res Int 2013; 2013: 1-10.
[http://dx.doi.org/10.1155/2013/952641] [PMID: 24027767]

[78]   Sakthiselvan P, Madhumathi R. Kinetic evaluation on cell growth and biosynthesis of polyhydroxybutyrate (PHB) by *Bacillus safensis* EBT1 from sugarcane bagasse. Eng Agric Environ Food 2018; 11(3): 145-52.
[http://dx.doi.org/10.1016/j.eaef.2018.03.003]

[79]   Singhaboot P, Kaewkannetra P. A higher in value biopolymer product of polyhydroxyalkanoates (PHAs) synthesized by *Alcaligenes latus* in batch/repeated batch fermentation processes of sugar cane

juice. Ann Microbiol 2015; 65(4): 2081-9.
[http://dx.doi.org/10.1007/s13213-015-1046-9]

[80]   Abdeshahian P, Ascencio JJ, Philippini RR, Antunes FAF, dos Santos JC, da Silva SS. Utilization of sugarcane straw for production of β-glucan biopolymer by *Lasiodiplodia theobromae* CCT 3966 in batch fermentation process. Bioresour Technol 2020; 314123716.
[http://dx.doi.org/10.1016/j.biortech.2020.123716] [PMID: 32650262]

[81]   Paiva JMF, Frollini E. Sugarcane bagasse reinforced phenolic and lignophenolic composites. J Appl Polym Sci 2002; 83(4): 880-8.
[http://dx.doi.org/10.1002/app.10085]

[82]   Kocabas A, Ogel ZB, Bakir U. Xylanase and itaconic acid production by *Aspergillus terreus* NRRL 1960 within a biorefinery concept. Ann Microbiol 2014; 64(1): 75-84.
[http://dx.doi.org/10.1007/s13213-013-0634-9]

[83]   R P, S K, K S. Bioprocessing of sugarcane factory waste to production of Itaconic acid. Afr J Microbiol Res 2014; 8(16): 1672-5.
[http://dx.doi.org/10.5897/AJMR2013.6554]

[84]   Nieder-Heitmann M, Haigh KF, Görgens JF. Process design and economic analysis of a biorefinery co-producing itaconic acid and electricity from sugarcane bagasse and trash lignocelluloses. Bioresour Technol 2018; 262: 159-68.
[http://dx.doi.org/10.1016/j.biortech.2018.04.075] [PMID: 29704763]

[85]   Borges ER, Pereira N Jr. Succinic acid production from sugarcane bagasse hemicellulose hydrolysate by *Actinobacillus succinogenes.* J Ind Microbiol Biotechnol 2011; 38(8): 1001-11.
[http://dx.doi.org/10.1007/s10295-010-0874-7] [PMID: 20882312]

[86]   Xi Y, Dai W, Xu R, *et al.* Ultrasonic pretreatment and acid hydrolysis of sugarcane bagasse for succinic acid production using *Actinobacillus succinogenes.* Bioprocess Biosyst Eng 2013; 36(11): 1779-85.
[http://dx.doi.org/10.1007/s00449-013-0953-z] [PMID: 23649828]

[87]   Ong KL, Li C, Li X, Zhang Y, Xu J, Lin CSK. Co-fermentation of glucose and xylose from sugarcane bagasse into succinic acid by *Yarrowia lipolytica.* Biochem Eng J 2019; 148: 108-15.
[http://dx.doi.org/10.1016/j.bej.2019.05.004]

[88]   Campanhol BS, Silveira GC, Castro MC, Ceccato-Antonini SR, Bastos RG. Effect of the nutrient solution in the microbial production of citric acid from sugarcane bagasse and vinasse. Biocatal Agric Biotechnol 2019; 19101147.
[http://dx.doi.org/10.1016/j.bcab.2019.101147]

[89]   Calabia BP, Tokiwa Y. Production of d-lactic acid from sugarcane molasses, sugarcane juice and sugar beet juice by *Lactobacillus delbrueckii.* Biotechnol Lett 2007; 29(9): 1329-32.
[http://dx.doi.org/10.1007/s10529-007-9408-4] [PMID: 17541505]

[90]   Wei D, Liu X, Yang ST. Butyric acid production from sugarcane bagasse hydrolysate by *Clostridium tyrobutyricum* immobilized in a fibrous-bed bioreactor. Bioresour Technol 2013; 129: 553-60.
[http://dx.doi.org/10.1016/j.biortech.2012.11.065] [PMID: 23270719]

[91]   Zhu L, Wei P, Cai J, *et al.* Improving the productivity of propionic acid with FBB-immobilized cells of an adapted acid-tolerant *Propionibacterium acidipropionici.* Bioresour Technol 2012; 112: 248-53.
[http://dx.doi.org/10.1016/j.biortech.2012.01.055] [PMID: 22406066]

[92]   Chen F, Feng X, Xu H, Zhang D, Ouyang P. Propionic acid production in a plant fibrous-bed bioreactor with immobilized *Propionibacterium freudenreichii* CCTCC M207015. J Biotechnol 2013; 164(2): 202-10.
[http://dx.doi.org/10.1016/j.jbiotec.2012.08.025] [PMID: 22982366]

[93]   Hoornweg D, Bhada-Tata P. What a Waste : A Global Review of Solid Waste Management https://openknowledge.worldbank.org/handle/10986/173882012.

[94] Patel H. Environmental valorisation of bagasse fly ash: a review. RSC Advances 2020; 10(52): 31611-21.
[http://dx.doi.org/10.1039/D0RA06422J] [PMID: 35520640]

[95] Srivastava VC, Mall ID, Mishra IM. Treatment of pulp and paper mill wastewaters with poly aluminium chloride and bagasse fly ash. Colloids Surf A Physicochem Eng Asp 2005; 260(1-3): 17-28.
[http://dx.doi.org/10.1016/j.colsurfa.2005.02.027]

[96] Mall ID, Srivastava VC, Agarwal NK, Mishra IM. Adsorptive removal of malachite green dye from aqueous solution by bagasse fly ash and activated carbon-kinetic study and equilibrium isotherm analyses. Colloids Surf A Physicochem Eng Asp 2005; 264(1-3): 17-28.
[http://dx.doi.org/10.1016/j.colsurfa.2005.03.027]

[97] Jiménez-Quero VG, León-Martínez FM, Montes-García P, Gaona-Tiburcio C, Chacón-Nava JG. Influence of sugar-cane bagasse ash and fly ash on the rheological behavior of cement pastes and mortars. Constr Build Mater 2013; 40: 691-701.
[http://dx.doi.org/10.1016/j.conbuildmat.2012.11.023]

[98] Yang SD, Liu JX, Wu J, Tan H-W, Li Y-R. Effects of vinasse and press mud application on the biological properties of soils and productivity of sugarcane. Sugar Tech 2013; 15(2): 152-8.
[http://dx.doi.org/10.1007/s12355-012-0200-y]

[99] Jauhri KS. Modified sugarcane pressmud: a potential carrier for commercial production of bacterial inoculants. Indian J Agric Res 1990; 24: 189-97.

[100] Rondina DJG, Ymbong DV, Cadutdut MJM, *et al.* Utilization of a novel activated carbon adsorbent from press mud of sugarcane industry for the optimized removal of methyl orange dye in aqueous solution. Appl Water Sci 2019; 9(8): 181.
[http://dx.doi.org/10.1007/s13201-019-1063-0]

[101] Shankaranand VS, Lonsane BK. Sugarcane-pressmud as a novel substrate for production of citric acid by solid-state fermentation. World J Microbiol Biotechnol 1993; 9(3): 377-80.
[http://dx.doi.org/10.1007/BF00383084] [PMID: 24420047]

[102] Kumar R, Saha SK, Mendiratta SK. Effect of feeding sugarcane press mud on carcass traits and meat quality characteristics of lambs. Vet World 2015; 8(6): 793-7.
[http://dx.doi.org/10.14202/vetworld.2015.793-797] [PMID: 27065649]

[103] López González LM, Pereda Reyes I, Romero Romero O. Anaerobic co-digestion of sugarcane press mud with vinasse on methane yield. Waste Manag 2017; 68: 139-45.
[http://dx.doi.org/10.1016/j.wasman.2017.07.016] [PMID: 28733111]

[104] Nimbalkar PR, Khedkar MA, Gaikwad SG, Chavan PV, Bankar SB. New Insight into sugarcane industry waste utilization (press mud) for cleaner biobutanol production by using *C. acetobutylicum* NRRL B-527. Appl Biochem Biotechnol 2017; 183(3): 1008-25.
[http://dx.doi.org/10.1007/s12010-017-2479-3] [PMID: 28474218]

[105] Sanchez N, Ruiz R, Plazas A, Vasquez J, Cobo M. Effect of pretreatment on the ethanol and fusel alcohol production during fermentation of sugarcane press-mud. Biochem Eng J 2020; 161107668.
[http://dx.doi.org/10.1016/j.bej.2020.107668]

[106] Dalsasso RR, Pavan FA, Bordignon SE, Aragão GMF, Poletto P. Polyhydroxybutyrate (PHB) production by *Cupriavidus necator* from sugarcane vinasse and molasses as mixed substrate. Process Biochem 2019; 85: 12-8.
[http://dx.doi.org/10.1016/j.procbio.2019.07.007]

[107] Sen KY, Hussin MH, Baidurah S. Biosynthesis of poly(3-hydroxybutyrate) (PHB) by *Cupriavidus necator* from various pretreated molasses as carbon source. Biocatal Agric Biotechnol 2019; 17: 51-9.
[http://dx.doi.org/10.1016/j.bcab.2018.11.006]

[108] Zakeri A, Pazouki M, Vossoughi M. Use of response surface methodology analysis for xanthan

biopolymer production by *Xanthomonas campestris*: focus on agitation rate, carbon source, and temperature. Iran J Chem Chem Eng 2017; 36: 173-83. [IJCCE].

[109] Huang G, Chen X, Wang C, *et al.* Photoluminescent carbon dots derived from sugarcane molasses: synthesis, properties, and applications. RSC Advances 2017; 7(75): 47840-7.
[http://dx.doi.org/10.1039/C7RA09002A]

[110] Sreńscek-Nazzal J, Kamińska W, Michalkiewicz B, Koren ZC. Production, characterization and methane storage potential of KOH-activated carbon from sugarcane molasses. Ind Crops Prod 2013; 47: 153-9.
[http://dx.doi.org/10.1016/j.indcrop.2013.03.004]

[111] Bento HBS, Carvalho AKF, Reis CER, De Castro HF. Single cell oil production and modification for fuel and food applications: Assessing the potential of sugarcane molasses as culture medium for filamentous fungus. Ind Crops Prod 2020; 145112141.
[http://dx.doi.org/10.1016/j.indcrop.2020.112141]

[112] Takahashi M, Morita T, Wada K, *et al.* Production of sophorolipid glycolipid biosurfactants from sugarcane molasses using *Starmerella bombicola* NBRC 10243. J Oleo Sci 2011; 60(5): 267-73.
[http://dx.doi.org/10.5650/jos.60.267] [PMID: 21502725]

[113] Pan NC, Pereira HCB, da Silva MLC, Vasconcelos AFD, Celligoi MAPC. Improvement production of hyaluronic acid by *Streptococcus zooepidemicus* in sugarcane molasses. Appl Biochem Biotechnol 2017; 182(1): 276-93.
[http://dx.doi.org/10.1007/s12010-016-2326-y] [PMID: 27900664]

[114] Ai H, Liu M, Yu P, *et al.* Improved welan gum production by *Alcaligenes* sp. ATCC31555 from pretreated cane molasses. Carbohydr Polym 2015; 129: 35-43.
[http://dx.doi.org/10.1016/j.carbpol.2015.04.033] [PMID: 26050885]

[115] Ruiz SP, Martinez CO, Noce AS, Sampaio AR, Baesso ML, Matioli G. Biosynthesis of succinoglycan by *Agrobacterium radiobacter* NBRC 12665 immobilized on loofa sponge and cultivated in sugar cane molasses. Structural and rheological characterization of biopolymer. J Mol Catal, B Enzym 2015; 122: 15-28.
[http://dx.doi.org/10.1016/j.molcatb.2015.08.016]

[116] de Oliveira MR, da Silva RSSF, Buzato JB, Celligoi MAPC. Study of levan production by *Zymomonas mobilis* using regional low-cost carbohydrate sources. Biochem Eng J 2007; 37(2): 177-83.
[http://dx.doi.org/10.1016/j.bej.2007.04.009]

[117] Treichel H, Mazutti MA, Filho FM, Rodrigues MI. Technical viability of the production, partial purification and characterisation of inulinase using pretreated agroindustrial residues. Bioprocess Biosyst Eng 2009; 32(4): 425-33.
[http://dx.doi.org/10.1007/s00449-008-0262-0] [PMID: 18820951]

[118] Sguarezi C, Longo C, Ceni G, *et al.* Inulinase production by agro-industrial residues: optimization of pretreatment of substrates and production medium. Food Bioprocess Technol 2009; 2(4): 409-14.
[http://dx.doi.org/10.1007/s11947-007-0042-x]

[119] De Azeredo LAI, Castilho LR, Leite SGF, Coelho RRR, Freire DMG. Protease production by Streptomyces sp. isolated from Brazilian Cerrado soil: optimization of culture medium employing statistical experimental design. Appl Biochem Biotechnol 2003; 108(1-3): 749-56.
[http://dx.doi.org/10.1385/ABAB:108:1-3:749] [PMID: 12721412]

[120] Sheoran A, Yadav BS, Nigam P, Singh D. Continuous ethanol production from sugarcane molasses using a column reactor of immobilized*Saccharomyces cerevisiae* HAU-1. J Basic Microbiol 1998; 38(2): 123-8.
[http://dx.doi.org/10.1002/(SICI)1521-4028(199805)38:2<123::AID-JOBM123>3.0.CO;2-9] [PMID: 9637012]

[121] Cazetta ML, Celligoi MAPC, Buzato JB, Scarmino IS. Fermentation of molasses by *Zymomonas*

mobilis: Effects of temperature and sugar concentration on ethanol production. Bioresour Technol 2007; 98(15): 2824-8.
[http://dx.doi.org/10.1016/j.biortech.2006.08.026] [PMID: 17420121]

[122]   Maiti B, Rathore A, Srivastava S, Shekhawat M, Srivastava P. Optimization of process parameters for ethanol production from sugar cane molasses by *Zymomonas mobilis* using response surface methodology and genetic algorithm. Appl Microbiol Biotechnol 2011; 90(1): 385-95.
[http://dx.doi.org/10.1007/s00253-011-3158-x] [PMID: 21336926]

[123]   Inaba T, Watanabe D, Yoshiyama Y, *et al.* An organic acid-tolerant HAA1-overexpression mutant of an industrial bioethanol strain of *Saccharomyces cerevisiae* and its application to the production of bioethanol from sugarcane molasses. AMB Express 2013; 3(1): 74.
[http://dx.doi.org/10.1186/2191-0855-3-74] [PMID: 24373204]

[124]   Gabisa EW, Bessou C, Gheewala SH. Life cycle environmental performance and energy balance of ethanol production based on sugarcane molasses in Ethiopia. J Clean Prod 2019; 234: 43-53.
[http://dx.doi.org/10.1016/j.jclepro.2019.06.199]

[125]   Deshmukh AN, Nipanikar-Gokhale P, Jain R. Engineering of *Bacillus subtilis* for the production of 2,3-Butanediol from sugarcane molasses. Appl Biochem Biotechnol 2016; 179(2): 321-31.
[http://dx.doi.org/10.1007/s12010-016-1996-9] [PMID: 26825987]

[126]   Chan S, Kanchanatawee S, Jantama K. Production of succinic acid from sucrose and sugarcane molasses by metabolically engineered *Escherichia coli*. Bioresour Technol 2012; 103(1): 329-36.
[http://dx.doi.org/10.1016/j.biortech.2011.09.096] [PMID: 22023966]

[127]   Sharma A, Vivekanand V, Singh RP. Solid-state fermentation for gluconic acid production from sugarcane molasses by *Aspergillus niger* ARNU-4 employing tea waste as the novel solid support. Bioresour Technol 2008; 99(9): 3444-50.
[http://dx.doi.org/10.1016/j.biortech.2007.08.006] [PMID: 17881224]

[128]   Feng J, Yang J, Yang W, Chen J, Jiang M, Zou X. Metabolome- and genome-scale model analyses for engineering of *Aureobasidium pullulans* to enhance polymalic acid and malic acid production from sugarcane molasses. Biotechnol Biofuels 2018; 11(1): 94.
[http://dx.doi.org/10.1186/s13068-018-1099-7] [PMID: 29632554]

[129]   Dorta C, Cruz R, de Oliva-Neto P, Moura DJC. Sugarcane molasses and yeast powder used in the Fructooligosaccharides production by *Aspergillus japonicus*-FCL 119T and *Aspergillus niger* ATCC 20611. J Ind Microbiol Biotechnol 2006; 33(12): 1003-9.
[http://dx.doi.org/10.1007/s10295-006-0152-x] [PMID: 16835781]

[130]   Xie Y, Zhou H, Liu C, *et al.* A molasses habitat-derived fungus *Aspergillus tubingensis* XG21 with high β-fructofuranosidase activity and its potential use for fructooligosaccharides production. AMB Express 2017; 7(1): 128.
[http://dx.doi.org/10.1186/s13568-017-0428-8] [PMID: 28641403]

[131]   Boviatsi E, Papadaki A, Efthymiou MN, *et al.* Valorisation of sugarcane molasses for the production of microbial lipids via fermentation of two *Rhodosporidium* strains for enzymatic synthesis of polyol esters. J Chem Technol Biotechnol 2020; 95(2): 402-7.
[http://dx.doi.org/10.1002/jctb.5985]

[132]   Do Valle JS, Vandenberghe LP de S, Santana TT, *et al.* Optimization of *Agaricus blazei* laccase production by submerged cultivation with sugarcane molasses. Am J Ment Retard 2014; 8: 939-46.

[133]   Marim RA, Oliveira ACC, Marquezoni RS, *et al.* Use of sugarcane molasses by *Pycnoporus sanguineus* for the production of laccase for dye decolorization. Genet Mol Res 2016; 15(4)gmr15048972.
[http://dx.doi.org/10.4238/gmr15048972] [PMID: 27813609]

[134]   Furlan SA, Schneider ALS, Merkle R, de Fátima Carvalho-Jonas M, Jonas R. Formulation of a lactose-free, low-cost culture medium for the production of β- D-galactosidase by *Kluyveromyces marxianus*. Biotechnol Lett 2000; 22(7): 589-93.

[http://dx.doi.org/10.1023/A:1005629127532]

[135] Schneider ALS, Merkle R, de Fátima Carvalho-Jonas M, Jonas R, Furlan S. Oxygen transfer on β- D-galactosidase production by *Kluyveromyces marxianus* using sugar cane molasses as carbon source. Biotechnol Lett 2001; 23(7): 547-50.
[http://dx.doi.org/10.1023/A:1010338904870]

[136] Neto DC, Buzato JB, Borsato D. L-asparaginase production by *Zymomonas mobilis* during molasses fermentation: optimization of culture conditions using factorial design. Acta Sci Technol 2006; 28: 151-3.

[137] Paterson-Beedle M, Kennedy JF, Melo FAD, Lloyd LL, Medeiros V. A cellulosic exopolysaccharide produced from sugarcane molasses by a *Zoogloea* sp. Carbohydr Polym 2000; 42(4): 375-83.
[http://dx.doi.org/10.1016/S0144-8617(99)00179-4]

[138] Ortiz ME, Fornaguera MJ, Raya RR, Mozzi F. *Lactobacillus reuteri* CRL 1101 highly produces mannitol from sugarcane molasses as carbon source. Appl Microbiol Biotechnol 2012; 95(4): 991-9.
[http://dx.doi.org/10.1007/s00253-012-3945-z] [PMID: 22350320]

[139] Bhosale P, Gadre RV. ? -Carotene production in sugarcane molasses by a Rhodotorula glutinis mutant. J Ind Microbiol Biotechnol 2001; 26(6): 327-32.
[http://dx.doi.org/10.1038/sj.jim.7000138] [PMID: 11571614]

[140] Fuess LT, Rodrigues IJ, Garcia ML. Fertirrigation with sugarcane vinasse: Foreseeing potential impacts on soil and water resources through vinasse characterization. J Environ Sci Health Part A Tox Hazard Subst Environ Eng 2017; 52(11): 1063-72.
[http://dx.doi.org/10.1080/10934529.2017.1338892] [PMID: 28737443]

[141] Bettani SR, de Oliveira Ragazzo G, Leal Santos N, *et al.* Sugarcane vinasse and microalgal biomass in the production of pectin particles as an alternative soil fertilizer. Carbohydr Polym 2019; 203: 322-30.
[http://dx.doi.org/10.1016/j.carbpol.2018.09.041] [PMID: 30318219]

[142] dos Reis KC, Coimbra JM, Duarte WF, Schwan RF, Silva CF. Biological treatment of vinasse with yeast and simultaneous production of single-cell protein for feed supplementation. Int J Environ Sci Technol 2019; 16(2): 763-74.
[http://dx.doi.org/10.1007/s13762-018-1709-8]

[143] Magrini FE, de Almeida GM, da Maia Soares D, *et al.* Effect of different heat treatments of inoculum on the production of hydrogen and volatile fatty acids by dark fermentation of sugarcane vinasse. Biomass Convers Bioref 2020; pp. 1-14.

[144] Iltchenco J, Almeida LG, Beal LL, *et al.* Microbial consortia composition on the production of methane from sugarcane vinasse. Biomass Convers Bioref 2019; pp. 1-11.

[145] Crivelaro SHR, Mariano AP, Furlan LT, Gonçalves RA, Seabra PN, Angelis DF. Evaluation of the use of vinasse as a biostimulation agent for the biodegradation of oily sludge in soil. Braz Arch Biol Technol 2010; 53(5): 1217-24.
[http://dx.doi.org/10.1590/S1516-89132010000500027]

[146] Marques SSI, Nascimento IA, de Almeida PF, Chinalia FA. Growth of *Chlorella vulgaris* on sugarcane vinasse: the effect of anaerobic digestion pretreatment. Appl Biochem Biotechnol 2013; 171(8): 1933-43.
[http://dx.doi.org/10.1007/s12010-013-0481-y] [PMID: 24013860]

[147] Tasic MB, Bonon A de J. Rocha Barbosa Schiavon MI, et al. Cultivation of Chlamydomonas reinhardtii in Anaerobically Digested Vinasse for Bioethanol Production. Waste Biomass Valor 2020; 1-9.

[148] Awasthi MK, Kumar V, Yadav V, *et al.* Current state of the art biotechnological strategies for conversion of watermelon wastes residues to biopolymers production: A review. Chemosphere 2022; 290133310.
[http://dx.doi.org/10.1016/j.chemosphere.2021.133310] [PMID: 34919909]

[149]   Liu H, Kumar V, Jia L, *et al.* Biopolymer poly-hydroxyalkanoates (PHA) production from apple industrial waste residues: A review. Chemosphere 2021; 284131427.
[http://dx.doi.org/10.1016/j.chemosphere.2021.131427] [PMID: 34323796]

[150]   Awasthi SK, Kumar M, Kumar V, *et al.* A comprehensive review on recent advancements in biodegradation and sustainable management of biopolymers. Environ Pollut 2022; 307119600.
[http://dx.doi.org/10.1016/j.envpol.2022.119600] [PMID: 35691442]

[151]   Duan Y, Tarafdar A, Kumar V, *et al.* Sustainable biorefinery approaches towards circular economy for conversion of biowaste to value added materials and future perspectives. Fuel 2022; 325124846.
[http://dx.doi.org/10.1016/j.fuel.2022.124846]

[152]   Kumar V, Sharma N, Umesh M, *et al.* Emerging challenges for the agro-industrial food waste utilization: A review on food waste biorefinery. Bioresour Technol 2022; 362127790.
[http://dx.doi.org/10.1016/j.biortech.2022.127790] [PMID: 35973569]

[153]   Awasthi SK, Sarsaiya S, Kumar V, *et al.* Processing of municipal solid waste resources for a circular economy in China: An overview. Fuel 2022; 317123478.
[http://dx.doi.org/10.1016/j.fuel.2022.123478]

[154]   Kumar V, Sharma N, Maitra SS. *In vitro* and *in vivo* toxicity assessment of nanoparticles. Int Nano Lett 2017; 7(4): 243-56.
[http://dx.doi.org/10.1007/s40089-017-0221-3]

[155]   Vinay K, Neha S, Maitra S. Protein and Peptide Nanoparticles: Preparation and Surface Modification, in Functionalized Nanomaterials I. CRC Press 2020; pp. 191-204.

[156]   Vallinayagam S, *et al.* Recent developments in magnetic nanoparticles and nano-composites for wastewater treatment. J Environ Chem Eng 2021; 9(6)106553.
[http://dx.doi.org/10.1016/j.jece.2021.106553]

[157]   Kumar V, Sharma N, Lakkaboyana SK, Maitra SS. Silver nanoparticles in poultry health: Applications and toxicokinetic effects, in Silver Nanomaterials for Agri-Food Applications. Elsevier 2021; pp. 685-704.
[http://dx.doi.org/10.1016/B978-0-12-823528-7.00005-6]

[158]   Egbosiuba T C. Biochar and bio-oil fuel properties from nickel nanoparticles assisted pyrolysis of cassava peel Heliyon, vol 8, no 8, p e10114 2022.
[http://dx.doi.org/10.1016/j.heliyon.2022.e10114]

[159]   Aguiar MM, Ferreira LFR, Monteiro RTR. Use of vinasse and sugarcane bagasse for the production of enzymes by lignocellulolytic fungi. Braz Arch Biol Technol 2010; 53(5): 1245-54.
[http://dx.doi.org/10.1590/S1516-89132010000500030]

[160]   Silvat D. Biodegradation of vinasse: fungal lignolytic enzymes and their application in the bioethanol industry. Fungal Enzy 2013; p. 65.

[161]   Bastos RG, Morais DV, Volpi MPC, *et al.* Influence of solid moisture and bed height on cultivation of *Aspergillus niger* from sugarcane bagasse with vinasse. Braz J Chem Eng 2015; 32: 377-84.
[http://dx.doi.org/10.1590/0104-6632.20150322s00003423]

SUBJECT INDEX

feruloyl esterase 47
genes 44
glucose-fructose oxidoreductase 132
hydrolytic 8, 278
immobilization methods 66
maleic isomerase 117
microbial pathway degradation 206
nitrogenase 95
oxidoreductase 132
Essential oil (EO) 115, 129, 161, 164, 165,
   166, 169, 170, 171, 174, 176

F

Factors 45, 46, 79, 81, 208, 217, 225, 249,
   250, 252, 254, 260, 266
   allosteric transcription 45
   aromatic-responsive transcription 46
Fatty acids 48, 69, 70, 80, 81, 95, 133, 134,
   149
   essential 80
Fermentable sugars 2, 13, 17, 64, 165, 210,
   282
Fermentation 9, 13, 14, 15, 19, 20, 37, 65, 71,
   115, 123, 133, 154, 166, 167, 168, 173,
   175, 177, 205, 216, 221, 264, 278, 280,
   283
   anaerobic 264, 283
   biofuel 205
   industry 133
   microbial 13, 14, 37, 71, 123, 173, 216, 221,
   280
   yeast-mediated 166
Fermentation processes 5, 10, 14, 16, 20, 173
   microbial 5, 16
Fermentative 2, 134
   microorganisms 2
   production 134
Fertilizers 63, 64, 81, 217, 220, 227, 253, 256,
   262, 281, 285
   commercial 256
Fibers 42, 66, 83, 274, 280
   coconut 83
Filters, coffee 62
Fluidized-bed gasification (FBG) 194
Fluorescence-activated cell sorting (FACS) 46
Food 62, 63, 69, 81, 131, 156, 164, 176, 227,
   280
   composition 63
   industries 62, 69, 81, 131, 164, 227, 280

nutritious 156, 176
Food waste 61, 62, 63, 64, 66, 67, 68, 69, 70,
   216, 217, 219, 220, 221, 222, 228, 229,
   230, 232, 234
   valorizing 216
Fruit wastes 112, 179
Fuel consumption 68
Fuel conversion 67, 189, 191, 194, 206
   oxygenates 67
   process 189
   technologies 191, 194
   renewable 206
Fumaric acid production 117, 118
Fungal 69, 115, 125
   autolysis 115
   hydrolysis of food waste 69
   metabolism 125

G

Galactose 16, 17, 65, 67, 68, 121, 127, 228
Gases 1, 12, 63, 69, 70, 96, 147, 155, 165,
   170, 188, 192, 195, 203, 254, 259, 264,
   265, 272
   greenhouse 63, 70, 165, 259, 272
   petroleum 69
   pyrolytic 265
   toxic 188
Gasification 96, 187, 191, 192, 194, 195, 197,
   203, 209, 265
   catalytic de-polymerization 191
   fluidized-bed 194
   system 192
Genes 17, 50
   transporting 17
   vanillin dehydrogenase 50
Genome 47, 49, 52
   editing tools 47, 52
   sequencing data 49
Global waste cleaning network 267
Glucoamylase 68
Gluconobacter metabolism 121
Gluconolactone 118, 132
Glucose 15, 16, 17, 18, 67, 68, 114, 116, 119,
   121, 123, 124, 127, 130, 132, 168, 177
   cellulosic 15
   metabolize 119
   oxidase 119
Glucose isomerase 65
   immobilizing enzyme 65

www.ingramcontent.com/pod-product-compliance
Lightning Source LLC
Chambersburg PA
CBHW050810220326
41598CB00006B/165